Conserving Humanity at the Dawn of Posthuman Technology

"From *Homo sapiens* to *Homo futuro*, Carvalko tracks our posthuman trajectory beginning with everything that defines us as 'being.' *Who* or *what* will remain given our insatiable desire to keep innovating, even at the expense of our very selves? He asks that vital question: 'is this path what we claim as our humanity?' A gripping depiction of a technological roadmap based on confluence incorporating societal repercussions, and the role of policy and ethics. Carvalko is an engineer and lawyer who interweaves his practice with wisdom and storytelling. A must read."

—Katina Michael, *Professor in the School for the Future of Innovation in Society and the School of Computing, Informatics and Decision Systems Engineering at Arizona State University, USA, and Founding Editor-in-Chief of* IEEE Transactions on Technology and Society

"Joseph Carvalko offers the reader a masterfully presented monograph, bringing together philosophical, historical, scientific and technological perspectives of 'post-human' in an engaging and illuminating fashion. Through a multi-disciplinary treatment of the subject matter, the author guides the reader through the changing nature of 'creativity' and technology's effects on human dignity as we inch toward a trans- and post-human existence that is akin to a science fiction novel, in which trans-humans, cyborgs and humanoid robots are commonplace. This book is a must-read for practitioners, scholars, students and those interested in the tensions between technological progress, creativity and ethics. Highly recommended!"

—Roba Abbas, *Lecturer in Operations and Digital Business, School of Management, and Operations and Marketing at the University of Wollongong, Australia, and Associate Editor of the* IEEE Technology and Society Magazine

"This is a deeply inspiring book, written by a technologist who is truly a humanist, a professor who talks simply like your dear friend drinking coffee with you, and a lawyer who doesn't let laws muddy your mind, making the legal concepts simple. But most importantly Carvalko's astonishing writerly hands gift us a book about major issues of our time, ones that can be easily understood in one sentence: 'My son, the answer to that question is in your hands.'"

—Da Chen, New York Times *bestselling author of* Colors of the Mountain, *a memoir (1999)*

"This is an important book. The author, Joseph Carvalko, understands the technical, legal and cultural aspects of our burgeoning high technology based on his experience as an engineer, an attorney, and a musician of uncommon talent. He brings his polymathic perspective to issues that are more commonly addressed from only a single point of view. He wrestles with deep questions of our identity as human beings, and our future when our capacity to change our genome, our brains, and our bodies have increased beyond anything conceivable in prior eras. The future may be unknowable in detail, but it is certainly something we need to prepare for, so that we may steer in the direction of human freedom and flourishing. This book will help you ponder how we might accomplish that."

—Michael LaTorra, *former Professor of English at New Mexico State University, USA, member of the Board of Directors of the Institute for Ethics and Emerging Technologies, and former Chair of Humanity Plus*

"Carvalko brings a pragmatic look at the co-shaping entanglement of people and technologies characteristic of our age. In this timely contribution, he considers the various implications of this productive entanglement, including the ethical ones: what 'personhood' means, what it means 'to be human' is redefined by the technologies around. Carvalko's message is cautiously optimistic: technologies will never take away our humanity. But it is only in reflecting on the productive impact of technologies on people and drawing on our creativity that we can continuously rediscover the value of humanity in the technological age."

—Olya Kudina, *postdoctoral researcher at Delft University of Technology, The Netherlands*

"'Your scientists were so preoccupied with whether or not they could that they didn't stop to think if they should' (Jurassic Park). Jeff Goldblum's famous line precisely sums up the premise of Carvalko's book: scientists never consider the second and third order of effects that arise from their discoveries. But new technologies always present a double-edged sword of promise and danger. His well-researched book tackles the dilemma of how genetic engineering and artificial intelligence may fundamentally change what it means to be considered human, and whether creativity is a trait exclusive to the non-technically enhanced human."

—Deborah Carr, *DVM, Virginia-Maryland Regional College of Veterinary Medicine, M.P.H., Uniformed Services University of the Health Sciences, and M.F.A., Fairfield University, USA*

Joseph R. Carvalko Jr.

Conserving Humanity at the Dawn of Posthuman Technology

palgrave
macmillan

Joseph R. Carvalko Jr.
Interdisciplinary Center for Bioethics
Yale University
New Haven, CT, USA

ISBN 978-3-030-26406-2 ISBN 978-3-030-26407-9 (eBook)
https://doi.org/10.1007/978-3-030-26407-9

This Palgrave Macmillan imprint is published by the registered company Springer Nature Switzerland AG
The registered company address is: Gewerbestrasse 11, 6330 Cham, Switzerland

Preface

According to scientists the world is home to over 8.7 million species, which in large numbers, from the tiniest protozoa to the largest whale, have been and are exploited for a human purpose rendering them perilously at risk of extinction. Intentionally or through ignorance, humans decimate species at the rate of 100–1000 per million each year, mostly due to human-caused climate change and habitat destruction.[1] Overtime some species will survive largely as a consequence of their genetic malleability. Ironically, for *Homo sapiens*, the fact of surviving the present environment may be coming to an end, not through climate change or habitat destruction, but through genetic engineering.

For humans, the act of survival has seemed always to include more than genetic adaptation. History tells the story that humans endowed with self-awareness survived by clustering into families and tribes, while inventing tools, cultivation, and language that helped construct complex civilizations. As our planet silently continues spinning through space, we

[1]"Human actions have already driven at least 680 vertebrate species to extinction since 1500…. Habitat loss and deterioration, largely caused by human actions, have reduced global terrestrial habitat integrity by 30 per cent relative to an unimpacted baseline; combining that with the longstanding relationship between habitat area and species numbers suggests that around 9 per cent of the world's estimated 5.9 million terrestrial species—more than 500,000 species—have insufficient habitat for long-term survival, are committed to extinction, many within decades, unless their habitats are restored." See, IPBES Global Assessment Summary for Policymakers, 6 May 2019. https://www.ipbes.net/news/ipbes-global-assessment-summary-policymakers-pdf.

move ahead, adopting norms, values, and the quest for "what it means to be human," in a real sense—, our essence. At times this notion of essence has appeared absolute, some indomitable spirit that carried us from eon to eon. However, as a species living through a meteoric culture change, largely as a consequence of technology, our essence is being called into question like never before. In the past we resorted to familial connections, religion, ethnic solidarity, and patriotic duty to find meaning. But we cannot ignore that our evolution itself plays a role in our essence, regardless of our beliefs, affiliations and social constructs. So what can we anticipate if we put our thumb on evolution by contriving to dominate and manipulate the very kernel of who we are?

Our genome is agnostic; among us 99% is shared, with 1% left to Darwinian chance. Whether consciously or not, we place great value in those traits that got us here, the ones in the making for over 3.5 billion years, especially the ones that express our physicality, well-being, skills, talents, creativity, intellect, emotions, and empathy. Few would quarrel with the claim that the creative idea, defined simply as one that is both novel and useful or influential in a particular social setting, has been the key driver in the progress of human civilization.[2] It's largely creativity and the closely aligned trait, referred to as the intellect, that distinguishes us from other species and from one another. If either or both of these were to be enhanced in any extant population, beyond what we regard as the normal distribution of these traits, the consequences for humanity would be cataclysmic.

The human qualities of intellect and creativity bear on a vast array of behaviors, some which chiefly concern invention, music and storytelling or more narrowly our narratives, which will in different ways delineate the coming posthuman. I will examine these three kinds of creativity because, in addition to our capacity to show compassion or love, our inventions, music and narratives serve as metaphors for our inimitability as a creative species.

As a modified human species, posthumans will not all collect on one side of a bright line, but overlap with their existing human contemporaries in expressing humanistic values, at least during a transition phase, that is prior to becoming a principal planetary species. As explained in the

[2]The terms "creative" and "creativity" as will be developed throughout, more generally apply to: (1) a person, (2) a process, or (3) a product, whether manifested via (a) an idea; (b) an observable performance; (c) or an artifact.

text, for metaphysical reasons, posthumans will be endowed with perceptions and intellectual capacities that will lead them to create in ways the human soul has yet to plumb.

The work that follows is divided into: (1) The Introduction; (2) Biology and Brains; (3) Technology of Creativity; (4) Creative Psychology & Essence; (5) Posthuman Humanities; (6) Societal Repercussions; (7) Policy and Ethics; (8) Final Thoughts.

New Haven, USA

Joseph R. Carvalko Jr.

Acknowledgements

I want to express my gratitude to those individuals who have helped make this book a success through their inspiration and suggestions, particularly to my wife, Susie, for her insights and her patience for projects postponed. Also, I wish to thank discerning reader/critics, who provided invaluable feedback: Roba Abbas, Ph.D., Lecturer in Operations and Digital Business, School of Management, Operations and Marketing at the University of Wollongong, Australia, and Associate Editor of the IEEE Technology and Society Magazine; Natalie Kofler, Ph.D., cellular, molecular, and medical biosciences, founder of Editing Nature and scholar at the Yale Interdisciplinary Center for Bioethics; Sheeva Azma, Science writer, M.S. Neuroscience, B.S. Brain and Cognitive Science; Lynn Hargrove, Ph.D., Psychology; Stephen McGraw, Ph.D., Psychology; Monica List, Ph.D., Philosophy; Cara Morris, B.S. and Patent Attorney; and Samantha Halkowicz, B.S., Social Work. I also want to express my appreciation to the dedicated team at Palgrave Macmillan, particularly Rachel Daniel and Madison Allums. I also want to acknowledge my collaborators, who inspire my interest in technology and ethics, Katina Michael, Ph.D., currently Professor in School for the Future of Innovation in Society and School of Computing, Informatics and Decision Systems Engineering, at Arizona State University; and Lori Bruce, M.S., Bioethics/Medical Ethics, currently Director, Sherwin B. Nuland Summer Institute in Bioethics, Associate Director, Interdisciplinary Center for Bioethics, Yale University, Chair, Community Bioethics Forum, Yale School of Medicine. And, not lastly my mentors, who encouraged me every step of the way: Richard Paul, SSgt U.S.A.F., Richard Kasper, M.D.; Kendall Preston, M.Sc., Physics; M.J.E. Golay, Ph.D.; Emil Bolsey, B.S.; and especially my parents, Joseph and Lucille, who never stopped believing in the power of the creative mind.

Contents

About the Author

Joseph R. Carvalko Jr. is an American technologist, academic, patent lawyer, and writer. As an inventor and engineer, he has been awarded sixteen U.S. patents in various fields. He has authored academic books, articles, and fiction throughout his career. Currently he is an Adjunct Professor of Law at Quinnipiac University, School of Law, teaching Law, Science and Technology; Chairman, Technology and Ethics Working Research Group, Interdisciplinary Center for Bioethics, Yale University; member, IEEE, Society on Social Implications of Technology; summer faculty member, Interdisciplinary Center for Bioethics, Yale University.

List of Figures

PART I

The Inevitable Path Ahead

CHAPTER 1

Introduction

At this moment in history a virtual cornucopia of creativity in all spheres of human endeavor blossoms on a scale never before witnessed. But as science and the arts ascend into an amorphous digital cloud for all to access, this new age of enlightenment conceals an ironic twist in human evolution, a twist which threatens to set in motion a vortex into which our social existence, as we know it, vanishes.

With advances in genomic modification and computer-based prosthetics, our species tumbles headlong into a future world that will be populated by societies of transhumans—what I shall call *Homo futuro*—technological offshoots of *Homo sapiens* exhibiting greater intellect and creative capacities, who will have longer, healthier lives, and off-load much of the world's mundane, arduous, and dangerous work to ever improving robots.[1]

[1] Plato spoke about the need for every human to be acknowledged by others as human, and deserving of respect. Some attribute recognition, by others, as the driving force leading to creation of works of art, discovery of new lands and the founding of empires. And although transhumans may exhibit enhanced intellects, as will be argued below, Francis Fukuyama, in The End of History and the Last Man (1992), suggested, a need for hyper-recognition can lead to domination and oppression of others, and devolve into a life of hedonistic, materialistic pleasure and self-obsession.

J. R. Carvalko Jr., *Conserving Humanity at the Dawn of Posthuman Technology*, https://doi.org/10.1007/978-3-030-26407-9_1

This is our facticity—a truth about our existence—one that will, for better or worse, profoundly influence the lives of our children and our children's children and bear on our choices, our transcendence.[2] But, when we will engage *Homo futuro* in its full panoply remains uncertain, as do many techno-scientific predictions. However, like the dawn of a new day, this Cimmerian event radiates below the horizon, waiting to reveal to we *Homo sapiens* that our time honored human trek through history may soon chart a new trail into the future.[3]

Until now, the world's advanced nations have nearly unanimously said "no" to allowing germline modifications of the human genome—that is those that get passed on to our offspring—fearing its incalculable and irreversible consequences for the human race.[4] But it's difficult to ignore the power of genome editing when it can be used for correcting birth defects, building resistance to disease, increasing tolerance to environmental conditions or enhancing senses or abilities. And, as history demonstrates time and again, it only takes the slightest crevice for the genie to escape the laboratory's Petri dish and become our new reality.

Particularly at this moment we cannot ignore evidence that we have reached a historic threshold between two different evolutionary states.

[2]Facticity is the third person viewpoint that others have about me, e.g., weight, gender, skin color, as well as demographic and psychological characteristics. But as a first person I can choose, who and what I am. Other senses of transcendence refer to exceeding the limits of ordinary experience. Jean-Paul Sartre speaks of transcendence in describing the relation of the self to the object oriented world and to others. I leave it to the reader to pick up on the various connotations of the word throughout the book.

[3]Cimmerians are mythical peoples described by Homer as dwelling in a remote realm of mist and gloom.

[4]In 1866, the German biologist August Weissman provided the basis for our understanding of two types of genes, germ line and soma line. Germ line genes are comprised of sex cells and passed along to subsequent generations, while the soma line cells are a result of environment, determining physical and behavioral characteristics. According to the National Academies of Sciences, Engineering, and Medicine, 2017, "Germline editing" refers to all manipulations of germline cells (primordial germ cells, gamete progenitors, gametes, zygotes, and embryos). "Heritable genome editing" is a form of germline editing that includes transfer of edited material for gestation, with the intent to generate a new human being possessing the potential to transmit the genetic "edit" to future generations. See, *Human Genome Editing: Science, Ethics, and Governance.* Washington, DC: The National Academies Press. https://doi.org/10.17226/24623. Somatic cell editing is underway for purposes of treating genetic diseases pertaining to the various tissues of the body, but in contrast to heritable germline editing the effects of changes made to somatic cells are limited to the treated individual and would not be inherited by future generations.

In November 2018, geneticist He Jiankui announced details about the two babies he claims to have genetically modified as embryos using the gene editing technology known as CRISPR.[5] Dr. He managed a genetics lab at the Southern University of Science and Technology in China, where he'd modified the CCR5 gene in what is believed to be the first heritable edit to our species—, *Homo sapiens.* The CCR5 gene is known for its role in Human Immunodeficiency Virus, or HIV, although it also has a less well-understood role as a modulator of learning, memory, and neuroplasticity.[6] If the doctor's claims bear out, this event marks an inflection point and radical departure from the historically imperceptible changes over the course of human evolution.

Throughout this book, I raise the suggestion that future designs instantiated in transhumans will compete with the initiative of the modern human to discover truth and beauty, and that these offshoots of the *Homo sapiens* will, in their own way, flourish in love, compassion, and we would hope, peace and humility. Yet, I am discomfited by the prospect that we and certainly our progeny may someday coexist with beings of extraordinarily superior intellect and physicality. And, if this portends our future, it has the potential to change the essence of who we are, the "*I*" that defines our transcendent core.

Recognizing that we stand at the cusp of an evolutionary departure from the modern human, what should we choose to do?[7] Stop progress at all costs or manage the revolution as best we can? G.E. Moore (1873–1958) in his influential work *Principia Ethica* (1903), wrote that assessing what's good or bad may be indefinable,

[5] Clustered Regularly Interspaced Palindromic Repeats (CRISPR/Cas9) is a breakthrough technology enabling the correction of errors in the genome. With CRISPR, scientists can turn on or off genes in cells and organisms quickly, cheaply and with relative ease to fix diseases such as HIV, cancer, and other genetically based diseases (see, https://www.ncbi.nlm.nih.gov/pmc/articles/PMC4975809).

[6] Reducing CCR5 activity may improve learning and memory according to one study (see, https://elifesciences.org/articles/20985).

[7] Neo-Darwinian theory holds evolution is based upon a population of organisms that (a) vary in heritable traits influencing their chances for reproduction and survival; (b) have off-spring resembling their parents more than they resemble randomly chosen members of the population; and (c) produce more offspring on average than are needed to replace members removed from the population by emigration and death.

but if revised life forms come to exist, how should we respond if we aim to preserve what is fundamentally human?[8]

Modern humans have been endowed with biochemical consciousness yielding a set of emotions and thoughts, which combines with other like minds to add to the world's output: its products, processes and social constructs, all of which bear on our humanity for better or worse. The origin of these productions seat deeply in neurological structures and processes that have evolved to advance civilization to ever increasing levels of knowledge and norms that convey riches in expression of love, art, music, literature—, and wisdom. As Seneca wrote: *Homo sum: humani nihil a me alienum puto*; thus, "I am a human being: and I deem nothing pertaining to humanity is foreign to me." So, will it be that as our *Homo sapiens* descendants come upon the successors of our species, that they will greet them with equanimity?

Over the course of history humankind has defended interests, whether territorial, economic, social or political—and pursued, at bottom, goals for survival, love, power, and self-actualization. Against this backdrop, it's our need for recognition that motivates us to pursue our potential as individuals largely through life experience, learned knowledge, language, art, and technology. But, now we face the brink of dramatic change. As we venture into a world of *Homo futuro*, the real possibility exists that ways in which humans have heretofore imagined and actualized as modern humans, will be in the best case augmented or, in the worst case, replaced by species of higher intellect and creative potential of a kind.

If we embrace the idea that enhanced human intelligence in the long run will be developed through advances in technology, we also must consider that *Homo sapiens* risk obsolescence, the kind of outmodedness that led to Neanderthal extinction, our last hominid rival, forty thousand years ago. Before extinction, some humans will culturally and biologically assimilate with *Homo futuro*. This premise follows from the idea that initially transhumans may not be significantly distinct from the *Homo sapiens*. Increasingly, over time, *Homo futuro* will become intellectually

[8]Joel Garreau, argues that we are engineering the next stage of human evolution through advances in genetic, robotic, information and nanotechnologies, by altering our minds, memories, metabolisms and where unrestrained technology may bring about the ultimate destruction of our entire species. Joel Garreau, Radical Evolution: The Promise and Peril of Enhancing Our Minds, Our Bodies—And What It Means to Be Human (Doubleday, 2005).

exceptionally gifted compared to *Homo sapiens* and likely will assume positions of power and achievement in every walk of life. Eventually, compared to its successors, the *Homo sapiens'* apparently inferior brain power will lead to its extinction. When the new speciation begins, we should anticipate its greatest effect will be in the classes least able to intellectually compete.

A lack of access to these breakthrough technologies will put populations in less developed nations at an intellectual disadvantage. As to affluent countries that can afford these technologies, it will drive a veritable "arms race" toward a better brain/physicality. For example, China is widely regarded as a developing nation, but strong in science, and could successfully compete with other technologically advanced nations.

As behavioral geneticist Robert Plomin reminds us, genetics do not tell us about how or what we think, or how our brain works; instead, this study of our molecular makeup gets to the core of who we are—it is about individual differences, or in other words, what makes us unique [1]. Humans have anywhere between about 19,000 and 23,000 genes distributed on the two sets of their 23 chromosomes, of which an astonishing 6000 genes are expressed in the brain. How many of these encode for traits and biological markers or indicators of intelligence is unknown. And importantly, as between human beings, there are no scientifically verified differences in intelligence that favors or accretes positively to any social class, ethnic origin, phenotypical distinction, or sex.

It does appear that the number of genes implicated in intelligence may be large. This vast collection of intelligence-related genes may even work in combination, such that any particular gene set has a small influence on overall IQ. Despite this vastness, it may not be unreasonable to assume that by manipulating a relatively small percentage of discrete differences, a burgeoning population of individuals with marginally enhanced cognitive traits would emerge. Enhanced characteristics could include increased working memory, augmented visuospatial processing, and sheer mental processing speed. Once we concede that humans will be subjected to genetic modification to increase cognitive capacities, we are led to consider its implication—that is, what end will this enhancement serve, and what unintended consequences shall we face, in terms of human productivity, creativity, and values.[9]

[9] Monozygotic or identical twin studies combined with imaging of brain structures and features confirm a genetic basis for particular forms of intelligence.

Gene flow, the concept that explains how a gene version is carried to a population where it previously did not exist, may be an important source of genetic variation. If a genetic trait is advantageous and helps the individual to adapt and thus improve their odds of surviving and reproducing, the genetic variation will be more likely inherited through the process known as natural selection. Humans without this genetic variation will have lower survival rates, while humans possessing this genetic code will live longer, producing more offspring with greater survival rates. Over time, as subsequent generations with the favorable trait continue to reproduce, the trait will become increasingly more common in the gene pool, making the population different from its original ancestral one, and resulting in what we'd consider a new species.[10] Of course, the rate of evolutionary change that brings about this effect may be fast or slow, depending on the degree to which the enhancement regimen would be adopted.

To carry forward this hypothesis, two or more new dissimilar classes or species, caused by genetically engineered traits that express for intelligence, would only add pressure to existing societies, many of which have been unable to integrate and assimilate disparate ethnic, religious, and racial immigrants. History has shown with remarkable constancy that when an invading ethos collides with an indigenous culture, the latter goes through stages of resisting, capitulating, and abandoning its aboriginal fabric, culturally, racially, and on an evolutionary scale, physiologically.

The idea that we face an inevitable posthuman co-evolution assumes that average differences between *Homo sapiens* and *Homo futuro* genomes will result from human motivated selection of genes based on breakthroughs in identifying how intelligence and creativity function within the brain. Leaving aside the myriad ways in which intelligence can be defined, current research is deep into the study of intelligence genes, gene complexes, and brain structures. Hundreds of scientists are engaged in genome-wide association studies (GWAS), seeking to link genetics to not only intelligence, but also a variety of specific disorders, both inherited conditions and those whose genetic/environmental disorders are less certain.

[10] See, https://ghr.nlm.nih.gov/primer/mutationsanddisorders/evolution.

Four traits with especially large published and replicated GWAS results are height, body mass index, educational attainment, and depression, the last two implicating intelligence and a linkage to creativity itself [2]. And using some of the most recent GWAS data, scientists have identified specific alleles, or variations in genes, that account for at least 20% of the 50% heritability of intelligence [3].

Our posthuman future does not only lie in what scientists will learn about where biological roots of intelligence are located, but also to where artificial intelligence (AI) can be engineered and interfaced, not only in computers directly, but the human brain. The best example thus far is the implementation of a prosthetic system that was able to use an individual's own memory patterns to facilitate the brain's ability to encode and recall memory, showing a short-term memory improvement of 35–37% over baseline measurements [4, 5, 6, 7].[11]

In part, the idea of enhanced intelligence goes beyond genetic engineering and prosthetics—it includes seemingly futuristic technological innovation such as AI, utilizing silicon-based semiconductors and associated software, as well as synthetic biological devices for control and computation in the human brain [8].

Synthetic biology has opened up a world of genomic editing that can be used to modify or design genetic sequences at the level of individual base pairs and, potentially, at multiple sites found in a given gene or an organism's entire genome. Someday, synthetic biology, coupled with the inorganic chemistries of AI wet-ware, that is, biological equivalents of computer silicon chips, loaded with AI, will transform the science of genetic editing and other applications for physical implants.[12] Biological AI constructs, unlike conventional computer chips, will have the potential to affect the germline, and thus the speciation of future generations. We see evidence in recent successes to manufacture bacteria out of whole cloth; through the technology of 3-D printing for creating anatomical organs; and, in a more profound way, using gene technology to create

[11]For example, a hippocampal electronic prosthesis, implanted to improve memory or replace the function of damaged brain tissue.

[12]Synthetic biology is a scientific discipline that relies on chemically synthesized DNA to create organisms with generally novel, enhanced characteristics or traits. The subject combines disciplines from biotechnology, genetic engineering, molecular biology, molecular engineering, systems biology, membrane science, biophysics, electrical engineering, computer engineering, control engineering and evolutionary biology.

complex computer-like circuits, such as AND, OR, and NOR gates capable of computation [9].

It's axiomatic that useful technologies evolve and spawn new technologies, and with each innovation, we see a convergence with other technologies.[13] In part, this is because technologies have utility, which services complex, progressing systems: for example, social, biological, and mechanistic systems, such as banking, brains and computers, respectively. The same holds for human enhancement that has been assimilated into various technologies for reasons of survival or good health—for example, pacemakers to keep the heart beating, or Prozac to maintain mental wellness.

Likewise, transhumans will not depend on a single technology, but multiple technologies distributed within their anatomies. The acronym, Nano-Bio-Info-Cogno refers to nanotechnology, bioengineering, information systems, and cognitive technology that may join to boost human performance, physiologically and psychologically. In various ways, these classifications of technology will be manifested in synthetic DNA molecular computers, micro-sized miniaturization of silicon computers, communication networks within the body, micro-mechanical drug delivery vehicles, and brain networks, which use AI to optimize and work better [10]. In my earlier book, *The Technohuman Shell—A Jump in the Evolutionary Gap*, I detailed illustrative electronic devices that already supplement and in the future will further enhance physiology. However, here the emphasis is on technology that will augment perception, intuition, thinking and creativity in a posthuman world.

Artificial intelligence is dangerously reductionist, especially when used for decisions that require human judgments, such as matters of life or death. Ethicist Wendell Wallach wonders if robotic war machines will reach complete autonomy and if so, "Will they pick their own targets and pull the trigger without the approval of a human?" [11]. Consider an amplification of this concern, when robotic autonomy invades human autonomy, and compromises principles to do no harm and fairly mete out justice. If we yield our place on earth to decision-making technology, we surrender humanity's role in decision-making to systems that lack any sense for the meaning of life.

[13] Any patented invention illustrates the point of convergence, which will be taken up in chapters that follow.

To gauge whether society is ready to accept genetic engineering enhancements when these become available, we need only turn attention to the market for in vitro fertilization (IVF), a practice well-entrenched in most modern societies.[14] For several generations the use of surrogacy, donor eggs, cryogenic technology and testing embryos for genetic markers have allowed infertile parents, straight or gay, married or single—to have children. The in vitro embryonic screening technology involves extracting cells from embryos and determining relevant genomic markers before choosing which embryos, based on a specification, might be implanted into a uterus.

A proposed offshoot of IVF technology uses stem cell-derived gametes and iterates embryo selection (IES), which could accelerate enhanced intelligence [12].[15] Shulman and Bostrom, from the Future of Humanity Institute, Oxford University, argue that IES in vitro, through the compression of multiple generational selections, will lead to improved IQ [13]. Furthermore, they hypothesize that expanses in IQ gain will depend on the number of embryos used in selection. For example, 1 in 2 embryo selections would yield a 4.2 point IQ gain, and 1 in 10 embryo selections would result in an 11.5 point gain. Over five generations, a 1 in 10 embryo selection protocol would max out at a 65 point gain, due to diminishing returns; and a 10 generation protocol would likewise max out at a 130 point gain. I again emphasize that the Shulman and Bostrom idea focuses on a biological solution for increasing IQ; however, other nonbiological technologies, as AI computational devices that provide enhanced memory, may also result in an apparent increase in IQ.

Needless to say, adding gains of between 5 and 130 IQ points to any population, even to a few individuals within a demographic, could effectively empower intellectual capital. Such capital could be employed for social, political, economic, or scientific advances. The world largely suffers from a class based mode of distribution, and if the matter of improved IQ were feasible, it likely would be made available to the more affluent in society.

IVF, IES, or CRISPR-type gene editing for increasing intelligence fits squarely within the realm of scientific possibility, but whether these or

[14]In 1978, Louise Brown was the first child conceived with IVF and according to a 2012 report by the Society for Assisted Reproductive Technology, more than 61,000 babies were conceived with the help of IVF. It costs about $12–15,000 for an IVF procedure.

[15]IES has been used to produce fertile offspring in mice and gamete-like cells in humans.

similar technologies would be accepted by societies at-large or, on the other hand, legislatively prohibited, remains uncertain.[16] Because of these genetic editing technologies' significance in determining global economic competitiveness through improving human intellectual capital productivity, developed countries would likely look favorably at these advances. BGI, a Chinese company possessing the largest gene-sequencing facility in the world, launched a project to study the basis of intelligence in 2012 [14]. Whether BGI will succeed in its endeavor is an open question. Some geneticists have expressed doubt and others optimism. The jury is still out inasmuch as no study to date has positively identified the gene set for enhanced intelligence [15, 16]. That said, numerous studies involving hundreds of geneticists are underway, precisely to discover where the genetic determinates are located within the genome [17].

Understanding *whether* a genetic basis for intelligence exists, and if such a basis can be manipulated, requires delving into the multiple disciplines of: evolution and population genetics; genes and genetic epidemiology; molecular biology; the brain and cognitive neuroscience; and the architecture of cognitive psychology. This is a tall order, and one which, in overview, provides an introduction to what the future may have in store. I draw upon and cite much of the latest research in these fields, in addition to literary works discussing these topics, such as: *Blueprint* (Plomin, 2018); *Gödel, Escher, Bach: An Eternal Golden Braid* (Hofstadter, 1979); *An Interaction, Creativity, Cognition, and Knowledge* (Dartnall, Ed., 2002); and, *The Cambridge Handbook of Creativity* (Kaufman and Sternberg, Eds., 2010) [18, 19, 20, 21].

Contemporary research in genetics, artificial intelligence, human intelligence, and creativity serves as a looking glass into aspects of our resourcefulness that are likely to change. These changes will be powered by the acceleration of technology, as the early adopters of these changes

[16] U.S. law does not limit the number of children a sperm donor may produce, while other countries do restrict the numbers (Anywhere from one in Taiwan to 25 in the Netherlands). No U.S. law prevents selecting embryos, providing federal funds are not used, so there'd be no prohibition to select embryos with "high IQ." Pre-implantation genetic diagnosis in the UK is regulated by the Human Fertilisation and Embryology Authority, which permits identification of embryos with a schedule of rare genetic diseases for which the parents are known carriers. And UK's restrictions allow prenatal screening for Down's syndrome, a disability that produces low to moderate intellectual disability, so in some sense, they have one foot in a eugenic practice that reduces the probability of lower IQ children being born.

begin to embrace CRISPR-like gene editing and artificial intelligence for themselves and their offspring.

I have added footnotes that explain ideas or define terms and concepts that may be familiar to those who are not experts in a particular field. I agree that where technology is central, we need rigor and objectivity. But in appreciation of this fact, much room exists for analogies and metaphors to gain a conceptual understanding of complexities that terse technical definitions may not well serve. On the other hand, aesthetics, found in art, poetry, music or prose, need to be dealt with on a human level, and I've granted myself license to employ all appropriate avenues, including straight definitions, rhetorical flourishes, metaphors, imagery, fiction, and poetry to explore a balanced perspective of what the future may hold in store.

Let me restate two of the more salient issues that we shall explore in the following chapters: (a) how will the technologies of genetic engineering and electronics, e.g., AI, improve cognition and influence human potential, particularly creativity; and (b) what questions should be raised and answered before we transform society into new divisions of *Homo sapiens* and *Homo futuro*, generally.

As Biologist and Naturalist, E. O. Wilson observed, the two great branches of learning, science and the humanities are complementary and "arise from the same creative process in the human brain." Science lays bare our potential and the humanities counsels us to tread wisely. In sum these truths light our way into an uncertain future [22].

Notes

1. Plomin, R.B. (2018). *How DNA Makes Us Who We Are*. Cambridge, MA: MIT Press.
2. Ware, E., et al. (2017). "Heterogeneity in Polygenic Scores for Common Human Traits." *bioRxiv* 106062. https://doi.org/10.1101/106062.
3. Plomin, R., et al. (2018). "The New Genetics of Intelligence." *Nature Reviews Genetics* 19: 148–159.
4. Developing a hippocampal neural prosthetic to facilitate human memory encoding and recall, *Journal of Neural Engineering* (2018). https://doi.org/10.1088/1741-2552/aaaed7, iopscience.iop.org/article/10....088/1741-2552/aaaed7.
5. Tsien, J.Z. (2016). "Principles of Intelligence: On Evolutionary Logic of the Brain." *Frontiers in Systems Neuroscience* 9. https://doi.org/10.3389/fnsys.2015.00186. ISSN 1662-5137.

6. Tsien, J.Z., pioneered Cre-loxP-mediated brain subregion- and cell type-specific genetic techniques in 1996, enabling researchers to manipulate or introduce any gene in a specific brain region or a given type of neuron. See, Tsien, J.Z. (2016). "Cre-Lox Neurogenetics: 20 Years of Versatile Applications in Brain Research and Counting…" *Frontiers in Genetics*. https://doi.org/10.3389/fgene.2016.00019, http://journal.frontiersin.org/article/10.3389/fgene.2016.00019/abstract.
7. Amara, A. (2013). "Electronic Hippocampal System Turns Long-Term Memory on and Off, Enhances Cognition, Kurzweil, R., AI." Kurzweil Accelerating Intelligence. http://www.kurzweilai.net/artificial-hippocampal-system-restores-long-term-memory. Earlier progress was reported in Brain-implantable biomimetic electronics as the next era in neural prosthetics, Proceedings of the IEEE, Volume: 89, Issue: 7, July 2001.
8. Nielsen, A.A.K., and Voigt, C.A. (2014). "Multi-input CRISPR/Cas Genetic Circuits That Interface Host Regulatory Networks." *Molecular Systems Biology* 10 (11): 763.
9. Charles, Q.C. (2015, October 23). "Organs on Demand? 3D Printers Could Build Hearts, Arteries." https://www.livescience.com/52571-3d-printers-could-build-organs.html.
10. Gurney, K. (1997). *An Introduction to Neural Networks*. London and NY: Routledge; Reyneri, L.M. (1999). "Theoretical and Implementation Aspects of Pulse Streams: An Overview." In: Proceedings of the Seventh International Conference on Microelectronics for Neural, Fuzzy and Bio-Inspired Systems, 1999. MicroNeuro'99. IEEE. Murray, A.F., et al. (1991). "Pulse-Stream VLSI Neural Networks Mixing Analog and Digital Techniques." *IEEE Transactions on Neural Networks* 2 (2): 193–204.
11. Wallach, W. (2015). *A Dangerous Master*. Basic Books.
12. Sparrow, R. (2013). "In Vitro Eugenics." *Journal of Medical Ethics* 40: 725–731. Published online first: 4 April 2013. https://doi.org/10.1136/medethics-2012-101200.
13. Shulman, C., et al. (2014). "Embryo Selection for Cognitive Enhancement: Curiosity or Game-Changer?" *Global Policy* 5 (1): 85–92. https://doi.org/10.1111/1758-5899.12123. ISSN 1758-5899.
14. Yong, E. (2013). "Chinese Project Probes the Genetics of Genius." *Nature* 497 (7449): 297–299. https://doi.org/10.1038/497297a.
15. Chabris, C.F., et al. (2012). *Psychological Science* 23: 1314–1323.
16. Davis, O.S., et al. (2010). *Behavior Genetics* 40: 759–767.
17. Deary, I.J., Johnson, W., and Houlihan, L.M. (2009). *Human Genetics* 126: 215–232.

18. Plomin, R.B. (2018). *How DNA Makes Us Who We Are*. Cambridge, MA: MIT Press.
19. Kaufman, J.C., and Sternberg, R.J. (eds.). (2010). *The Cambridge Handbook of Creativity*. Cambridge University Press.
20. Hofstadter, D.R. (1979). *Gödel, Escher, Bach: An Eternal Golden Braid*. Basic Books.
21. Dartnall, T. (ed.). (2002). *An Interaction, Creativity, Cognition, and Knowledge*. Westport, CT: Praeger Publishers.
22. Wilson, E.O. (2014). *The Meaning of Human Existence*. Liveright Publishing Corporation.

CHAPTER 2

A Bridge and Not an End

Sarouk: Isn't there something innate in our being, beyond the biological that keeps us evolving?

Mensa: Yes, I think there is, but merely evolving into another techno-creature, or form, has in and of itself proved that the entity we once regarded as human, lost the battle against its extinction.

Sarouk: How so?

Mensa: Mr. Sarouk, we have severed the modern world from all that preceded it, as if it never existed. We live in a culture that fractured into a million unrecoverable shards. Gone are traditions and a common wisdom. The ethos into which I was born, exposed me to beliefs, values, needs. I once aspired to leave at least a ripple of myself on the wake of greater humanity. I aspired to love. In this process of living life at its fullest, our forbearers brought to bear an inseparable blend of intention, feeling, judging and morality concerning what they did. Today's posthuman worships technical progress, not tradition. Common wisdom's been replaced by a world order of interconnected internalized technology.

Sarouk: Being a few generations ahead of you, I have no sense of what life was like… So, how far has this moved us from what had been our former trajectory?

Mensa: When I was born, I knew the connections to my biological and cultural predecessors. In form with them anatomically, intellectually, spiritually. We were the same species from whom we inherited a familial DNA,

J. R. Carvalko Jr., *Conserving Humanity at the Dawn of Posthuman Technology*, https://doi.org/10.1007/978-3-030-26407-9_2

even though we lived in diverse cultures. What changed me and made those that followed different was that we were co-opted by invention … newer genetic constructs, programs that helped us think faster.

Sarouk: I take it you are critical of this new reality?

Mensa: In ways yes, in other ways … no. We gained a greater potential for surviving, free from destructive passions that in my day spawned interminable wars. But, unlike past revolutions, political or social, which failed to bring conformity, the posthuman transformation has eradicated the individual's claim to his or her own essence. We are techno-human beings, one-hundred percent homogeneous.

Sarouk: You have to admit that we have not completely destroyed the human spirit. We continue to make strides in engineering and science.

Mensa: True enough, although my processors work hard to discourage my thinking about the past, in part because it evokes tender memories, which they've determined are illogical. Emotion is deemed inefficient. We now have a neat, reductive, techno-culture, where few mistakes are made… principles and laws are not influenced by human sentiment, but programs that objectively mediate and flatten the culture. We are no longer on some eye opening journey.

Sarouk: How so?

Mensa: Nietzsche, wrote that "What is great in man is that he is a bridge and not an end." That bridge kept going for our ancestors. They followed their dreams, while pushing against a never-ending assortment of obstacles we called the daily grind. But we no longer look back, we push forward against no apparent existential threat. Our imperative to survive and achieve significance has vanished, no longer serving as the drivers for procreation or self-expression.[1]

[1]This conversation was inspired by an earlier article by the author. See, Carvalko, J. (2016). "Crossing the Evolutionary Gap—IEEE Technology and Society." *IEEE Technology and Society Magazine*, Copyright, Joseph R. Carvalko.

CHAPTER 3

Irrepressible Creativity

No other living being creates as prolifically as humans, who now create a way out of their own essence.

…
We are resolved into the supreme air,
We are made one with what we touch and see,
With our heart's blood each crimson sun is fair,
With our young lives each spring-impassioned tree
Flames into green, the wildest beasts that range
The moor our kinsmen are, all life is one, and all is change.
With beat of systole and of diastole
One grand great life throbs through earth's giant heart,
And mighty waves of single Being roll
From nerve-less germ to man, for we are part
Of every rock and bird and beast and hill,
One with the things that prey on us, and one with what we kill [1].

In the Symposium, Plato explains that "beauty" is not found in "a creature or the earth or the heavens," but only "in itself and by itself." It's not clear how this "beauty" can be synthesized or mechanized, although thousands of technologists are attempting just that. Elsewhere in the Symposium, Socrates quotes Diotima who refers to beauty as the object of every love's yearning, the soul's progress toward ever-purer beauty, from one body to all, then through all beautiful souls, laws, and

J. R. Carvalko Jr., *Conserving Humanity at the Dawn of Posthuman Technology*, https://doi.org/10.1007/978-3-030-26407-9_3

kinds of knowledge, to arrive at beauty itself [2]. What happens if we create a society that designs away from what no one has successfully explained: love, beauty, the soul, the "wow" or eureka moment, itself? [3]. And, what kind of society will evolve, if through a form of eugenics, humans produce a species of superior intelligence, each creatively indistinguishable?

In exchange for living less strenuous lives, and succumbing to a one size fits all intellect, our most cherished values, significantly our independence or "free will," risk falling under the spell of the deterministic logic of cyber-machinery. Without free will, we lose the power to exercise: integrity, excellence, courage, compassion, and temperance. Knowing this, we must acknowledge that we face a future where anatomically embedded processors communicating with cloud computers will select what we see or hear. We face a future where autonomous agents track our moves, physically and psychologically, via the artificial intelligence of learning machines, facial recognition algorithms that curb options, training our behavior, millisecond by millisecond, communication packets to and fro, to order compliance with the likes of the vigilant eye of government bureaucracies and social media conglomerates. In a real sense, we risk losing autonomy, making it difficult at best to exert our preference for alternative social lives, decreasing our ability to engage face to face, not in an atmosphere, where pheromones and perfumes pass, but separated by video screens, enveloped by an electromagnetic spectrum, where through a world wide web, only electrons communicate. We mustn't ignore the onslaught of technology that rattles the foundation of what it has always meant to be human.

A number of AI futurists warn that when a computer's power of cognition surpasses general human intelligence, it stands to become the planet's dominant thought-form. The brain receives, organizes, retains, and uses information through various cognitive functions that include: acquiring information (perception); selecting what to focus on (attention); representing sensory input in terms of a biologically induced pattern (symbolically); expressing what a pattern represents (epistemological predicate); retaining information (memory); and using retrieval to direct and guide behavior. Computers have reached the potential to perform each and every one of these functions. But, to achieve the human dominant-thought form also demands that a computer have the potential to create an aesthetic palette and a moral fabric out of whole-clothe, ones that fits into the framework of human values, those which are widely considered immutable and those which must change as we journey in the direction of greater enlightenment.

Think about inventions, such as the electric light, radio, television, or the Internet. It seems much to ask a computer to achieve on this level of innovation. So, computers wouldn't likely contribute to particular creative domains. But, perhaps achieve creativity unique to their well-matched ends. As for we humans the motivation to invent is a reflection of particular values: to create, to produce something having utility, to disclose our knowledge-about something unknown and non-obvious. If we abrogate all spheres of human ingenuity, in the long run we risk a sundering of values, as our quest for excellence in Aristotelian terms, or the thirst to create in the humanistic sense, severing the practices that guided our journey from the African savannah to the village, from the metropolises, to what may become an unrecognizable Singularity.[1]

We can't avoid the role computers play in our lives, their superiority for decision making, especially as to national defense, economics or medical diagnosis. Other applications, some having to do with social networking, commercial enterprise, and transportation, and others having to do with security, involve new found applications for such things as artificial intelligence. For example, facial recognition has the power to identify criminals, one's likely IQ, political ideology, or with stunning accuracy sexual orientation [4].[2] Closely aligned to AI is virtual reality (VR), which immerses us into mind-wanderings that give new meaning to consciousness, unconsciousness, dreaming, and raises questions about personal versus sub-personal mental processes; the "*I*" which we regard as ourselves. Needless to say modern innovations engage us in ways that introduce us to new experiences, but which for many is addictive, witnessed by our dependence on smartphones and social media.

Nature has always defined the biological factors that produce life's animating force. Yet as conscious agents, we create our social dominions via the endless ways that we express our fears, aspirations and imaginative urges, including inventing the tools with which we achieve our aims. It's a pliable world, where, within limits, we influence it, ecologically and socially, and vice versa. Beginning with the Industrial Revolution,

[1]The hypothesis that artificial general intelligence will trigger runaway technological growth, resulting in unfathomable changes to human civilization.

[2]According to a Stanford University study, a neural network distinguished between gay and heterosexual with an accuracy rate of 81% for men and 74% for women. Stanford University researchers trained the AI on more than 35,000 public images of men and women sourced from an American dating site.

no other specie has modulated its social reality with the intensity and speed humans have. Humans are unique among creatures who invent pseudo-worlds, and which now boldly embark on a project to radically reinvent themselves. The transition to transhumanism is an inevitable extension of this long history of innovation, which here and now includes the species itself. Few limits have ever corralled our imaginations or ingenuity. Why should we limit ourselves at this point?[3]

Humans are neither zombies nor savants, but employ multiple intellectual stratagems in navigating the complexities of surviving, and unlike other life on the planet, to this end, largely exercise an extraordinarily complex brain in a process we blithely speak to as thinking. An alien observing the species would notice, first its conspicuous explicit cognitive ability, which is shorthand for intelligence, i.e., that feature of life we have come to associate with a conscious, deliberate and reflective learning process, or more succinctly critical thinking. It also includes implicit learning, characterized by automatic, associative, nonconscious, and unintentional learning processes, referred to as affecting intuition and the ability to automatically and implicitly detect complex and noisy regularities in the environment. We constantly seem busy expressing our ideas, searching for truths of a certain kind.

Yet, beyond our rational side, we express predilections based on what we often refer to as our feelings. Psychologists refer to this as affective engagement, analogous to Jung's feeling function that denotes the preference for evaluation based on emotions, feelings, and empathy [5]. And, apropos to our desire to create in an artistic sense, we exercise an aesthetic engagement, where we appreciate beauty, harmony, fantasy, and seek out cultural stimuli. For reasons to be addressed, it's doubtful that affective or aesthetic engagement can be engineered into cyborgs, or humanoid robots, at least not of the kind that has defined the modern human, but the AI technology currently under development, which seeks to create art, music and literature in the motifs of one's culture, may be employed to directly interface to the biological brain, thus influencing what originality may come to pass.

Throughout this book we will explore whether genetic engineering will favor a certain kind of intellectual enhancement, one that will narrow

[3]Think about flying, electronic communication, human travel to the Moon, 300,000 kilometers into a vacuous cosmos, or to the ocean floor, 300,000 centimeters into a pressure-fill amorphous mass.

the concept of divergent thinking, behavior that in large part influences imagination across a large swath of domains. And, we shall drill down into theories of creativity; first the role of representational thought; second, where the elements used in the conception of an object are roughly predetermined; and third, quite the opposite, where the crucial feature used in the creation of an inventive or aesthetic object may not be revealed in advance.[4] This last form takes us in a direction, where "what we know about" something matters, process for instance, forms of performance, music, writing, and art. In some cases it becomes impossible to separate the dance from the dancers.

Notes

1. Wilde, O., Panthea, Poems 1881.
2. Pappas, N. (Fall 2017 Edition). "Plato's Aesthetics." In: E.N. Zalta, ed., *The Stanford Encyclopedia of Philosophy.* https://plato.stanford.edu/archives/fall2017/entries/plato-aesthetics/.
3. Schickore, J. (Summer 2018 Edition). "Scientific Discovery." In: E.N. Zalta, ed., *The Stanford Encyclopedia of Philosophy.* https://plato.stanford.edu/archives/sum2018/entries/scientific-discovery/.
4. Wang, Y., et al. (2018). "Deep Neural Networks Are More Accurate Than Humans at Detecting Sexual Orientation from Facial Images." Retrieved from osf.io/zn79k.
5. Jung, C.G. (1971). *Psychological Types.* Princeton, NJ: Princeton University Press.

[4]Throughout what follows, word meaning may prove problematic, because we will be moving between subjects within science (e.g., psychology, neuroscience, computer science), mathematics, law (e.g., real property and intellectual property), ethics, and aesthetics where a word in one discipline may mean something different in another. Word meanings for "domain," "object," and "entity" for instance are sometimes dependent on discipline, context and the purpose used (e.g., philosophy, science). If a word needs a special definition I will state so, otherwise words have plain meaning, as for example, "object" which will mean anything that we can imagine, but manifested generally through its properties (e.g. its color, size, weight, smell, taste, and location).

CHAPTER 4

The Narrative

No different from all those creatures we share Earth, we have no privileged powers to ignore our condition. Like them, we are forced to engage a never-ending hodge-podge of obstacles that stand in the way of survival. Each of us as members of a species is limited in what we can do, or how far in time we can possibly go. No, we are not special, and at the end of days the last standing creature well may be the cockroach. Although, every living creature struggles much like Sisyphus portrayed it, we do differ from other life forms and from each other in our private actuality, each of us turning, twisting and pushing each day, repetitions up, down, in, out, hopes and dreams, some dashed, some fulfilled.[1] We live separate lives even from those closest to us. What we sacrifice for love, ego, what ambitions and vulnerabilities we shelter while we turn, how long we wait to turn and twist, pushing against the next barrier in the queue—what we believe as our limits (actual or assumed) and when we decide to quit—if anything this is what separates each of us, one from the other. While we live through our daily travails, forces of solidarity shore us up. The obvious are families, neighbors, countrymen, friends, in the flesh and cyberspace, a collection of interests, framed as compatriots and citizens. We seek out the values, norms that a culture symbolizes, our shared fates. Our past, place, and perspective in this web of circumstance is what writes our narrative.

[1] Sisyphus was a legendary king of Corinth condemned eternally roll a heavy rock up a hill in Hades only to have it roll down again as it neared the top.

J. R. Carvalko Jr., *Conserving Humanity at the Dawn of Posthuman Technology*, https://doi.org/10.1007/978-3-030-26407-9_4

The philosopher Ortega y Gasset famously pronounced, "*Yo soy yo y mi circunstancia*," that I cannot exist as "me" without things, and things cannot exist without me. In this modern era, that central thing *is* our technology, which exists because of us, and because of it we exist. We are self-referenced by and reflections of technology at the same time. Few of us would survive but weeks, if the world's industrial infrastructure stopped working in any significant way. On the other hand our dependence on technology undermines important human values, such as autonomy, stifling thought, subjecting us to waves of digitized "ones and zeros" that plug minds back into themselves, each byte fed back as an element in a closed system.[2] Each of us basically closes the loop, completes the circuit. Few of us pause to actually think how it is that technology comes out of a virtual cloud. How it invisibly fires electrons that strike at the center of imagination, wresting the power to choose, to farm, forage, fight or vote. In the maelstrom of intransigence we can't even come together to stop wars or save Earth from a thermal meltdown. Who among us does not feel vulnerable?

Now in an era of genetics and mechanistic manipulation, the moment arrives when we face a threat that goes to the teleological core of living things: survival. As each of us is an individual, each of us is integral to the species, for if just one of us becomes a carrier of a germline modification, the species potentially diverges from its natural arc. We need to ask: "Who am '*I*' in all this?"

Jean Paul Sartre maintained that existence precedes essence. Three-hundred years before him, Spinoza had already said that what surrounds us determines the essence of human existence, which in the modern era, we must include technology. Our physiology integrates itself into the environment, from the composition of every inseparable molecule we breathe and eat allowing the memes of our culture to infiltrate our minds.[3] As do these philosophers, I believe that the human "*I*," cannot

[2]Philosophers P.P. Verbeek and D. Ihde are proponents of technological mediation, an idea that technology is not neutral as it shapes who we are and by human actions and practices we in turn shape technology, in a feedback loop. Rather than approaching technologies as material objects or as extensions of human beings, our inventions are mediators of human-world relations. Mediation theory is rooted in the 'post-phenomenological' approach in philosophy of technology, which was founded by Don Ihde.

[3]A meme acts as a unit for carrying cultural ideas, symbols, or practices, that can be transmitted from one mind to another through writing, speech, gestures, rituals, or other imitable phenomena with a mimicked theme. See, Dawkins, R. (1989). *The Selfish Gene* (2 edition). Oxford University Press.

be detached from the world in which we live. So, as we transition to an era of post humanism, who we are, that is the sum and substance of our essence will change, step-wise, that is dramatically, once genetic enhancements kick in affecting memory, thoughts, and emotions.

One's ontological being shifts into another category of things to the degree we become inventions of ourselves, i.e., by the technology we incorporate into our physiology, not technology that simply achieves some therapeutic or prosthetic outcome, but permanently changes one's psychological potential, and with some measure of hubris passes those potentialities on to its descendants.

We live pre-reflectively integrating perceptions, thoughts, and judgments about the world we live in [1].[4] The memory of all that passes through our mind informs imagination and predilections, which manifest in art, music, literature, science, mathematics, inventions, and institutions that enrich and regulate our lives. The mind conforms to the way these objects seem in a "mind to world" direction of fit, melding into our beliefs.[5]

We carry preternatural objects, that is those beyond what is natural, into our states-of-consciousness, which ultimately *represent our reality*—that bundle of incorporeal processes, intangible entities, and states-of-affairs, collectively subsumed under the heading "social reality." At this point we could only speculate how *Homo futuro* will process these same objects in ways that may frame new social constructs, new realities, new narratives. We will endeavor to examine the details of these issues, ones that in our every day life, we rarely stop to consider.

Note

1. Searle, J. (1998). *Mind, Language and Society*. Basic Books.

[4]Antecedents comprise a world that exists independent of our experiences, thoughts, perceptions through our senses; where statements are true or false depending on whether they correspond to facts in the world.

[5]According to German Idealism, and the manner in which I employ the words "thing" and "object" are as follows: "thing" denotes a being existing independent of humans, and "object" denotes phenomena or appearances humans perceive. Within this framework, "things" can never be known by humans insofar as they only experience objects, phenomena, or appearances structured by human cognition.

PART II

Brains and Biology

CHAPTER 5

Thin Edge of the Wedge

Mensa: My parents, called Millennials, were born in the century before me, and were oblivious to things like cyborg technology, genetic enhancements, and even climate change. By the time I entered the workforce, late 2020s, technology had already been used widely for therapies for better than fifteen or twenty years, like stimulators for pain management, depression, regulating bioprocesses like the heart, skeletal prosthetics, and artificial organs, hearts, kidneys. This was the first wave. By 2025, genetic engineering was well on its way. Augmentation technology to survive the meltdown came in the early part of the 2030s, and then of course the digital brain enhancements of the mid-30s to increase mental acuity.

Sarouk: Fascinating, go on.

Mensa: Back then, any two of us looked the same, as far as form. Inside, we were diverging due to germline enhancements and bio computer technologies, all with the same objective: survival and superior intellect.

Sarouk: Are you suggesting that these differences created, for the lack of a better descriptor, different species?

Mensa: We did not think of it that way. Although the germline alteration business was proliferating, most procreation still resulted in aboriginal offspring. But anatomies were gradually being modified by different equipment, applications, installed at younger and younger ages. Between germline alterations and hard technology, we shifted into different social spheres, what our predecessors referred to as demographics.

J. R. Carvalko Jr., *Conserving Humanity at the Dawn of Posthuman Technology*, https://doi.org/10.1007/978-3-030-26407-9_5

Sarouk: But, the main point was to survive, increase longevity, right?

Mensa: Yes, therapeutic technology was headed there. Yet, it was impossible to deal with simply the physical body without wanting to improve the psychology of individuals, their outlooks, and of course treat emotional disturbances, like depression. Psychotropic medications had been around for a hundred years, both drugs and in the'20s, deep brain electronic stimulators.

Sarouk: Were they installed just to alleviate illness?

Mensa: No, these feel good technologies were available to anyone, subject to cost. Many designed to alter sensations, what we felt. Affectively.

Sarouk: Affectively?

Mensa: Yes, some backed technologies that reduced emotional barriers, so we could see things for what they were, impartially. Introversion, introspection, forms of self-awareness, neuroticism and narcissism, were suppressed. Systems that favored cognitive over affective thinking, over time, produced personalities with lesser degrees of openness, extraversion, and great levels of agreeableness.

Sarouk: Were there attempts to deal with attitudes, say empathy or indifference.

Mensa: They came at this a little differently, developing technology to influence arousal, curiosity and motivational intensity.

Sarouk: What do you mean, exactly?

Mensa: We've forgotten that early on some stimulators were prompted by the need to survive environmental changes. As such they were integrated into the sympathetic nervous system to slow down the production of adrenaline, and thus channel the affects of 'arousal.' For instance, while increasing our sensitivity to temperature change, it dampened 'flight response,' or the impulse to act impulsively. Once installed, these types of processors gave the network time to size-up threats, before transmitting to our processors, how to respond.

Sarouk: But, how did that change people? Did it alter their perceptions, their dispositions?

Mensa: I can say, I'm not the person today that I was a hundred or even fifty years, ago. My temperament, mental focus, and even socialization changed each time a new operating system was downloaded.

Sarouk: How did it affect socialization?

Mensa: Close friends and relatives eventually drifted away into their own space, we became different in how we each saw the world. And, then we began to lose touch, not because we were incorporating different operating systems, but because somehow the technology was de-socializing, partitioning us into electronic spaces where we could no longer effectively communicate. Protocols were not standard, inputs produced different outcomes.

CHAPTER 6

The Platform

Evolutionary biologists trace our beginnings back thirty-six million years ago to *Dryopithecus*, a small primate, who walked on all fours and had a brain one-eighth the size of modern humans. Over millenniums, *Dryopithecus* gave rise to the gorilla, orangutan, and a combination chimpanzee-humanlike creatures, which split off about 20 million years ago into two lines: chimpanzee and hominid.

Several versions of hominids flowered in the ensuing time period, stretching over a period 15 million to 2 million years ago [1]. *Homo habilis* lived approximately between 2.1 and 1.5 million years ago. Experts differ on exact intervals, and even modify the names of closely related species based on origins. Paleoanthropologists now use the term *Homo ergaster* for the non-Asian forms of *Homo erectus*, and reserve the latter only for those fossils found in Asia and meet certain skeletal and dental requirements, which differ slightly from *Homo* ergaster. *Homo erectus* lived between 1.9 million and 1.4 million years ago, and was a direct ancestor of various proto-human species including *Homo heidelbergensis*, *Homo antecessor*, *Homo neanderthalensis*, and *Homo denisova*. *Homo erectus* was also a direct ancestor of the primate species, to which *Homo sapiens* belong. Evidence of our emergence, i.e., the anatomically modern human, has been determined as 340,000 years ago, but it was not until 200,000 years ago that a culture first appeared in Africa, before disseminating into the far reaches of the planet, roughly 100,000 years ago [2]. That being noted, *Homo sapiens*, continue to naturally evolve (see, Fig. 6.1).

J. R. Carvalko Jr., *Conserving Humanity at the Dawn of Posthuman Technology*, https://doi.org/10.1007/978-3-030-26407-9_6

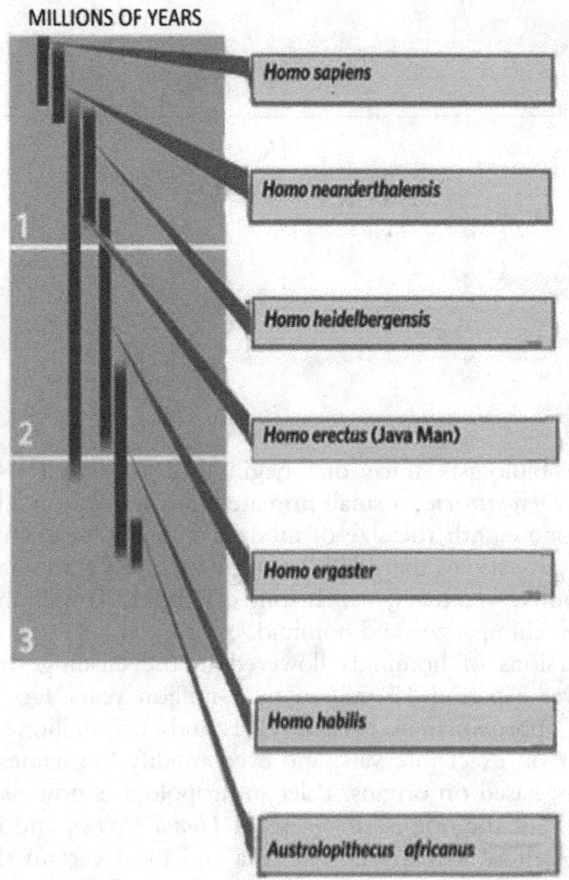

Fig. 6.1 Timeline of human evolution

Evolution, the change in the heritable characteristics of biological populations over successive generations, separates species and their respective genomes. To this end, DNA sequencing, and research into mitochondrial DNA (mtDNA), followed by Y-chromosome DNA (Y-DNA) has significantly advanced the understanding of human origins and heritability. Neo-Darwinism posits that adaptation is a combined result of Heredity + Variation + Selection. Variations within a species may manifest in terms of differences in cell structure, bone mass, trunk strength, wing length, maneuverability, and cunning. Other variations, such as an added toe caused by a chance environmental accident are not inherited by offspring.[1] Heritable traits that help organisms survive and reproduce become more common in a population over time. Said another way, traits which do not promote survival or cause early death will not be passed on in the species, because the organisms will die off. At the level of genes, evolution depends on changes in the frequency of variations in different genes within populations over time [3].

Original Modern Synthesis (OMS) is a theory in evolutionary biology that describes the merging of Mendelian genetics with Darwinian evolution to form a unifying theory. OMS holds that evolution is a process of adaptive change that occurs through the shifting frequencies of small-effect genetic variations at many different places in the genome, without the direct involvement of new mutations. Instead, OMS focuses on causal agents of evolution such as random genetic drift, gene flow, mutation pressure, and natural selection.

An alternative argument to OMS is that molecular evolution is dominated by selectively neutral evolution and genetic drift [4]. Recently, Huneman and Walsh pointed to other influences that allow new traits to arise and get passed on. Among other things, these factors include the Baldwin effect, where if an animal's environment suddenly changes, some respond by learning new behaviors or by ontogenetically adapting.[2] Huneman and Walsh further note that the foregoing all contribute to a developmental system, where evolution works at levels from genes and cells to organisms and cultures [5].

For Richard Dawkins, ethologist and evolutionary biologist, memes are to cultural progress, what genes are to biological progress [6]. It is a combination of (a) the ideas related to a developmental system and (b) Dawkins memes, to which I subscribe [7].[3] As we proceed, keep in

[1] In mammals, somatic cells give rise to internal organs: skin, bones, blood and connective tissue, while germ cells give rise to spermatozoa and ova, which fuse during fertilization to produce a cell called a zygote, which divides and differentiates into the cells of an embryo, and although somatic cells do not transmit their mutations to the next generation germ cells do.

[2] While developmental (i.e., ontogenetic) processes can influence subsequent evolutionary (e.g., phylogenetic) processes, individual organisms develop (ontogeny), while species themselves evolve (phylogeny).

[3] Memes are the cultural parallel to the biological genes, having the properties of objects, events, and processes, but instantiated in the brain, allowing individuals to coordinate their participation in respect to those entities.

mind that evolutionary biology is not governed by any master theory, and what theories do exist will likely undergo revision as the effect of genetic engineering plays a larger role in moving the genotype further and further from its historically relative stasis.

Speciation refers to the evolutionary processes through which new species arise naturally from existing organisms. Artificially induced speciation can occur by crossbreeding plants and in animal husbandry. Surprisingly, in 2006 researchers discovered that a gene has the potential to jump from one chromosome to another—contributing, in effect, to the establishment of a new species. Now we stand at the precipice of yet another form of speciation, one that is induced in the laboratory.

Artificially induced speciation has been practiced by humans for thousands of years through the selective breeding of plants and animals. In the new age of genetic manipulation, the terms "transgenic," "genetically engineered," and "genetically modified" and have taken on slightly different meanings, which have been largely driven by commercial and regulatory interests. For example, the acronym "GMO" refers to genetically engineered food products; and we refer to selectively bred animals as "genetically modified."

About 32,000 years ago, canines were probably the first animals to be genetically modified, resulting in the domestication of wolves. Over the course of thousands of years, the practice of genetic modification of canines has led to the hundreds of varieties of dogs. Similarly, selective breeding has taken place for all animals regarded as domesticated. In the same way, farmers have experimented with the cultivation of crops, such as wheat grasses bred for larger grain and overall hardier varieties.[4]

In a sense, the selective breeding occurring thousands of years ago began the long march toward what we now know as genetic science. In this journey, two important years in modern science stand out: 1865 and 1952. In 1865, Gregor Mendel published the results of his experiments to cross-pollinate varieties of peas. In 1952, Watson, Crick, and Franklin determined the structure of deoxyribonucleic acid, or DNA, as double helix, spiraled arrangement of molecules supported on a sugar-phosphate backbone with nucleotide base pairs.

[4]Examples are: corn from teosinte, a small eared wild grass with a few kernels, to what we serve with holiday dinner, a bright yellow ear of corn with hundreds of kernels.

The four DNA bases are: adenine (A), guanine (G), thymine (T), and cytosine (C). Scientists knew that in DNA the A and G nucleotides and the T and C nucleotides appeared in equal quantities. With this information, Watson and Crick hypothesized that the four nucleotide bases paired together in specific combinations, where A binds to T and G binds to C. For example, one strand of base materials CATC, results in the complementary strand of GTAG, as shown in Fig. 6.2.

In a simplified version of how DNA is involved in making proteins, again refer to Fig. 6.2. A process called transcription includes unwinding the DNA strand to create an mRNA strand, and which also substitutes the thymine T nucleic acid with uracil U nucleic acid. A molecular machine, called the ribosome, not illustrated here, translates the mRNA, taking three nucleotides in series, which specify an amino acid called a codon, to produce a protein.

To be clear, variations always exist between individuals and whole populations. Modern genetics has given rise to new understanding of variations between organisms. These are variations in the genotype, or the genetic constitution of an organism. Genotypes contain singular nucleotide polymorphisms, or SNPs, which pepper the average human genome. SNP reflect differences in the genome, which result from a wide variety of factors, such as environment or social conditions. Although there are upwards of 80 million SNPs throughout the world's population, any given genome may differ from the currently established referenced genome at 4–5 million sites. But, I'm calling attention to further nonnatural changes that will occur in the human genome, analogous to genetically modified animals.

As we transition through the technologies of genetic or cybernetic revision, both transhumans and modern humans will coincide for some uncertain period. A parallel in history occurred when *Homo sapiens* and *Homo neanderthalenis* coexisted for over 300,000 years, and although considered disparate species exhibited cultural and symbolic behavioral transmission, as well as interbreeding over a 5000 year period [8]. What might foreshadow a future society that will include more than one dominant species, especially given that territoriality and intolerance for racial, ethnic, or religious diversity, has plagued humankind reaching as far back as recorded history?

It's likely that *Homo futuro* would create a new social reality, a new pluralism. And, to that end, societal pressures would build as individuals of higher IQ to seek educational opportunities and careers appropriate

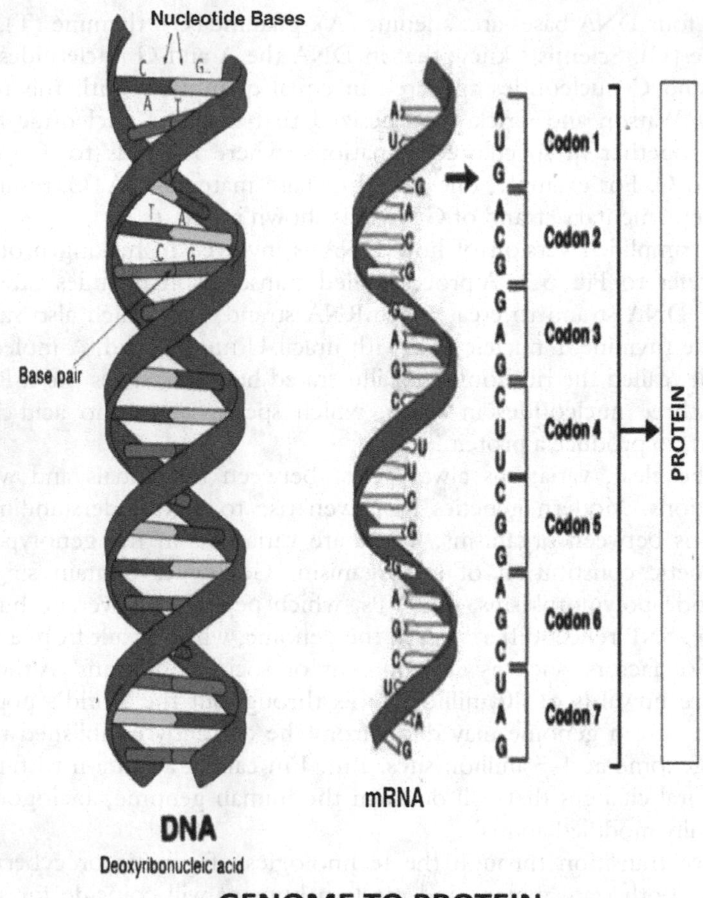

Fig. 6.2 Transformation from genome to protein

to their aptitudes. This could force a realignment of members of society who would not have the intelligence to meaningfully participate in the new world dominated by the newest, most advanced species.

Depending on how society assigns significance and status, over a wide spectrum of enhancements, *Homo futuro* may find itself in different modes of being, which will influence cultural modalities or status norms, such as where they fit into a particular commercial or social structure, such as the workplace, family unit, or change our sensibilities regarding tastes in art, music, or literature.[5] Joshua Spencer had it right when he wrote, "But, we don't think these things all exist in the same sense as cars and human beings. If they exist or have being at all, then they have different ways of being [9]."

Notes

1. Panati, C. (1984). *The Browser's Book of Beginnings.* Boston: Houghton Mifflin Company.
2. McDougall, I., Brown, F.H., and Fleagle, J.G. (2005). "Stratigraphic Placement and Age of Modern Humans from Kibish, Ethiopia." *Nature* 433, 733–736; Campbell and Tishkoff (2008); Conard, N.J. (2010). "Cultural Modernity: Consensus or Conundrum?" *Proceedings of the National Academy of Sciences of the United States of America* 107: 7621–7622.
3. Duret, L. (2008). "Neutral Theory: The Null Hypothesis of Molecular Evolution." *Nature Education* 1 (1): 218.
4. Kimura, M. (1968). "Evolutionary Rate at the Molecular Level." *Nature* 217 (5129): 624–626. https://doi.org/10.1038/217624a0.
5. Huneman, P., and Walsh, D. (2017). *Challenging the Modern Synthesis: Adaptation, Development, and Inheritance.* Oxford University Press. Introduction. ISBN 978-0-19-068145-6.
6. Dawkins, R. (1976). *The Selfish Gene.* Oxford University Press.
7. Jablonka, E., and Lamb, M. (2007, August). "Evolution in Four Dimensions." *The Behavioral and Brain Sciences* 30 (4): 353–392.
8. Neanderthals coexisted with anatomically modern humans for up to 5400 years long. Science News. (2014). http://www.sci-news.com/othersciences/anthropology/science-neanderthals-coexisted-humans-02111.html (Last visited 7/29/2018).
9. Spencer, J. (2012, November 12). "Ways of Being." *Philosophy Compass* 7 (12): 910–918. https://doi.org/10.1111/j.1747-9991.2012.00527.x.

[5] Status norms explain the widespread recognition of differences in social order and value, differentiating age, sex, physique, education, occupation, neighborhood, and as genetic engineering of physiology and psychology take hold, difference in those with extraordinary potentials to live longer, healthier lives, and exhibit superior intellect.

CHAPTER 7

The Making of Homo Futuro

What do the words "*Homo futuro*" and "transhuman" mean exactly? I treat them synonymously, but keep in mind that in some contexts, they signify a philosophical movement, rather than a physiological entity.[1] As a movement, "transhumanism" deals with science or technology related to the evolution of the human species, particularly vis-a-vis human-centric viewpoints [1]. The label "posthumanism" also considers similar ideas, but looks outward as to how these successors to the modern human will impact our planet and achieve harmony within the framework of ecologically complex networks [2]. Transhumanism and posthumanism each share the idea of technogenesis that asserts "that humans and technics have… coevolved" [3].[2]

In 1957, Julian Huxley, the famous evolutionary biologist, coined the word "transhumanism" [4]. Shortly afterward, New School futurology professor, FM-2030, identified people, who adopt technologies, lifestyles, and worldviews, and referred to them as "transhuman" [5, 6].

[1]There are at least seven definitions according to Francesca Ferrando, See, "Posthumanism, Transhumanism, Antihumanism, Metahumanism, and New Materialisms: Differences and Relations" (2013). http://www.existenz.us/volumes/Vol.8-2Ferrando.pdf.

[2]What transhumanism and posthumanism share is the notion of technogenesis, which as Katherine Hayles explains, the human species is defined by its co-evolution with various tools and technologies; or, that our subjectivity is always "contaminated by the outside—technics." See, https://ivc.lib.rochester.edu/how-we-think-digital-media-and-contemporary-technogenesis/.

J. R. Carvalko Jr., *Conserving Humanity at the Dawn of Posthuman Technology*, https://doi.org/10.1007/978-3-030-26407-9_7

Then, in the 1990s, a transhumanist movement introduced the idea of re-engineering humanity [7]. About this time, philosopher Max More, defined a transhuman as: "Someone in the transition stage from human to biologically, neurologically, and genetically posthuman. One who orients his/her thinking towards the future to prepare for coming changes and who seeks out and takes advantage of opportunities for self-advancement" [8]. The World Transhumanist Association (WTA), founded in 1998, expanded the idea this way:

> Posthumans could be completely synthetic artificial intelligences, or they could be enhanced uploads ... or they could be the result of making many smaller but cumulatively profound augmentations to a biological human. The latter alternative would probably require either the redesign of the human organism using advanced nanotechnology or its radical enhancement using some combination of technologies such as genetic engineering, psychopharmacology, anti-aging therapies, neural interfaces, advanced information management tools, memory enhancing drugs, wearable computers, and cognitive techniques. [9]

It's this more specific WTA meaning that's dealt with throughout this book, that is, a *Homo futuro* refers to a human entity, whose state at any given time reflects a human in transition, or transhuman. But as germline alterations become routine, our shared anatomical identity may also begin to move into other directions. Not only will germline alterations drive structural changes, but so will the hard technology of the computer and its lifeblood, software, serving to create hybrid beings that will have superior sensory access and intellect, e.g., visual, olfactory and auditory, and perhaps direct access to virtual realities, through powers of thought [10].[3]

Homo sapiens share physiological identity with one another: a torso from which two arms and two legs hang, topped off with a skull filled

[3]In May 2012, Cathy Hutchinson a 58-year-old woman paralyzed by a stroke and unable to move her arms or legs, sipped a cinnamon latte, assisted by a mind-controlled robotic arm, operating under the direction of an implanted sensor about the size of an 80 mg aspirin. The BrainGate device is a sensor implanted in the brain and an external decoder device, which bypasses nerve circuits and replaces them with wires outside Hutchinson's body. Also, the Biomechatronics group at the MIT Media Lab, develops bionic limbs using technology that integrates human physiology, such as nerve and muscle movement controlled by thought, to effectuate electromechanical processes that result in lifelike limb functionality.

with grey matter; 78 major organs; 206 bones; and 10 trillion nerve cells.[4] Transhumans and cyborgs may operate with nearly identical human structural elements, but underlying its externalities, its genome may be modified via synthetic neuronal bio-circuits or electronic circuits, to not only negotiate complex social patterns or process massive amounts of data and accelerate computational functionality, but better tolerate adverse climates.

In the near term the concentration of technology will be on the development of medical devices and genomic alteration to correct disabling or life threatening disease. One direction is in noninvasive gene therapy, illustrated in by the drug Zolgensma, which cures spinal muscular atrophy, an inherited disease, often fatal for patients before age two.[5] Another direction is medical device technologies, which will be down-sized, through advances in synthetic DNA, and molecular and nano-sized processors.[6] These will array within organs as nonorganic, internal adjuncts to our anatomy for use as: nano-prosthetics, nano-stimulators/suppressors, artificial organ processors, metabolic and cognitive enhancers, and diagnostic tools to ensure our physical and psychological well-being as we head toward a practically interminable lifetime.

The apparent embodiments of the transhuman may be revealed through positive externalities, of a better health and mental acuity, e.g., improvements to the immune system or cognitive functioning. Note that transhumans enhanced for reasons of excelling in activities, sports or warfare for instance, may be viewed, depending on one's ethical view, as negative externalities [11]. Taking into consideration these types of transformations, over time modifications may inwardly manifest as novel, aesthetic and intellectual ideas, enlightened emotional aptitudes and social engagements, which combined, revolutionize the essence or category of the experience as had been passed down since the dawn of *Homo sapiens* arrival.

[4]Physiology refers to a sub-discipline of biology, and how organisms, organ systems, organs, cells, and biomolecules carry out the chemical or physical functions that exist in a living system.

[5]See, A $2 Million Drug Is About to Hit the Market, Wall Street Journal, May 26, 2019.

[6]One of many examples are dendrimers, hyper branched, tree-like structures that contain a core moiety, and branching units, with an overall globular structure and encloses internal cavities. Its size is less than 10 nm. This technology is used for long circulatory, controlled delivery of bioactive material (e.g., liver targeted delivery) of bioactive particles to macrophages. See, https://www.omicsonline.org/open-access/nanotechnology-and-its-applications-in-medicine-2161-0444-1000247.php?aid=41535.

Homo futuro, for our purposes include genetically modified humans and cyborgs, distinguished by physiological function or enhanced abilities, due to the integration of artificial (mainly analog and digital electrical components), which may use some sort of feedback, as may be provided by a AI systems, locally or remotely. Apart from the cyborg is the humanoid robot (e.g., android, or biorobot), which is a fully artificial construct, having a body shape resembling a human.[7] There may be significant reasons to anticipate they may exceed the transhuman in their potential for a wide range of activities and even become the dominant creative force going forward. As MIT's Max Tegmark said,

> "We are made of carbon, that's why only we can be intelligent", and I think there is no scientific basis for this. We should be more humble and realize life is a beautiful thing that can occur with or without carbon and we're on a true trajectory towards perhaps building other kinds of life, but that's not something we should take lightly.[8]

Material composition of a humanoid robot does not seem relevant, but whether they acquire some form of pseudo-consciousness does, especially as it relates to creative potentialities. For example, would it know what it was creating independent from human intervention? On the other hand, *Homo futuro*, as a technological innovation, whether biological, electronic, nano-technological or a combination of these, will behave autonomously.[9]

Homo futuro, as agents of biological augmentation, will come about in-part through AI augmentation of brain function, something mainstream futurists agree on. Differences between commentators center on the actualization of superhuman levels of artificial general intelligence (AGI). A weight of authority believes that "full thinking machines" (FTMs) will inhabit the planet sometime during the next 55–60 years, although, estimates vary from 10 years to over a century. Futurists

[7] Ackerman, E. (2018) IHMC Teaches Atlas to Walk Like a Human. *IEEE Spectrum*, see, https://spectrum.ieee.org/automaton/robotics/humanoids/ihmc-teaches-atlas-to-walk-like-a-human.

[8] See, Tegmark, M. (2017) The Intelligence Explosion. https://www.52-insights.com/ai-max-tegmark-the-artificial-intelligence-explosion-science-interview-the-future/.

[9] Autonomous defined as: Denoting or performed by a device capable of operating without direct human control. See, https://en.oxforddictionaries.com/definition/autonomous.

Vernor Vinge and Ben Goertzel put it in the relatively short term, i.e., 2025–2030 timeframe, whereas Ray Kurzweil puts the arrival time in the vicinity of mid-century:

> 2029 is the consistent date I have predicted for when an AI will pass a valid Turing test and therefore achieve human levels of intelligence. I have set the date 2045 for the 'Singularity' which is when we will multiply our effective intelligence a billion fold by merging with the intelligence we have created.[10]

Max Tegmark reiterates his point that we must not focus on humans as we know them when we consider the matter of intelligence:

> … I respect people who think it's not going to happen in the next 100 years but there are also a lot of leading researchers from the top companies who are building it who think it will happen within decades, so I think it's a possibility. As I said, I think many people are stuck in this mindset that we need a special sauce to have intelligence and that somehow if you're not made of flesh and blood it's impossible to be smarter, because there is something magical about carbon atoms or the soul or something like that.[11]

But staying with the thought, whether AGI will be realized before the end of the century, Bostrom and Müller conducted a survey of 550 experts during the 2012/2013, timeframe and found [12]:

Median optimistic year (10% likelihood) → 2022
Median realistic year (50% likelihood) → 2040
Median pessimistic year (90% likelihood) → 2075

Essentially the 50% likelihood represents a flip of the coin that high-level machine intelligence develops between 2040 and 2050 and although this statistic is not especially impressive, the nine in ten chances, by 2075, is.

Most explanations of AGI suggest an adaptive intelligence, but nonetheless employing mechanistic problem solving in a problem space described symbolically. Computer solutions treat data structures, the

[10] See, https://futurism.com/kurzweil-claims-that-the-singularity-will-happen-by-2045/.

[11] See, https://www.52-insights.com/ai-max-tegmark-the-artificial-intelligence-explosion-science-interview-the-future/ (Last visited 8/1/2018).

underlying data, and interrelationships and apply objective problem solving algorithms to arrive, explicitly or inferentially at the intended solution, e.g., to recognize patterns and responses. In time, AGI may be indistinguishable from how we imagine a fully developed group of neurons will perform, ones born of a physio-chemical biological gene set. Discussion about the potential of AGI in all manner of human activity unsurprisingly fixes on the power to access information and to solve problems logically. And as we shall see, AI systems, having access to patterns and algorithms, already compose music, and author jokes and poetry. Nevertheless there are aesthetic limitations, outside the domain of logic that constrain AI's creative potential, a subject with which we will deal in later chapters.

The incorporation of technology per se, which modifies an anatomy, in and of itself does not produce a transhuman. Prosthetics have been used to aid individuals anatomically or medically challenged for some time. Electromechanical devices, such as an imbedded insulin pump or pacemaker, hardly affects the essence of an otherwise normal person. Although we cannot say the same about devices slated to change what we remember, or our outlooks and moods.

Essence changes when the materials of construction, or one's original form or purpose change to the extent that one no longer views reality as they did prior to the alteration. Being able to sense phenomena, beyond the context of normal human experience, could be an indication that one's essence has changed. One of the ways gene editing can affect essence is through the design of sensitivities that alter perception, awareness or memory.

A gene that extends human memory hasn't been discovered, although scientists have made progress in memory augmentation using computer chips that attach to a rat's hippocampus to recall forgotten memories.

The hippocampus has a limited capacity and deals primarily with short term memory. It is a cashew-shaped organ located in the temporal lobe, which records or memorizes information and experiences, before it's moved to the prefrontal cortex, where the event is interpreted (see, Figs. 7.1 and 7.2).[12] One success in enhancing memory, reported in 2011, dealt with a cortical neural prosthesis that was applied

[12] Credit for Fig. 7.1: NIH National Library of Medicine. https://ghr.nlm.nih.gov/condition/polymicrogyria; Credit for Fig. 7.2: Wikimedia Commons, File: Blausen 0102 Brain Motor & Sensory (flipped).png.

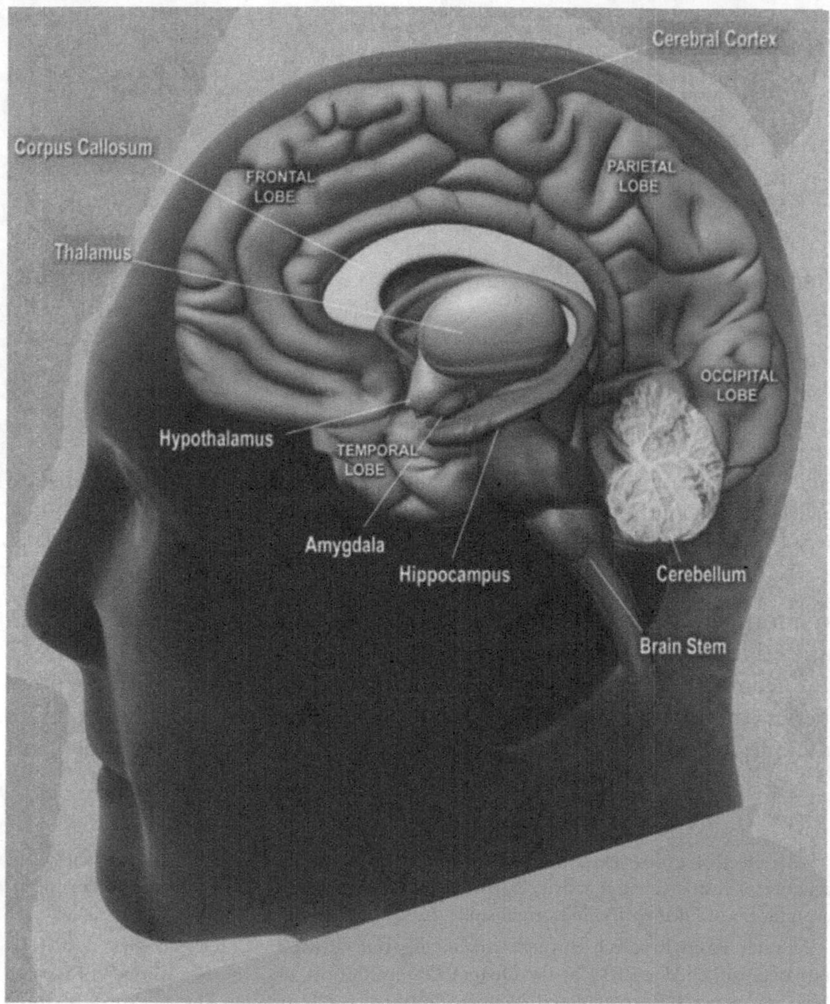

Fig. 7.1 Human brain left side, image courtesy of NIH National Library of Medicine. https://ghr.nlm.nih.gov/condition/polymicrogyria

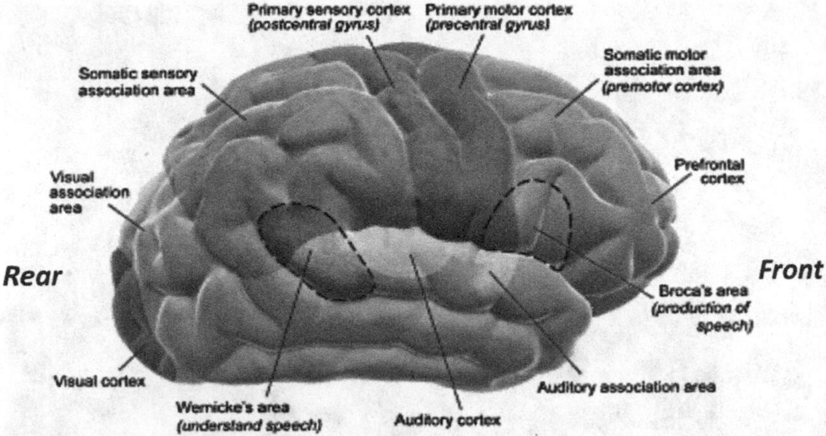

Fig. 7.2 Partial brain cortex functionality. Blausen.com staff (2014). "Medical Gallery of Blausen Medical 2014." *WikiJournal of Medicine* 1 (2). https://doi.org/10.15347/wjm/2014.010. ISSN 2002-4436

to information processing in two subregions of the hippocampus, which assisted long-term memory in rats [13].[13]

Recent advances have proven feasibility for memory storage prosthetics to improve recall, and which appear poised for human brain integration [14].[14] As with all prosthetics, these devices obviously can't

[13] An electronic prosthetic device was applied to a rodent hippocampus and shown to facilitate or restore mental activity impaired by events such as lack of attention to or misinterpretation of information in normal subjects.

[14] Under Brain Research through Advancing Innovative Neurotechnologies® (BRAIN) Initiative, launched in 2014 by the Obama Administration, the Defense Advanced Research Projects Agency's Systems-Based Neurotechnology for Emerging Therapies (SUBNETS) program is working to create implantable devices to help people with neuropsychiatric illnesses. Also, scientists implemented a prosthetic system that uses a person's own memory patterns to facilitate the brain's ability to encode and recall memory that showed short-term memory improved 35–37% over baseline measurements. See, Developing a hippocampal neural prosthetic to facilitate human memory encoding and recall, *Journal of Neural Engineering* (2018). https://doi.org/10.1088/1741-2552/aaaed7, iopscience.iop.org/article/10....088/1741-2552/aaaed7. Earlier progress was reported in Brain-implantable biomimetic electronics as the next era in neural prosthetics, Proceedings of the IEEE (Volume: 89, Issue: 7, July 2001).

be inherited. Nevertheless, this technology fits within a repertoire that meaningfully contributes to quicker thinking, future knowledge store, and creative expression. As products of human design, at some basic level, technologies that artificially supplement human experience reflect a designer's intention, which has ramifications for one's artistic freedom. The degree, quality, and content, to which future technology will enhance memories depends, not entirely on the individual in whom the prosthetic is applied, but the engineers, programmers, and scientists, who come from the ranks of academia, industry, and government.

Technology that does not render a permanent behavioral affect, but merely alters attitude, mood, or cognition in some transient way, is not what would be considered *Homo futuro*. *Homo futuro* more closely tracks what Max More described as: "…[moving] through the transhuman stage into post humanity, where our physical and intellectual capacities will exceed a human's as a human's capacities exceed an ape's."

Homo futuro, as used here, implies improved intellectual and creative potentials of a permanent nature, which has a specific direction and perhaps dimension, meaning access to domains typically inaccessible by modern humans. Access would be through one or more or a combination of pharmacologic, genomic, or cybernetic technology. Individuals, not enhanced, would consider their new fellow citizens as mentally quicker, more creative, emotionally stable, socially engaged, and if other health-related genes were in play, less afflicted by chronic or deadly disease.

Notes

1. Ferrando, F. (2013). "Posthumanism, Transhumanism, Antihumanism, Metahumanism, and New Materialisms: Differences and Relations." *Existenz* 8/2: 26–32. First posted 3-4-2014, rev. 3-13-2014.
2. Ibid.
3. Hayles, K. (2011). "Wrestling with Transhumanism." In: G.R. Hansell and W. Grassie, et al., eds., *H+: Transhumanism and Its Critics.* Philadelphia, PA: Metanexus Institute; also Hayles, K. (2012). *How We Think: Digital Media and Contemporary Technogenesis.* Chicago: University of Chicago Press.
4. Huxley, J. (1957). *New Bottles for New Wine: Essays.* London: Chatto & Windus, pp. 13–17.

5. Hughes, J. (2004). *Citizen Cyborg: Why Democratic Societies Must Respond to the Redesigned Human of the Future.* Westview Press. ISBN 0-8133-4198-1. OCLC 56632213.
6. FM-2030. (1989). *Are You a Transhuman?: Monitoring and Stimulating Your Personal Rate of Growth in a Rapidly Changing World.* New York, NY: Warner Books.
7. Dvorsky, G. (2009, June 22). "Transhumanism and the Intelligence Principle." *Sentient Developments, Science, Futurism, Life.* http://www.sentientdevelopments.com/2009/06/transhumanism-and-intelligence_22.html (Last visited 8/8/2012).
8. More, M. (1993, Winter/Spring). "Technological Self-Transformation, Expanding Personal Extropy." *Extropy* 10 (4): 2. http://www.maxmore.com/selftrns.htm (Last visited 7/26/2018).
9. Rifkin, J. (1983). *Algeny: A New Word—A New World.* Viking, writes about transhumanist goals for genetically modifying human embryos in order to create "designer babies". Also see, Kass, L.R. (1971). "Babies by Means of In Vitro Fertilization: Unethical Experiments on the Unborn?" *New England Journal of Medicine* 285: 1174–1175.
10. Nicolelis, M. (2012, September). "Mind in Motion." *Scientific American* 307 (3): 60–63.
11. VanNostrand, C. (2014). *Prosthetics: The Ethical Issues Surrounding Them.* http://www.pitt.edu/~cmv36/wa3.pdf (Last visited 5/26/2019).
12. Müller, V., et al. (2014, forthcoming). "Future Progress in Artificial Intelligence: A Survey of Expert Opinion." In: V.C. Müller, ed., *Fundamental Issues of Artificial Intelligence* (Synthese Library; Berlin: Springer).
13. Berger, T.W., et al. (2011). "A Cortical Neural Prosthesis for Restoring and Enhancing Memory." *Journal of Neural Engineering* 8 (4): 046017.
14. Amara, A. "Electronic Hippocampal System Turns Long-Term Memory on and Off, Enhances Cognition, KurzweilAI." Kurzweil Accelerating Intelligence. See, http://www.kurzweilai.net/artificial-hippocampal-system-restores-long-term-memory.

CHAPTER 8

Fashioning Life Forms

In the Origin of Species (1859), Darwin wrote: "Almost every part of every organic being is so beautifully related to its complex conditions of life that it seems as improbable that any part should have been suddenly produced perfect, as that a complex machine should have been invented by man in a perfect state" [1]. By the late 1960s, companies and universities had begun in earnest to change Darwin's perceived improbability by modifying the DNA of living organisms and setting the stage for genetically modified microbes, plants, and animals.

In 1972, biochemist Paul Berg of Stanford University created the first recombinant DNA molecules in the laboratory by splicing DNA from a bacterial virus into a monkey virus called SV40. The following year, Herbert Boyer, of the University of California at San Francisco, and Stanley Cohen, at Stanford University, used *E. coli* restriction enzymes to insert DNA into plasmids in the development of recombinant DNA (rDNA), to create novel genetic sequences not found in the genome of a living organism. A year earlier, a General Electric microbiologist named Ananda Chakrabarty genetically engineered a new bacterium by the addition of four DNA fragments to the bacteria *Pseudomonas putida* to create a living multiplasmid hydrocarbon-degrading *Pseudomonas*. The Chakrabarty organism had markedly different biological features from anything found in nature [2].[1]

[1] Claim one of the patent reads: A bacterium from the genus Pseudomonas containing therein at least two stable energy-generating plasmids, each of said plasmids providing a separate hydrocarbon degradative pathway.

J. R. Carvalko Jr., *Conserving Humanity at the Dawn of Posthuman Technology*, https://doi.org/10.1007/978-3-030-26407-9_8

General Electric filed a patent application claiming a new species capable of breaking down hydrocarbon components in crude oil [3]. The U.S. Patent and Trademark Office rejected the application, on the long established grounds that its impermissible to grant patents for "products of nature" [4].

Chief Justice Warren Burger, writing for a majority of the U.S. Supreme Court observed that the Commissioner of Patents also argued against patentability of genetic material on the basis of "grave risks that may be generated" [5]. The Court reviewed over a century of precedent on the long standing proscription against patenting natural phenomena, but that in this case, holding that no inherent limitation existed to patenting a live, ***human-made*** micro-organism under the patent statute, 35 U.S.C. §101 [6]. From Chakrabarty forward, the relevant distinction would be whether the biological entities for which patent protection was being sought were compositions of human ingenuity carried out as the Burger Court put it, by "the hand of man." Since the Chakrabarty decision, almost 40 years ago, over 286,518 patents have issued that cover the creation or manipulation of genetic material [7]. More about this in a later chapters that examine the history surrounding the patenting of genetically engineered products.

The motivations for genetically altering plants, insects, and animals are many ranging from improving the quality of a fruit or seafood, to manufacturing a medical product, such as insulin, or preventing a nonindigenous invasive insect from breeding and thus controlling the spread of disease. But, genetic engineering taken to another level, can be used to create transgenic nonhuman mammals. In some instances this results in hybrids and chimeras, particularly in the amalgamation of two species.[2] The obvious ethical barrier to human experimentation has led researches to use genetically modified animals. In 1975, two years after Berg's success adding DNA from a bacterial virus to that of a monkey virus, scientists created the first transgenic mammal by introducing foreign DNA

[2] In 1984, a chimeric geep was produced by combining embryos from a goat and a sheep, In 2003, Shanghai Second Medical University in China fused human skin cells and rabbit ova to create the first human chimeric embryos. In 2007, scientists at the University of Nevada School of Medicine created a sheep whose blood contained 15% human cells and 85% sheep cells.

into mouse embryos.[3] An early instance of patenting a genetically modified organism, the first of many thousands to follow, occurred in 1987, when the U.S. patent office ruled that a type of oyster, having been artificially treated to alter the number its chromosomes, was patentable [8].

In 1988, Harvard scientists Philip Leder and Timothy Stewart were granted a U.S. patent on a genetically altered mouse to increase its susceptibility to certain forms of cancer [9]. These inventions are transgenic nonhuman eukaryotic animals, whose germ and somatic cells contain an activated oncogene sequence introduced into the animal or its ancestor, at an embryonic stage, usually by injecting genes from another species into a fertilized animal egg and then implanting the egg into a host mother. The Leder-Stewart invention, referred to as the OncoMouse®, at the time raised moral and ethical issues about whether patents should be granted for genetically engineered animals [10]. Never before had a whole, living animal patent been allowed. That concern seems to have vanished as we approach the second decade of the twenty-first century.

Thanks to the explosion of genetic research, scientists can now breed "super" species with enhanced cognitive abilities compared to their non-transgenic counterparts. In 1999, researchers at Princeton created Doogie, a type of smart mice, which learned faster than ordinary mice. In particular, Doogie mice showed enhanced synaptic plasticity, memory retention, and flexibility in learning new patterns [11]. The rodent's brilliance is due to an enhancement to a brain receptor called N-Methyl-D-Aspartate (NMDA), found in the cortex, the center of consciousness, and in the hippocampus, the center of emotion, memory, and the autonomic nervous system.[4,5] The NMDA receptor, critical for learning and memory, also is implicated in a variety of brain disorders, which are typically treated through pharmacological products [12]. This engineered transgenic mouse over-expresses a component of the NMDA receptor,

[3]A hybrid is formed by combining the DNA from the same species, whereas a chimera is formed by combining the DNA of different species.

[4]NMDA receptors are glutamate-gated cation channels with high calcium permeability, critical for the development of the central nervous system, generation of rhythms for breathing and locomotion. They also underlie learning, memory, and neuroplasticity and as such enhancement of NMDA receptor activity has been proposed as a therapeutic strategy for conditions associated with altered cognitive function.

[5]In humans, the cerebral cortex takes up 77% of the available space in the brain and it's where the higher functions reside, such as the temporal lobe, which serves as the language center, or the frontal lobes that serve the executive functions.

which is called the NR2B subunit [13].[6] Research indicates that the discovery of NR2B represents a genetic factor for memory enhancement. This pioneering work led to the discovery of more than two dozen memory enhancement genes, many of which regulate the NR2B pathway [14].

Fast forward to 2009, when scientists working with the NR2B gene reported that they created transgenic rats, in which NR2B was overexpressed in the cortex and hippocampus, enhancing memory function in adulthood across multiple mammalian species. Overexpression of NR2B improves memory performance in a variety of experimental tests of memory—and different kinds of memory including spatial memory, recognition, and working memory in mice [15].

A sense of where these developments are taking us is stark. Natalie Kofler, molecular biologist, and scholar at the Yale Interdisciplinary Center for Bioethics, recently reviewed Philip Ball's 2019 work, How to Grow a Human: Adventures in Who We Are and How We Are Made. She writes, "[It] examines his 'mini-brain'; how scientific advances from genomics to assisted reproduction influence human identity… by introducing us to a collection of signaling neurons grown from his own reprogrammed skin cells…. His observation of 'part of himself' in a Petri dish begins a journey that spans centuries, giving context to a not-so distant future in which organs are grown to order and gene editing steers human evolution" [16].

Darwin believed that nature doesn't make leaps—but that's about to change as CRISPR/Cas9 biotechnology combines with digital technology to profoundly modify the epicenter of our anatomical core, the genome. What happens when the hubris of technology invents something so fundamental as whom we are? Will a surprisingly novel exterior reality emerge? Will the society in which these future beings emerge suddenly flip our culture on its virtual head? Or will the changes go

[6]In 1949, Donald Hebb proposed a learning rule that describes how the neuronal activities influence the connection between neurons, i.e., the synaptic plasticity, where we can assume that learning and memory are related to changes in synaptic strength among neurons simultaneously active. If the NMDA receptor, acts as a switch for memory formation, enhanced signal detection by NMDA receptors should enhance learning and memory. The Princeton lab results of the experiments on mice suggest that genetic enhancement of mental and cognitive attributes such as intelligence and memory in mammals is feasible.

unnoticed, like the daily vitamin, acting undetected in society, our psyche and physiology, on our social reality, all the while following the contours that point *Homo sapiens* to a preordained evolutionary graveyard?

Notes

1. Darwin, C. (1859). *The Origin of Species* (1st edition, Chapter 2).
2. U.S. Patent 4,259,444.
3. U.S. Patent 4,259,444, entitled “Microorganisms Having Multiple Compatible Degradative Energy-Generating Plasmids and Preparation Thereof”.
4. See, 35 U.S.C. §101.
5. Diamond v. Chakrabarty, 447 U.S. 303 (1980).
6. See, Parker v. Flook, 437 U.S. 584, 98 S.Ct. 2522, 57 L.Ed.2d 451 (1978); Gottschalk v. Benson, 409 U.S. 63, 67, 93 S.Ct. 253, 255, 34 L.Ed.2d 273 (1972); Funk Brothers Seed Co. v. Kalo Inoculant Co., 333 U.S. 127, 130, 68 S.Ct. 440, 441, 92 L.Ed. 588 (1948); O’Reilly v. Morse, 15 How. 62, 112–121, 14 L.Ed. 601 (1854); Le Roy v. Tatham, 14 How. 156, 175, 14 L.Ed. 367 (1853).
7. October 8, 2018, results of Search in U.S. Patent Collection db for: (((((ISD/19800101->20181001 AND SPEC/”gene sequence”) OR genetic) OR polynucleotide) OR DNA) OR plasmid): 286518 patents.
8. Ex parte Allen, 2 U.S.P.Q. 2d 1425 (Bd. Pat. App. & Interf. April 3, 1987).
9. U.S. Patent 4,736,866, entitled “Transgenic Non-human Mammals”.
10. The trademark ‘OncoMouse’ is owned by Dupont, Inc.
11. Tang, Y.P., Shimizu, E., Dube, G.R., Rampon, C., Kerchner, G.A., Zhuo, M., Liu, G., and Tsien, J.Z. (1999, September). “Genetic Enhancement of Learning and Memory in Mice.” *Nature* 401 (6748): 63–69. https://doi.org/10.1038/43432. PMID 10485705.
12. U.S. Patent 8,822,462 (2014), entitled “Subunit Selective NMDA Receptor Potentiators for the Treatment of Neurological Conditions” provides pharmaceutical compositions and methods of treating or preventing disorders associated with NMDA receptor activity, including schizophrenia, Parkinson’s disease, cognitive disorders, depression, neuropathic pain, stroke, traumatic brain injury, epilepsy, and related neurologic events or neurodegeneration.
13. Ibid.
14. Lehrer, J. (2009, October 14). “Neuroscience: Small, Furry … and Smart.” *Nature News* 461 (7266): 862–864. https://doi.org/10.1038/461862a.

15. Genetic Enhancement of Memory and Long-Term Potentiation but Not CA1 Long-Term Depression in NR2B Transgenic Rats, Wang, D., Cui, Z., Zeng, Q., Kuang, H., Wang, L.P., et al. (2009). "Genetic Enhancement of Memory and Long-Term Potentiation but Not CA1 Long-Term Depression in NR2B Transgenic Rats." *PLOS ONE* 4 (10): e7486. https://doi.org/10.1371/journal.pone.0007486.
16. Kofler, N. (2019). "Dished Brains and Designed Babies." *Nature* 569.

CHAPTER 9

Confluence of Technologies

Geneticists have dedicated years looking for solutions to DNA errors that cause diseases, such as schizophrenia, bipolar disorders, dementia, and Alzheimer's. These labors are undoubtedly paying off as they rewrite new chapters in the field of genetics and mental disease. But, it is not just the melding of the science behind genetics and mental disease that is making enormous strides, the same can be said for hundreds of syndromes, ailments, and sicknesses that have plagued humanity for thousands of years. At this moment in history virtually every anatomical organ, and what might afflict them, is being looked at through the lens of genetics. It's an enormous undertaking involving scientists, whom we might envision sitting behind microscopes, test tubes, and computers, day after day. But, this conventional view of the practice of science goes far beyond men and women in lab coats, and more fittingly should be viewed as thousands of institutions employing tens of thousands of people working with an enormous number of cutting edge technologies to, yes reduce suffering, but also to enhance humans in ways that will improve their physical and mental potential.[1]

A view from 50,000 feet shows just a few of the technologies behind this extraordinary time in history. First, big data analytics' platforms

[1] Buckminster Fuller created the "Knowledge Doubling Curve" that showed until 1900 human knowledge doubled every century. Today average human knowledge doubles every 13 months. According to IBM, the build out of the "internet of things" will lead to the doubling of knowledge every 12 hours.

J. R. Carvalko Jr., *Conserving Humanity at the Dawn of Posthuman Technology*, https://doi.org/10.1007/978-3-030-26407-9_9

have directly aided GWAS analysis of DNA sequencing, using powerful sequencers as well as the newer sophisticated DNA microarrays costing less than $100. Comparative studies of brain imaging, also are routine, which now reveal gene complexes that are associated with cognitive activity as well as neuronal wiring and firing. Genetics have informed effectively every medical sub specialty, e.g., psychiatry, oncology, neurology, pediatrics, and pathology. And, given the power brought about by the confluence of these technologies, we now add one of the most significant advances in modern science, gene editing.

New gene-editing methods with nucleotide-level precision work by introducing a double-stranded DNA break at a specific location, which the cell then repairs. The differences between the different DNA editing methods lie in the ease with which new gene sequences can be targeted or found, how the systems achieve the double helix break, and the declining cost of the procedure. Zinc fingers were the first genomic-editing method, and employed custom DNA endonucleases called zinc-finger nucleases, hence the acronym ZFN. ZFN has shown itself to be expensive and not always reliable. These problems led to a newer method, known TALENs, which targets a single nucleotide, allowing researchers to reach nearly any sequence they choose, but cheaper, faster, and more reliably than ZFN [1].

In 1987, Yoshizumi Ishino discovered CRISPR, a molecule that can be programmed to aim at a specific sequence in a genome.[2] This technology has the obvious application of removing a medically undesirable sequence, or repairing or altering genetic code that suggests the risk of inheritable diseases like albinism, cystic fibrosis, muscular dystrophy, hemophilia, Tay-Sacks disease, and Huntington's disease [2]. Related to Ishino's discovery, six years later, researchers in the Netherlands and Spain published articles about a cluster of interrupted direct repeats in the causative bacteria of *Mycobacterium tuberculosis* in the archaeal organisms of *Haloferax* and *Haloarcula* species. By 2000, researchers had identified interrupted repeats in 20 species of microbes.

In 2012, researchers Jennifer Doudna, at the University of California, Berkeley, and Emmanuelle Charpentier, at the Hannover Medical School in Germany, demonstrated how CRISPR could be used in conjunction with the enzyme Cas9. With CRISPR/Cas9, the two molecules work

[2]The acronym stands for Clustered Regularly Interspaced Short Palindromic Repeats.

to cut out and/or modify genes at specific locations. The Cas9 enzyme enables the removal, addition, or other control of gene expression at the genetic locus or loci of interest. Researchers are discovering other enzymes: Cpfl, CasX, and CasY, which promise to make CRISPR even more versatile and exacting.

In a related development, the U.S. Court of Appeals for the Federal Circuit, in September 2018, awarded the patent for CRISPR to the Broad Institute of MIT and Harvard in Cambridge, Massachusetts, upholding a previous decision by the U.S. Patent and Trademark Office. The decision was a defeat for University of California, Berkeley (UC), and molecular biologist Jennifer Doudna, who remains a principal investigator in the forefront of this technology. But, the ownership fight only relates to who owns the technology, rather than the more profound question about the categories of genetically engineered technology that will be allowed to be patented. This will be covered in a latter chapter.

With the arrival of Dr. He's genetically engineered twins as a result of CRISPR technology, the modern laboratory will no longer simply serve as a metaphor for medical progress. The reality is that gene-editing laboratories will be where humans will be transformed into *Homo futuro*. The profound significance of gene editing cannot be overstated, as it rivals medical revolutions brought about through inventions such as the antibiotic, organ transplant, and pacemaker. But, this development goes well beyond its medical impact, as it will alter future generations at their core, changing forever the anatomical and physiological embodiment of the human race, the one in existence for the past 350,000 years.

Notes

1. Gaj, T., Gersbach, C.A., and Barbas III, C.F. (2013, July). "ZFN, TALEN, and CRISPR/Cas-Based Methods for genome Engineering." *Trends Biotechnol*, 31: 397–405.
2. CRISPR/Cas9, system of genome editing was named the 2015 Breakthrough of the Year by *Science*.

CHAPTER 10

A River of Humanity

… I'm less than a firefly in a galaxy that runs up uncountable numbers of solar systems that spawn planets, ones that divide between sea and land, between lowland, highland, fashioning germ-lines that fertilize the vacancies of mothers' wombs, giving rise to clans whose unremarkable beginnings repeat in offspring born in a cosmos of galaxies, so that I, less than a firefly, flicker.

With advances in genomics, interest abounds in discovering who we are genetically and psychologically. To this end, we probe our most prominent features, intelligence and creativity. Our metaphorical flicker comes packaged in a genome comprising 3 billion nucleotide base pairs inherited from each parent, spread out over 46 chromosomes. The order of the base pairs is the genetic code we each carry. One person may differ from another by only approximately 30 million base pairs, or about 1% of the total genetic makeup. A gene, within the larger genome, represents a sequence of base pairs that encodes for a molecule. Genes can have both a structural and operational function within the human body. In humans, roughly 20,000 genes serve as the building blocks for organ systems that carry out particular physical and, in the brain, mental processes. Some mental processes encoded in our DNA are conscious, but others are far below the level of consciousness.

Our shared genome as humans—the 99% of our genetic composition that makes us similar to one another—gives us what we all have in common. One heart, two eyes, two feet, and the learning and mechanisms

J. R. Carvalko Jr., *Conserving Humanity at the Dawn of Posthuman Technology*, https://doi.org/10.1007/978-3-030-26407-9_10

for learning acquired as a species. The 1.0% of the genes we do not share makes us unique, providing diverse biological leverages, personality, habits, intelligence, or physical appearance.

Single nucleotide polymorphisms, also known as SNPs, represent an example of biological diversity in living organisms. A SNP represents a difference in a single DNA base pair, represented by one of the four molecules adenine (A), thymine (T), guanine (G), and cytosine (C). For example, in a certain stretch of DNA, an A nucleotide may appear at a specific base position in most individuals, but in a minority of individuals, the position may be occupied by a C nucleotide. This means that there are two potential versions of this gene due to a SNP at this specific position. In other words, the gene has two possible nucleotide variations at the given genetic locus—A or C. This SNP, or variation, creates what are said to be two different versions of the given gene.[1]

According to convention, approximately 1% of the population must contain a particular nucleotide variation before it's deemed a SNP. That being said, there are an estimated 84 million different SNPs in the world. SNPs occur normally throughout a person's DNA and a typical genome differs from a reference human genome at 4 to 5 million sites.[2]

If we think of DNA as a ladder possessing a number of rungs, we can visualize each nucleotide pairing as a numbered rung in the overall sequence. Assume that on the same rung location of the DNA ladder, say at rung number 48, 256, or 345, the allele contributed by your mother is "A" and the paired chromosome contributed by your father "T," the genotype would be defined as AT. At any particular allele, different people may have different genotypes, such as AA, AT, or TT. A change in the sequence of one or more DNA nucleotides can alter the structure of proteins or the regulation of its production, or result in mutations increasing or decreasing the chance of survival.

The genotype is the genetic composition of an organism, that is the set of genes in our DNA, which is responsible for a particular trait. The phenotype determines the physical appearance of an organism, that is its

[1] An allele refers to different versions of the same gene.

[2] A reference genome is a nucleic acid sequence database used as a representative example of a species' set of genes, but do not accurately represent the set of genes of any one person.

physical expression, or characteristics for that trait.[3] SNPs function for forming part of a gene represent the most common type of genetic variation. These variations are associated with traits like height, weight or an inheritable disease (schizophrenia, and bipolar disorder) or a psychological trait, like extraversion, neuroticism, and even empathy [1].[4] Some traits are discrete (albinism, cystic fibrosis, Huntington's disease), and relatively easy to identify, yet others exhibit quasi-continuous features, as is the case with height, skin color, and learning ability [2].

Recall that a heritability of 40% or 50% for any one trait means that percentage of the difference, between individuals, which can be attributed to differences between them in an inherited DNA sequence. Over the last few decades, scientists have determined that 80% of our height and 70% of our weight is determined by the genes we inherit. But it's not just outward physical appearances, so are 50% of the general intelligence trait, 40% of a personality trait, 20–60% of temperament, and the heritability for remembering faces, 60%, with spatial ability, 70%.

Using the latest high-density SNP microarrays, genotyping technology now analyzes genomic variations across populations in large numbers. These populations are made available through the GWAS referred to earlier, and are proving invaluable for uncovering the genetic root of differences associated with a variety of traits related to working memory capacity or one's risk of developing Alzheimer's disease. But, these root differences are complicated by the fact that many traits, which underlie most birth defects and common diseases (e.g., coronary heart disease, diabetes mellitus, and obesity), combine into what is referred to as multifactorial and polygenic inheritance. GWAS data and multifactorial polygenic scoring are used to assess this type of inheritance, traits which depend on multiple genes each having only small effects [3]. By looking at factors contributed by multiple individual genes through GWAS,

[3]Susan Lindquist, former Director of the Whitehead Institute make the observation that: "Most people think of genetics as being only about DNA. But genetics is about the inheritance of traits. While most traits are inherited through transmission of DNA, the traits we study are inherited through proteins. Thus, these proteins can be every bit as much a genetic element as DNA. After all, they are heritable entities that span generations." Available: http://www.hhmi.org/news/lindquist2.html. Accessed 14 May 2019.

[4]In 2015 a $50 million project was launched under the auspices of The National Institute of Mental Health to investigate the complete spectrum of genomic elements active within the human brain and to expand on their role in development, evolution, and neuropsychiatric disorders.

investigators surprisingly have seen a correlation between gene sets and academic achievement levels [4, 5].

Gene associations with normally varying intelligence differences in adults show weak effects for any one gene [6]. However, where an overall polygenic score could be determined and then used to genetically engineer the more significant markers even a small increase in IQ, within a select population, could prove socially significant. That having been said, let's put polygenic scoring and this point particularly into perspective. If an SNP association is found, say for IQ on the order of 0.01% (i.e., a 0.01 change in the average IQ of 100), it means that to account for heritability in the 50% range, theoretically requires at least 5000 SNPs be found to show the sought after trait [7]. Some traits are well below 0.01%, by orders of magnitude, and therefore require enormous numbers of associations be found, in existing databases, which we will delve into below.

Some traits are pleiotropic, such that one gene maps to multiple phenotypic traits. This adds complexity to finding the influence genes have on the etiology of diseases or psychological conditions. So, traits often do not depend on one gene. In fact, it's frequently the case that thousands of DNA differences and their regulatory and epigenomic elements contribute to psychological traits for intelligence and mental disorders.[5]

Today, scientists have scant evidence that shows how genes bear on creativity, and therefore how creativity bears on intelligence or vice versa, but some researchers believe the two are related in ways not wholly understood [8]. Presumably, genes that express greater intelligence increase a species chances of surviving. *Homo futuro* will doubtlessly not differ in these ways, but will have accessed a gene set based on behavioral objectives or enhancing intelligence. If and when a gene or set of genes that encode for superior intelligence is found, it will be within that 1.0% of the genes, somewhere among the 20,000 that make us individual, the ones that make us flicker.

[5] About 360 genes have been identified for schizophrenia.

Notes

1. Lo, M., et al. (2017). "Genome-Wide Analyses for Personality Traits Identify Six Genomic Loci and Show Correlations with Psychiatric Disorders." *Nature Genetics* 49: 152–156. https://doi.org/10.1038/ng.3766; Luciano, M. (2017). "116 Independent Genetic Variants Influence the Neuroticism Personality Trait in on Over 329,000 UK Biobank Individuals." *bioRxiv*. https://doi.org/10.1101/168906.
2. Griffiths, A., et al. (eds.). (2000). "Quantifying Heritability." In: *An Introduction to Genetic Analysis* (7th edition). New York: W. H. Freeman. ISBN 978-0-7167-3520-5.
3. Dudbridge, F. (2013, March 21). "Power and Predictive Accuracy of Polygenic Risk Scores." *PLOS Genetics* 9 (3): e1003348. https://doi.org/10.1371/journal.pgen.1003348. ISSN 1553-7404. PMC 3605113. PMID 23555274.
4. Lee, J.J., et al. (2018, August). "Gene Discovery and Polygenic Prediction from a Genome-Wide Association Study of Educational Attainment in 1.1 Million Individuals." *Nature Genetics* 50 (8).
5. Lee, J.J., et al. (2016, July). "Predicting Educational Achievement from DNA." *Molecular Psychiatry*. https://doi.org/10.1038/mp.2016.107. ISSN 1476-5578. PMID 27431296.
6. Davies, G., et al. (2011). "Genome-Wide Association Studies Establish That Human Intelligence Is Highly Heritable and Polygenic." *Molecular Psychiatry* 16 (10): 996–1005. https://doi.org/10.1038/mp.2011.85. PMC 3182557. PMID 21826061.
7. Plomin, R. (2018). *Blueprint: How DNA Makes Us Who We Are*. Cambridge, MA: MIT Press.
8. Jung, R.E., and Haier, R.J. (2008). "Brain Imaging Studies of Intelligence and Creativity: What Is the Picture for Education?" *Roeper Review* 30 (3): 171–180.

CHAPTER 11

Revising the Bell Curve

Man's yesterday may ne'er be like his morrow; Nought may endure buy mutability!—Frankenstein—Mary W. Shelly

Intelligence may be defined as the competency of a system to adapt its behavior to meet goals in a range of environments [1]. The American Psychological Association Task Force on Intelligence writes: "Individuals differ from one another in their ability to understand complex ideas, to adapt effectively to the environment, to learn from experience, to engage in various forms of reasoning, to overcome obstacles by taking thought" [2]. The task force also concluded that "[d]ifferences in genetic endowment contribute substantially to individual differences in (psychometric) intelligence, but the pathway by which genes produce their effects is still unknown."

Related to intelligence is an intelligence quotient, popularly known as IQ.[1] The intelligence quotient is a score derived from several standardized tests of abstract-reasoning to arithmetic, vocabulary, or general knowledge. IQ scores are widely acknowledged as a metric representing intelligence. The heritability of IQ, according to one American

[1]IQ scores have a normal distribution with a mean of 100 and a standard deviation of 15; 50% of individuals score above 100, and 50% score below 100. The average IQ is 85–114, and 68% fall into this range. Only 16% have scores of 85 or lower, and 14% of have scores of 115–129. The top 2% of individuals have IQ scores of 130–145; these scores are higher than 98% of the population. IQ scores over 145 indicate the top 0.1%.

J. R. Carvalko Jr., *Conserving Humanity at the Dawn of Posthuman Technology*, https://doi.org/10.1007/978-3-030-26407-9_11

Psychological Association report is 45% for children and rises to 75% for late adolescents and adults [3, 4]. Today, thanks to the work of Howard Gardner and others, we know that intelligence is a multifaceted concept, and perhaps one that cannot be measured by IQ alone.[2] However, the more widely considered ways of determined measures of intelligence use multiple, different IQ tests, and a measure referred to as "g-factor" or simply "g."[3] The g-factor includes subsets of mental abilities related to fluid intelligence (Gf), crystallized intelligence (Gc), visuospatial processing (Gv), working memory (Gwm), and quantitative reasoning (Gq).[4] Fluid intelligence reflects the ability to reason, to form concepts, and to solve problems using new information. Crystallized intelligence reflects acquired knowledge, skills, and quantitative reasoning, including specific domain activities, as exemplified by well-practiced techniques for solving math problems.

Geneticist Robert Plomin believes that "The most far-reaching implications for science, and perhaps for society, will come from identifying genes responsible for the heritability of the g-factor. Despite the formidable challenges of trying to find genes of small effect, I predict that most of the heritability of g will be accounted for eventually by specific genes, even if hundreds of genes are needed to do it" [5]. Several studies now confirm that intelligence shows a substantial heritability of 54% [6, 7]. It's also confirmed that intelligence, to the extent that it is polygenic, means that more than one gene has an influence on the magnitude of the correlation.

In 2017, a study published in Nature Genetics, performed a detailed meta-analysis of intelligence involving 78,308 individuals, looking for associations between intelligence and specific genes [8]. This study identified 336 associated SNPs in 18 genomic loci, of which 15 were new. It confirmed 40 novel genes for intelligence. Approximately half of the SNPs implicated 22 genes that expressed for intelligence. The identified

[2]For example, a person may have a high level of musical intelligence, but a low level of logical/mathematical intelligence. Gardner identified eight intelligences: linguistic, logical-mathematical, musical, spatial, bodily/kinesthetic, interpersonal, intrapersonal, and naturalistic.

[3]The g-factor can be thought of as a subset of IQ, which in turn is a subset of intelligence, which in turn is a subset of the entirety of mental abilities.

[4]I list five of ten broad categories of g, which can be subdivided into seventy more narrow abilities.

genes were predominantly found in brain tissue, with one gene involving neuron growth. Overall the SNPs accounted for a non-negligible 5.0% difference in intelligence.

The scientists also reported that the study correlated findings of the 336 SNPs to findings from educational achievement, a trait linked to genes, of approximately 200,000 individuals, who had undergone DNA testing. Overall, studies confirm that performance on school achievement tests has a heritability of 60% [9]. An earlier study in 2014, also based on a GWAS population, found an individual's educational attainment implicates four specific genes, KNCMA1, NRXN1, POU2F3, and SCRT. All of these genes are associated with a particular neurotransmitter pathway, a glutamate neurotransmitter, related to brain plasticity, learning and memory. The pathway itself involves an NMDA receptor, glutamate binding, and synaptic changes. It's clear that science is unwrapping the genetic construction of intelligence [10].

GWAS studies confirm twin adoption studies that have long been used to determine whether behavior is a function of genetics or environment [11].[5] Studies of what is and what is not inherited on a wide range of traits and behavioral patterns have tracked people from childhood into adulthood. These kinds of studies are carried on throughout the world [12]. The subjects are twins raised in familial arrangements, as for example: (1) non-related adoptees from different parents, yet raised by a single family, or (2) related adoptees raised by different families, as well as (3) other twins raised by different families [13, 14]. These studies have consistently revealed genetic links to the psychological indicators of intelligence, autism, and schizophrenia, as well as susceptible to obesity, stomach ulcers, or various cancers.

Monozygotic (MZ) and dizygotic (DZ) twins, i.e., identical and non-identical twins, especially have yielded insights into the nature or nurture influence of traits on personality, intelligence, and psychopathology.

[5]The Colorado Adoption Project is an ongoing genetically-informative longitudinal study of behavioral development that began in the mid-1970s. The experiment accumulated data from both parent-offspring and related- versus unrelated- sibling comparisons to estimate the importance of genetic and shared environmental influences for resemblance among family members.

MZ twins have identical genomes, which because of a separation at birth have been especially helpful in bifurcating genetic influences.[6]

Current research on the heritability of intelligence confirms the early twentieth century work of Sir Cyril Burt, who administered intelligence tests to pairs of MZ twins and found an intelligence correlation of 0.771, suggesting a noteworthy genetic component to intelligence. In 2000, researchers at the University of Minnesota tested 139 identical twin pairs that were reared apart and determined that heritability of the g-factor was in the range of 70%—close to the worldwide average of 75%. Meta-analyses of family, twin, and adoption studies over the years have further shown that inherited differences in DNA sequences account for roughly half of the variance in measures of intelligence [15].[7] Twin studies have also found that heritability is greater for fluid intelligence than for crystallized intelligence [16].

Brain imaging technology and DNA sequencing now combine to routinely analyze subjects to determine the link between various g-factors, brain structure and function. In MZ twin studies, where the subjects have high and average IQs, a strong correlation has emerged, once again validating the inheritability of intelligence, and this time revealing a connection to brain features, such as white matter [17].

Researchers have found that genes do influence cognitive ability, but a specific "intelligence gene" has yet to be discovered, so at this point one cannot jump to any conclusion as to how scientists could edit the genome to bring about enhanced cognitive or creative potentiality. However, in a 2012 study, scientists identified networks of genetic variants that influence brain fiber integrity and intellectual performance in

[6] Monozygotic, or identical twin, studies combined with imaging of brain structures and features, consistently confirm a genetic basis for particular forms of intelligence. Until recently, because identical twins develop from a single fertilized egg, they were thought to have the identical genome. But this has been revised. Any differences between twins may be due to, as previously understood, their environments, but a new discovery has shown that twins from the same embryo can differ in phenotype, which of course is genetic. Part of the answer is that specific changes identified are found where a gene exists in multiple copies, or a set of coding letters where DNA is missing. Researchers are not sure if these changes in identical twins occur at the embryonic level, as the twins age or both. See, O'Connormarch, A. (2008). "The Claim: Identical Twins Have Identical DNA," *New York Times.* https://www.nytimes.com/2008/03/11/health/11real.html.

[7] Variance is a measure of how dispersed or spread-out data are in a statistical distribution, calculated as square of the standard deviation from the mean.

472 young healthy adults [18]. In 2015, Plomin and Deary reported on the current state of what we know generally about the heritability of intelligence making the observation that: (i) heritability of intelligence increases from about 20% in infancy to perhaps 80% in later adulthood, and (ii) intelligence captures genetic effects on diverse cognitive and learning abilities, which correlate phenotypically about 0.30 on average, but correlate genetically about 0.60 or higher [19, 20]. They point out further that among other studies, at least two GWAS investigations have indicated that inherited traits are polygenetic, that is caused by many genes of small effect. The largest GWAS of intelligence differences, which included nearly 18,000 children, found no genome-wide significant associations, with the largest effect sizes accounting for 0.2% of the variance of intelligence scores, and in smaller sample size of 1500 children the effect accounted for 0.5% of the variance of intelligence scores.

Many believe that IQ and creativity are synonymous, but, bear in mind that study after study has shown low correlations between intelligence and creativity [21, 22, 23]. Others believe in the threshold theory, which holds that only the top 10% of the population, with an IQ above 120, can be creative. These views may be a consequence that intelligence and creativity are different constructs [24, 25]. And, to this point, one study shows that too much knowledge, as associated with Gc for instance, may well hamper the conception of new ideas [26].

One theory is that if one were to genetically modify a brain to dramatically increase intelligence, it may well impede creativity. K. H. Kim, Professor of Creativity and Innovation at the College of William & Mary, writes: "People with high IQs exhibit adequate inbox thinking, but they are often conformists who exhibit inadequate outbox thinking. Creativity is making something unique and useful, therefore outbox thinking is necessary for unique ideas (in addition to inbox thinking for useful ideas)."[8]

But, it also seems reasonable to assume that to perform gene editing to effectuate cognitive or creativity enrichment using implantable natural or synthetic genes, would first require the identification of the gene complex that may be implicated in a particular kind of natural cognitive or creative process. Studies have now identified thousands of genetic

[8] H.K. Kim, The Creativity—Post Does Science Say Smart People Are Creative? See, https://www.creativitypost.com/article/does_science_say_smartest_people_are_creative (Last visited 5/23/2019).

variants that contribute to autism and psychiatric risk, personality traits, and intelligence. However, as of yet, although many neuropsychiatric conditions are heritable, no single genetic variant has been identified as contributing to the inheritance risk.[9]

Notes

1. Legg, S., and Hutter, M. (2007). *A Collection of Definitions of Intelligence*, Vol. 157, pp. 17–24. ISBN 9781586037581.
2. Neisser, U., Boodoo, G., Bouchard, T.J., Boykin, A.W., Brody, N., Ceci, S.J., Halpern, D.F., Loehlin, J.C., Perloff, R., Sternberg, R.J., and Urbina, S. (1996). "Intelligence: Knowns and Unknowns." *American Psychologist* 51: 77–101.
3. Neisser, U., et al. (1996). "Intelligence: Knowns and Unknowns." *American Psychologist* 51 (2): 77–101. https://doi.org/10.1037/0003-066x.51.2.77.
4. Devlin, B., Daniels, M., and Roeder, K. (1997, July 31). "The Heritability of IQ." *Nature* 388: 468–471. http://dx.doi.org/10.1038/41319, https://doi.org/10.1038/41319.
5. Plomin, R. (1999). "Genetics and General Cognitive Ability." *Nature* 402: C25–C29.
6. Sniekers, S., et al. (2017). "Genome-Wide Association Meta-Analysis of 78,308 Individuals Identifies New Loci and Genes Influencing Human Intelligence." *Nature Genetics 49* (7): 1107–1112. https://doi.org/10.1038/ng.3869.
7. Davies, G., et al. (2011, October). Genome-Wide Association Studies Establish That Human Intelligence Is Highly Heritable and Polygenic." *Molecular Psychiatry* 16 (10): 996–1005. https://doi.org/10.1038/mp.2011.85.
8. Ibid., see note 6 above.
9. Plomin, R. (2018). *Blueprint: How DNA Makes Us Who We Are.* Cambridge, MA: MIT Press.
10. Davies, G., et al. (2015). "Genetic Contributions to Variation in General Cognitive Function: A Meta-Analysis of Genome-Wide Association Studies

[9]SHANK3, DIXDC1, DISC1, C4, are gene candidates for autism, schizophrenia, and GRIN1, can lessen its gene expression and impair learning in children. A variation in the gene COMT can increase dopamine levels by four-fold in the frontal cortex, which can increase concentration.

in the CHARGE Consortium (N = 53,949)." *Molecular Psychiatry* 20: 183–192.
Bouchard, T.J., Lykken, D.T., McGue, M., Segal, N.L., and Tellegen, A. (1990). "Sources of Human Psychological Differences: The Minnesota Study of Twins Reared Apart." *Science* 250, 223–228; Polderman, T.J.C., et al. (2015). "Meta-Analysis of the Heritability of Human Traits Based on Fifty Years of Twin Studies." *Nature Genetics* 47: 702–709; Savage, J.E., et al. (2017). "GWAS Meta-Analysis (N = 279,930) Identifies New Genes and Functional Links to Intelligence." *bioRxiv*. http://biorxiv.org/content/early/2017/09/06/184853.1.abstract.

11. Rhea, S.A., Bricker, J.B., Wadsworth, S.J., and Corley, R.P. (2012). "The Colorado Adoption Project." *Twin Research and Human Genetics: The Official Journal of the International Society for Twin Studies* 16 (1): 358–365.
12. Sahu, M., and Prasuna, J.G. (2016). "Twin Studies: A Unique Epidemiological Tool." *Indian Journal of Community Medicine: Official Publication of Indian Association of Preventive & Social Medicine* 41 (3): 177–182. https://doi.org/10.4103/0970-0218.183593.
13. Baker, L.A., et al. (2013). "The Southern California Twin Register at the University of Southern California: III." *Twin Research and Human Genetics: The Official Journal of the International Society for Twin Studies* 16 (1): 336–343.
14. Petrill, S., et al. (2003). *Nature, Nurture, and the Transisiton to Early Adolescence*. Oxford University Press.
15. Knopik, V.S., et al. (2017). *Behavioral Genetics* (7th edition). Worth.
16. The Heritability of Intelligence: Not What You Think. https://blogs.scientificamerican.com/beautiful-minds/the-heritability-of-intelligence-not-what-you-think/ (Last visited 8/7/2018).
17. In a 2012 study, scientists identified networks of genetic variants that influence brain fiber integrity and intellectual performance in 472 young healthy adults. See, Chiang, et al. (2012, June 20). "Relating Gene Networks to Brain Integrity and Cognition." *Journal of Neuroscience* 32 (25): 8732–8745.
18. Ibid.
19. Plomin, R., et al. (2014). "Genetics and Intelligence Differences: Five Special Findings." *Molecular Psychiatry* 20 (1): 98–108. https://doi.org/10.1038/mp.2014.105.
20. Haworth, C.M., et al. (2010). "The Heritability of General Cognitive Ability Increases Linearly from Childhood to Young Adulthood." *Molecular Psychiatry* 15: 1112–1120.
21. Jauk, E., and Benedek, M., et al. (2013). "The Relationship Between Intelligence and Creativity: New Support for the Threshold Hypothesis

by Means of Empirical Breakpoint Detection." *Intelligence* 41 (4): 212–221.

22. Kaufman, J.C., et al. (2011). "Intelligence and Creativity." In: R.J. Sternberg, and S.B. Kaufman, eds., *The Cambridge Handbook of Intelligence*. Cambridge: Cambridge University Press.
23. Sternberg, R.J., and O'Hara, L.A. (1999). "Creativity and Intelligence." In: R.J. Sternberg, ed., *Handbook of Creativity*. Cambridge: Cambridge University Press, pp. 251–272.
24. Guilford, J.P. (1967). *The Nature of Human Intelligence*. New York: McGraw-Hill.
25. Kim, K.H. (2005). "Can only Intelligent People Be Creative?" *Journal of Secondary Gifted Education* 16: 57–66.
26. Batey, M., and Furnham A. (2006). "Creativity, Intelligence, and Personality: A Critical Review of the Scattered Literature." *Genetic, Social, and General Psychology Monographs* 132: 355–429.

CHAPTER 12

Neurobiological Correlates

Direct your eye right inward, and you'll find. A thousand regions in your mind.
—Walden—Henry David Thoreau

Creative impulses stream into sensory pathways. Some perceived and responded to, while others go unheeded only to re-emerge years later in something said or felt, a spontaneity that may create our world afresh. In the vastness of this inner brain space called the cortex, most vibes fall still, buried, limiting what the wisest and most among us can fathom. The greater amount of what does not get lost must navigate the depths of uncharted gray matter as it carves out new caverns of knowledge, predilections, and expectations. But, every so often this biological bundle of our creative genius sees a momentary flash that adds to what some call the soul. Can we build such a machine that operates within the folds of gray matter, and if we can, how does that change who we are, what is essentially—us?

On the surface, "thinking creatively" expresses itself in the products of our imagination, but we know that its origin comes from the depths of an enormously complex neurobiological substrate practically impossible to plumb [1].[1] At a superficial level we do know that hard work, culture,

[1]According to Peter Carruthers, "there is a very great deal of evidence from across many different levels in biology to the effect that complex functional systems are built up out of assemblies of sub-components, and that this sort of modular organization is in fact a pre-requisite of evolvability. This is true for the operations of genes, of cells, of cellular assemblies, of whole organs, of organ assemblies (like the respiratory system), ... it is

J. R. Carvalko Jr., *Conserving Humanity at the Dawn of Posthuman Technology*, https://doi.org/10.1007/978-3-030-26407-9_12

and environment play a role in the level, quality and amount of creative output. Along these lines, professional levels of creativity, such as exhibited by artists, writers, and musicians, who earn a living at practicing their art forms, take at least 10 years working in the domain of relevant knowledge [2]. However, genetic factors do contribute to defining brain structure and intelligence.

In 2013, researchers in Finland uncovered evidence for the molecular genetic background of music related phenotypes such as composing, improvising, and arranging music. Creativity in music was found to co-segregate with a duplication covering glucose mutarotase gene (GALM), at chromosome band 2p22, which influences serotonin release in the brain, and is associated with membrane trafficking of the serotonin transporter. Genes related to serotonergic systems have been shown to be associated, not only with psychiatric disorders, but also with creativity and music perception [3].

Genetic markers, specifically the alpha-synuclein gene (SNCA), were located in the region for musical aptitude, and found to be over-expressed when listening and performing music. The GATA-binding protein 2 gene (GATA2) was considered best associated with a region of musical aptitude as it regulates SNCA in dopaminergic neurons, thus linking DNA-based and RNA-based studies of music-related traits. Regions showing positive selection for musical aptitude also contained genes affecting auditory perception, cognitive performance, memory, human language development, and song perception [4].[2]

Biological mechanisms that account for variations in individual g-factors range from brain size and density to the synchrony of neural activity and overall connectivity within the cortex. Numerous studies of acknowledged geniuses, over the course of over more than two hundred years, has disclosed particular irregularities, but which are difficult or impossible to trace back to a particular kind of genius. Einstein's brain revealed

appropriate to think of cognitive systems as biological ones, which have been subject to natural selection."

[2]For those who wish to go deeper in the subject, Richard Haier's Neuroscience of Intelligence is an excellent history and topical summary on genetics, DNA, and imaging of brain connectivity and function. Haier, R. Neuroscience of Intelligence, (2016), Cambridge University Press (freely downloadable at: https://archive.org/details/TheNeuroscienceOfIntelligence/page/n1

the absence of the parietal operculum, and part of his Sylvian sulcus. He had an expanded inferior parietal lobule and larger than normal corpus callosum. Lenin had an unusually large prefrontal cortex, and an expanded temporal-lobe. In terms of size, Lord Byron, and Oliver Cromwell had enormous brains weighing upwards of 4.9 lbs. (the normal brain is 2.8 to 3.0 lbs.) [5, 6].

The mind constitutes a process influenced by physiology, society, and culture. The physiological part falls within the department of one's brain, which has inherited constituent materials, which in turn cause an associated functionality and behavior. The architectural development of the brain and its neurological function also depend on spatiotemporal regulation of the transcriptome, which varies considerably by brain region and cell type.[3] Assuming *Homo futuro* results in a superior being having enhanced creative abilities, it is probable that it would be a consequence of these inherited materials, which in some way alter overall brain volume, gray and white matter volume, white matter integrity, cortical thickness, cortical gyrification or convolution (observed as folds throughout the cortex), brain connectivity, dopamine and serotonin signaling mechanisms [7, 8, 9, 10, 11, 12, 13, 14].[4,5]

The brain contains the circuits through which sensations transform into perceptions, memories, and actions. It's been hypothesized that the steady increase of *Homo sapiens*' productive output as a species is either due to an increase in the number of cortical connections or other changes that have occurred within these connections [15]. Following along these lines, scientists have embarked on a project to map neuronal circuits or connectomes to better understand neural interactions. To date, researchers have only constructed the full atlas for the roundworm *Caenorhabditis elegans* and partial connectomes of a mouse retina and

[3]The transcriptome is the set of all RNA molecules in one cell or a population of cells and unlike the genome, largely fixed for a given cell line (excluding mutations). The transcriptome also can vary with external environmental conditions.

[4]MRI scans of 1583 children showed that SNP plasticity relates to intelligence and cortical thickness.

[5]As to cortical convolutions, statistically significant positive correlations intelligence scores were positively associated with the degree of folding within the left temporo-occipital lobe, in the outermost section of the posterior cingulate gyrus (retrosplenial areas). See, Luders, E., et al. (2007). Mapping the relationship between cortical convolution and intelligence: effects of gender. Cerebral cortex (New York, NY: 1991), 18 (9), 2019–2026. https://doi.org/10.1093/cercor/bhm227.

mouse primary visual cortex. When the human brain has been completely mapped it will involve a hundred billion neurons and over a quadrillion connections.[6]

The effects of the millions of interconnected tributaries represented by the receptor-nerve synapse combination, ultimately flow into that vast ocean of our brain where we form our interpretations, take actions, and commit our experiences to memory.[7] The neuron is a cell which comes in many varieties of complex shapes and electro-chemical properties. Such properties are determined largely by ion channels and interconnections to other neurons via synapses, which are often positioned on dendrites, somata, and axons.

Neurotransmitters are bioactive chemical agents that are released at the axon end of neurons. When a neurotransmitter fires, it transmits an electrical impulse down a nerve fiber, generally terminating at a synapse. When the voltage exceeds a certain threshold the synapse fires, further propagating impulses into one or many nerve fibers, which in turn, terminates into one or more neurons or tissue. A neurotransmitter can either "excite" neighboring neurons and thus increase activity, or "inhibit" neighboring neurons, thus suppressing activity. Excitatory neurotransmitters include glutamate, acetylcholine, norepinephrine, and nitric oxide. Inhibitory neurotransmitters on the other hand include acetylcholine, norepinephrine, glycine, GABA, dopamine, and serotonin.

Dopamine is a neurotransmitter in the extrapyramidal system of the brain that regulates movement, acting as an inhibitory transmitter controlling arousal levels, e.g., vital for physical drive often associated with fixing attention, stimulating learning, and providing for feelings of energy, arousal, and in some cases timelessness.

[6]Genetic brain maps may reveal how genes determine individual differences, and may shed light on the heritability of cognitive diseases, as well as other brain function, such as linguistic skills. The National Institute of Mental Health has funded a multidisciplinary team of investigators across 15 research institutes, referred to as the PsychENCODE Consortium, to create an integrative atlas of the human brain by analyzing transcriptomic, epigenomic, and genomic data of postmortem adult and developing brains at both the tissue and single-cells, across 2000 samples.

[7]The synapse transfers electric activity (information) from one cell to another. The transfer takes place from nerve to nerve or nerve to muscle. The region between the pre-and postsynaptic membrane is 30–50 nanometers.

Serotonin is a neurotransmitter that controls mood and acts to inhibit pain pathways in the spinal cord. High levels of serotonin, or sensitivity to this neurotransmitter, are associated with serenity and optimism. Overall, serotonin plays a role in modulating cognition, learning, memory, anxiety, sleep, appetite, and sexuality.

As scientists learn more about how the brain works, they are making inroads regarding the role certain genes have on cognition. In 2018, a group of scientists reported a study where 1475 adolescents, part of the IMaging and GENetics (IMAGEN) sample, showed that general IQ (gIQ) is associated with (1) polygenic scores for intelligence, (2) epigenetic modification of DRD2 gene, (3) gray matter density in striatum, and (4) functional striatal activation elicited by temporarily surprising reward-predicting cues [16]. They concluded that neurobiological correlates show both the malleability of gIQ and equally the "importance of genetic variance, epigenetic modification of DRD2 receptor gene, as well as functional striatal activation, known to influence dopamine neurotransmission."[8]

At this stage of neurological science, the functional significance of how all this is organized remains unclear. Consequently, we do not know if or how genetic augmentation will affect cognition, and hence creativity in a transhuman. What is clear is that scientists, backed by private and public funding, around the world in large numbers are working to advance the simulation of the mammalian brain. In 2018, Hewlett Packard Enterprise announced that it had been chosen by EPFL Blue Brain Project to build a next-generation supercomputer for modeling and simulation of the mammalian brain.[9]

We will reserve discussion about whether a computer that simulates the brain's connectivity could exhibit the functionality of a human brain. Let me simply suggest at this point that computers simulate many physical systems, for example, atomic explosions, but no one would imagine that a computer actually experiences the atomic forces of a detonation. Similarly, what is it about the brain, in addition to its circuitry, that prevents it from being represented by a computer? Perhaps this extra something is consciousness, which like the impossibility of actualizing actual

[8] Low dopamine also affects mental stasis.

[9] The Blue Brain, research initiative, aims to create a digital reconstruction of the brain by reverse-engineering mammalian brain circuitry. It was founded in May, 2005, by the Brain and Mind Institute of the École Polytechnique Fédérale de Lausanne (EPFL).

atomic forces in a computer, is also impossible to actualize. Whether the lack of consciousness also prevents a computer from actualizing forms of creativity is a question that will be addressed in later chapters. For now, we turn our attention to parts of the brain and the extent to which compared to other primates, hefty human brains, and their neocortices explain some of the extraordinary cognitive abilities observed in *Homo sapiens* [17].

Notes

1. Carruthers, P. (2006). *The Architecture of the Mind: Massive Modularity and the Flexibility of Thought*. Oxford University Press.
2. Simonton, D.K. (1997). "Creative Productivity: A Predictive and Explanatory Model of Career Trajectories and Landmarks." *Psychological Review* 104: 66–89.
3. Ukkola-Vuoti, L., Kanduri, C., Oikkonen, J., Buck, G., Blancher, C., Raijas, P., Karma, K., Lähdesmäki, H., and Järvelä, I. (2013). "Genome-Wide Copy Number Variation Analysis in Extended Families and Unrelated Individuals Characterized for Musical Aptitude and Creativity in Music." *PLoS One* 8 (2): e56356. https://doi.org/10.1371/journal.pone.0056356.
4. Järvelä, I. (2018). "Genomics Studies on Musical Aptitude, Music Perception, and Practice." Annals of the New York Academy of Sciences. https://doi.org/10.1111/nyas.13620.
5. DeFelipe, J. (2011, May 16). "The Evolution of the Brain, the Human Nature of Cortical Circuits, and Intellectual Creativity." *Frontiers in Neuroanatomy*. https://doi.org/10.3389/fnana.2011.00029.
6. Goldberg, E. (2018). *Creativity: The Human Brain in the Age of Innovation*. Oxford University Press.
7. Haier, R.J., Siegel, B.V., Nuechterlein, K.H., Hazlett, E., Wu, J.C., Paek, J., and Buchsbaum, M.S. (1988). "Cortical Glucose Metabolic Rate Correlates of Abstract Reasoning and Attention Studied with Positron Emission Tomography." *Intelligence* 12 (2): 199–217. https://doi.org/10.1016/0160-2896(88)90016-5.
8. Desrivières, S., et al. (2015). "Single Nucleotide Polymorphism in the Neuroplastin Locus Associates with Cortical Thickness and Intellectual Ability in Adolescents." *Molecular Psychiatry* 20: 263–274. https://doi.org/10.1038/mp.2013.197; published online 11 February 2014.
9. Witelson, S.F., Beresh, H., Kigar, D.L. (2006). "Intelligence and Brain Size in 100 Postmortem Brains: Sex, Lateralization and Age Factors."

Brain 129 (2): 386–398. https://doi.org/10.1093/brain/awh696. PMID 16339797.

10. Narr, K.L., Woods, R.P., Thompson, P.M., Szeszko, P., Robinson, D., Dimtcheva, T., and Bilder, R.M. (2007). "Relationships Between IQ and Regional Cortical Gray Matter Thickness in Healthy Adults." *Cerebral Cortex* 17 (9): 2163–2171. https://doi.org/10.1093/cercor/bhl125. PMID 17118969.
11. Gur, R.C., Turetsky, B.I., Matsui, M., Yan, M., Bilker, W., Hughett, P., and Gur, R.E. (1999). "Sex Differences in Brain Gray and White Matter in Healthy Young Adults: Correlations with Cognitive Performance." *Journal of Neuroscience* 19 (10): 4065–4072.
12. Penke, L., Maniega, S.M., Bastin, M.E., Hernandez, M.V., Murray, C., Royle, N.A., and Deary, I.J. (2012). "Brain White Matter Tract Integrity as a Neural Foundation for General Intelligence." *Molecular Psychiatry* 17 (10): 1026–1030. https://doi.org/10.1038/mp.2012.66. PMID 22614288.
13. Kaminski, et al. (2018). "Epigenetic Variance in Dopamine D2 Receptor: A Marker of IQ Malleability?" *Translational Psychiatry* 8: 169. https://doi.org/10.1038/s41398-018-0222-7.
14. Santarnecchi, E., and Rossi, S. (2016). "Advances in the Neuroscience of Intelligence: From Brain Connectivity to Brain Perturbation." *The Spanish Journal of Psychology* 19: E94.
15. Ibid., see note 5 above.
16. Ibid., see note 13 above.
17. Pietschnig, J., et al. (2015). "Meta-Analysis of Associations Between Human Brain Volume and Intelligence Differences: How Strong Are They and What Do They Mean?" *Neuroscience & Biobehavioral Reviews* 57: 411–432. https://doi.org/10.1016/j.neubiorev. PMID 26449760.

CHAPTER 13

The Form of the World

Creativity depends on biology, biography, environment, and circumstance, the same determinants upon which life depends. According to Nobel prizewinning biologist, E.O. Wilson, "Creativity is the defining trait of our species; and its ultimate goal, self-understanding: What we are, how we came to be, and what destiny, if any, will determine our future historical trajectory" [1]. It's expressed improvisationally—no scripts, scores, or stage directions. It cannot be programmed. This was true in the Stone Age and it remains true today. It requires an inner eye. That having been said, what is this inner eye, that something that helps us see things the way we do?

What's key in understanding our inner eye is as Isabel Myers wrote in Gifts Differing: "We often see different perspectives because each of us looks at the territory from different orientations." These differing perspectives foster creativity often through the cross-fertilization of ideas, outlooks and even a kind of spiritual awareness, which is impossible to explain through some formulaic recipe. Over 100 years ago, psychologist Carl Jung suggested that the form of the world into which we are born is innate in each of us as a virtual image. He labeled this the collective unconscious, and theorized that we are different in fundamental ways, in spite of this image, because our individual consciousness prefers to operate in unique ways. Individuals may think, feel, sense and intuit the same event differently. Jung observed that people either tend to be outgoing or reserved. He labeled these two types extraverts or introverts. The extrovert is oriented toward the external, objective, and is drawn to people and things. The introvert is oriented toward the internal and

J. R. Carvalko Jr., *Conserving Humanity at the Dawn of Posthuman Technology*, https://doi.org/10.1007/978-3-030-26407-9_13

subjective, being drawn to ideas and abstractions. Each of us prefers looking in one or the other direction, and thus our viewpoint.

Jung also had ideas about how we use our cognitive facilities to evaluate that which we perceive through our senses. The evaluative function for cognition or "thinking function" and its polar opposite the affective or "feeling function," according to Jung, determines how we assess what has been perceived. Just as a preference exists for the passage of time or how we take in facts, a predilection exists for how we interpret the meaning of what's been perceived. The "thinker" evaluates based on an impersonal analysis of facts and logical processes, perhaps influenced by how something is structured. On the other hand, affective thinkers, or "feelers," decide by valuing the experiences that life has wrought. They amass a tacit understanding about the essential qualities of events, objects, and things. Their sense, reason, and judgment come from what they learned from life. Jung did not have the benefit of modern science against which to test his ideas, but they do provide a frame of reference as borne out by more modern views of how personality and temperament affect our creative impulses.

Most psychologists now agree that human behavior is related to both genetics and socialization. Nonetheless there remains disagreement on which is the greater determinant of behavior. Freud and Jung stressed genetics. Adler and Fromm stressed environment. Twentieth-century biotechnology and neuropsychology carry Freud's argument from clinical and philosophic speculation to hard science. It would appear that both camps are correct. Our attitudes and our behaviors are innately linked to inherited traits and learned responses.

Carl Jung is largely credited with building on Hippocrates' notion that personality can be differentiated on the basis of how we perceive and evaluate our world. Jung's most important work, Psychological Types, first appeared in 1921, organizing over twenty years of study on the individual's relationship with the environment. The book's premise is that one facet of personality leans into a direction determining the manner and the limits of a person's judgment. Jung was supported by contemporaries such as Erich Adickes and Ernst Kretschmer, each whom also believed that personality could differentiate on the basis of how we perceive and evaluate our surroundings.

The more modern neuroscientific theory of creativity looks at a hemispheric lateralization model, which began in earnest during the 1970s, when a neurosurgical team lead by Philip Vogel and Joseph Bogen performed commissurotomies on epileptics.[1] After surgery patients were subjected to a series of tests to measure both the psychological and psychomotor effects of having two independent cortical hemispheres. The scientists later found subtle yet astonishing changes in the patient's abilities to perceive and evaluate information. It was in that era that researchers postulated that the nondominant hemisphere is specialized for creative activity, such as holistic pattern recognition, art, and music.

To survive, both humans and animals, which incidentally also have lateralized brains, developed two kinds of attention—the first is specific, narrow, sustained, broad, open, and focused. We associate these with the left cortical hemisphere as discussed above. But we also need attention for a broader awareness: that is, a scanning ability, to see the many parts of something, which represents yet more than or different from the combination of parts, or a gestalt. This fits with what some scientists have observed about the functioning of the right cortical hemisphere. It seems generally accepted, at least in the neurolinguistics research community, that right hemisphere is more holistic.

In 1984, David Kolb published a model of his theory of learning styles, which posited four distinct ways we learn, consistent with a left/right hemisphere partiality: concrete experience, reflective observation, abstract conceptualization, and active experimentation. The elements of concrete experience and abstract conceptualization constitute polar pairs, and are mutually exclusive. Reflective observation and active experimentation are also polar pairs and mutually exclusive. Most of us are predisposed to use one of the modes more than the other when learning. Some of us prefer to emphasize an integration of our experience which occurs during reflective observation.

Temperament has more to do with the emotional activity of a person, while personality is characteristic of the person's patterns of behavior, feelings, and thoughts. Scientists estimate that 20%–60% of temperament is determined by genetics, but specific genes that confer temperamental traits have not been located.[2] Personality has a strong genetic basis.

[1]A commissurotomy is a surgical incision or in this case bisecting the connection between the two hemispheres of the brain to treat certain psychiatric disorders.

[2]Gene variations may contribute to particular traits related to temperament e.g., variants in the DRD2 and DRD4 genes, linked to a desire to seek out new experiences; KATNAL2 linked to self-discipline and carefulness; the PCDH15 and WSCD2 genes linked to sociability; and the MAOA gene linked to introversion. See, https://ghr.nlm.nih.gov/primer/traits/temperament.

And, in December 1986, the New York Times reported on a study which found that personality traits were significantly inherited. The paper cited a long term project at the University of Minnesota which, dating back to 1979, studied more than 350 pairs of twins, who had gone nearly a week of extensive testing that included analysis of blood, brain waves, intelligence, and allergies. Social potency, a trait associated with forceful leadership and extroversion, was estimated to be roughly 55%–61% heritable. Social closeness, a trait associated with a preference for emotional intimacy, was determined 33% heritable and by as much as 67% due to environmental factors.

We now identify the personality traits that have a heritable basis as the Big Five: Openness to Experience, Conscientiousness, Extraversion, Agreeableness, and Neuroticism [2, 3]. Combined, these characteristics form our personality, the one that takes us through life's ever-changing events.

Openness to Experience, sometimes abbreviated as OTE, seems especially related to creativity, so let's focus a bit on this dimension. OTE can be viewed as having two separate components, Openness and Intellect, which combined captures what we regard as insight, critical thinking, curiosity, and innovation. When OTE is partitioned this way, research indicates that Openness predicts creative achievement, but not necessarily Intellect. However, intellect has been found to predict fluid reasoning [4]. Several self-reported studies of OTE indicate that it relates to high levels of imagination, where subjects performed well on cognitive tasks that measure creative thinking [5].

Along these lines, OTE correlates to a number of subsidiary factors, such as Explicit Cognitive Ability, which ties into intellectual conditioning that we associate with IQ tests [6, 7]. This conditioning includes automatic, associative, nonconscious, and unintentional learning processes. Another factor, Affective Engagement, refers to a process that tends toward a preferential evaluation based on emotions, feelings, and empathy as well as interest in aesthetics, fantasy, and our emotional absorption into our search for beauty [8].

Research into the genetic basis for these traits has included the type of twin studies mentioned earlier, which suggest that heritability and environmental factors both influence all five factors, equally [9]. In four twin studies, the mean percentage for heritability was calculated for each personality trait and the researchers concluded that genetic influence among the five factors were: extraversion 54%, conscientiousness 49%, neuroticism 48%, and agreeableness 42% [10]. In this same study, OTE was estimated to have a 57% heritability influence. This trait relates to intellectual curiosity, intellectual interests, perceived intelligence, imagination, divergent thinking, creative achievement, preference for novelty, inquisitiveness, aesthetic interests, fantasy richness, and unconventionality [11].[3]

In the next chapter we turn to the biological and neural mechanisms that may link intellect and personality with the elusive feature of creativity.

Notes

1. Wilson, E.O. (2017). *The Origins of Creativity*. Liveright Publishing.
2. Digman, J.M. (1990). "Personality Structure: Emergence of the Five-Factor Model." *Annual Review of Psychology* 41: 417–440. https://doi.org/10.1146/annurev.ps.41.020190.002221.
3. McCrae, R.R., and Costa, P.T., Jr. (1987). "Validation of the Five-Factor Model of Personality Across Instruments and Observers." *Journal of Personality and Social Psychology* 52 (1).
4. Nusbaum, E.C., and Silvia, P.J. (2011). "Are Openness and Intellect Distinct Aspects of Openness to Experience? A Test of the O/I Model." *Personality and Individual Differences* 51: 571–574.
5. Beaty, R.E., et al. (2015). Personality and Complex Brain Networks: The Role of Openness to Experience in Default Network Efficiency. *Human Brain Mapping* 37 (2): 773–779.
6. DeYoung, C.G., et al. (2005). Sources of Openness/Intellect: Cognitive and Neuropsychological Correlates of the Fifth Factor of Personality. *Journal of Personality* 73: 825–858.
7. Feist, G.J. (1998). A Meta-Analysis of the Impact of Personality on Scientific and Artistic Creativity. *Personality and Social Psychological Review* 2: 290–309.
8. Kaufman, S.B. (2013). Opening Up Openness to Experience: A Four-Factor Model and Relations to Creative Achievement in the Arts and Sciences. *The Journal of Creative Behavior* 47 (4): 233–255. https://doi.org/10.1002/jocb.33.
9. Jang, K.L., Livesley, W.J., and Vernon, P.A. (1996, September). "Heritability of the Big Five Personality Dimensions and Their Facets: A Twin Study." *Journal of Personality* 64 (3): 577–591. https://doi.org/10.1111/j.1467-6494.1996.tb00522.x. PMID 8776880.

[3] Peter Carruthers argues that the evolutionary function of childhood pretending is to practice creativity. See, *The Architecture of the Mind: Massive Modularity and the Flexibility of Thought* (2006). Oxford University Press.

10. Bouchard, T.J., and McGue, M. (2003, January). "Genetic and Environmental Influences on Human Psychological Differences." *Journal of Neurobiology* 54 (1): 4–45. https://doi.org/10.1002/neu.10160. PMID 12486697.
11. Ibid., see note 8 above.

CHAPTER 14

An Astonishing Specification

The human mind is that bend in the evolutionary curve that reaches for the stars from which we came.

During the past seven million years, the *Hominid* brain has tripled in size, most of it over the past two million years [1]. The increase in size appears coincident with our divergence from apes.[1] But, an even more remarkable increase occurred about 3.5 million years ago, beginning with *Australopithecine*, an early forbearer of ours, who lived over 2 million years ago, when cranial capacity measured about 450 cubic centimeters. *Homo habilis*, followed between 2.4 and 1.6 million years ago, which had a cranial capacity of between 729 and 824 cubic centimeters [2]. *Homo habilis* was followed by *Homo erectus*, who appeared 1.8 million years later, having a capacity of 800–1200 cubic centimeters [3, 4]. Then, we *Homo sapiens* as recently as 100,000 years ago, were up to 1274 cubic centimeters, which is where cranial capacity stands today [5, 6].

Encephalization is an evolutionary increase in the complexity or relative size of the brain, involving a shift of function from noncortical parts of the brain to the cortex, which in the genus *Homo* may have been over the past 2 million years caused by relatively simple genetic mechanisms [7, 8]. Some researchers believe that genetic enhancers—those that turn

[1] Allometry is the study of the relationship of body size to shape, anatomy, physiology and finally behavior, first outlined by Otto Snell in 1892, then by D'Arcy Thompson in 1917 in *On Growth and Form*, and by Julian Huxley in 1932.

J. R. Carvalko Jr., *Conserving Humanity at the Dawn of Posthuman Technology*, https://doi.org/10.1007/978-3-030-26407-9_14

on genes in certain cells—may have caused evolutionary changes in the DNA sequences that triggered changes both in brain size and hand modifications (digit and limb patterning) [9].[2]

And, it stands to reason that in addition to size, much of the changes in the brain occurred within its neuronal network. The human brain is estimated to have approximately 86 billion neuronal cells, and when we add supporting material such as scaffolding and insulation for brain cells, it brings the cell count to one trillion. Neurons also connect to thousands of synapses, bringing that total to over 86 trillion connections. In the case of memory alone, these networks exponentially increase storage capacity to about around 10^{15} or 2.5 petabytes, which is enough to hold three million hours of video in a typical computer [10]. The enormous complexity of this 2.8 pound bundle of white matter, gray matter and neuronal connections are products of genomic structure and processes, which at the level of consciousness bear on mental ability, such as thinking and creativity, which in turn are measured by the g-factor.[3]

The cerebrum and cerebral cortex are two prominent regions of the brain and comprise the bulk of the forebrain.[4] Many exceptions among species exist, but for those that have a neocortex (mammals), it divides into six layers, which are involved in higher-order brain functions, such as sensory perception, cognition, generation of motor commands, spatial reasoning, and language [11, 12]. These layers also feature such things as our ability to express empathy and compassion, essentially where our humanity may be found to materialize.

The cerebral cortex's two hemispheric parts, as mentioned above, are connected by five commissures or bundles of nerves that span a longitudinal fissure. The largest of these we refer to as the corpus callosum,

[2]A range of theories have been proposed to explain mammalian variation in relative encephalization. These include energetic considerations (e.g., diet or metabolism), ecological variables (e.g., visual acuity, arboreality), and most prominently diverse measures of species "intelligence" (e.g., tool use, social learning, tactical deception). See, Halley, A.C., and Deacon, T.W. (2017). "The Developmental Basis of Evolutionary Trends in Primate Encephalization." https://doi.org/10.1016/b978-0-12-804042-3.00135-4. Elsevier Inc.

[3]The average adult brain is about 2% of body weight, with 78% constituting water, 12% fat and 8% protein.

[4]The neocortex is the largest part of the cerebral cortex which is the outer layer of the cerebrum, with the allocortex making up the rest.

which connects more than 200 million fibers. According to some, these connections reflect the capacity for inter-hemispheric processes that modulate intellectual abilities, i.e., which relate to creativity [13, 14].[5]

The left hemisphere contains two important language sites called Broca's area for verbal articulation and Wernicke's area for the meaning of language. The left side of the cortex is associated with our logical, analytical, and verbal abilities. The right side of the cortex makes connections and is considered our artistic side. But, it's not here that we want to focus, just yet. We want to turn to the subject of whether a genetic basis exists for creativity within the two hemispheres of the frontal cortex. If a genetic basis does exist, and it can be enhanced in some yet to be determined way, we could conjecture a future where changes to our genetic paradigms could influence any number of human generated activity: general acquisition of knowledge, cultural change, scientific advancement, and artistic production.

Beneath the cerebral cortex important limbic structures include the amygdala that mediates fear and other emotions, as well as the hippocampus, which merges information from short-term and long-term memory into spatial memory to enable navigation.[6] Emotions are believed to originate from the amygdala to influence the upper cortical centers, where thinking occurs.

While our ancestral hominids may have bashed shells against rocks as the way to access the food inside, *Homo sapiens* applied a kind of intuition and innovation that ultimately led to the concept of the tool. Note that no tools predate 2.5 million years ago. Successful tool making is related to eye-hand coordination, and humans excel here. Studies of prehistoric tool making and artistry reveal strong evidence humankind has

[5] The corpus callosum is a wide flat bundle of fibers, beneath the cerebral cortex. It connects the left and right cerebral hemispheres, and enables communication between the hemispheres.

[6] Alice Flaherty has hypothesized a three-factor anatomical model of human idea generation and creative drive that focuses on interactions between the temporal lobes, frontal lobes, and limbic system. See, Flaherty A.W. (2005). "Frontotemporal and Dopaminergic Control of Idea Generation and Creative Drive." *The Journal of comparative Neurology* 493 (1): 147–153. https://doi.org/10.1002/cne.20768. The limbic system, is a set of brain structures located on both sides of the thalamus beneath the medial temporal lobe of the cerebrum that supports such functions as emotion, motivation, long-term memory, and olfaction. The amygdala is an almond-shaped node of gray matter, located deep and medially within the temporal lobes, plays a central role in our perception of fear and other emotions.

been predominantly right handed. Although primates have hands similar to ours, human hands have a finer control over muscle pressure, and are able to control objects with greater precision. This advantage is neurological and structural, and may extend as far back as the separation of the last common ancestor of the chimpanzee–human LCA, or Pan and Homo lineages 5–7 million years ago.[7]

The part of the brain that dominates our ability to coordinate eye and hand movement is the cerebellum. As shown in Fig. 7.1, the cerebellum resides beneath the cortex, at the back and bottom of the brain, behind the brainstem. It includes sensors that detect shifts in balance and movement, sending signals for coordination of multiple muscle groups so they can move smoothly. In some ways the cerebellum acts like an analog computer, calculating complex inputs. Imagine a ballistics computer receiving inputs from a fast ball it's been designed to react to. The equations of motion used to solve this trajectory would be expressed as complex differential equations, which would be quite difficult to solve, in fact. But the analogy to this fast ball coming at a computer is a batter who knows precisely when, how hard and where to swing the bat. The cerebellum computes much of this, controlling eye movement, as it estimates the trajectory of the ball, preparing the batter to swing the bat to meet the ball, while engaging the eye, back, leg, hip, arm, and hand muscles. We are learning much about the physiology of how this all occurs through the technology of bioimaging.

Fifty years ago, understanding how the brain functioned was limited to testing blood and urine for by-products of neurotransmitter activity, while measuring electrical activity using electroencephalography (EEG) recordings. The first deep look into the brain began in 1980, when the positron emission tomography (PET) scanner came on the scene. It used radioactive glucose injected into the blood stream. The radiolabeled

[7]HACNS1 (also known as Human Accelerated Region 2) is a gene enhancer "that may have contributed to the evolution of the uniquely opposable human thumb, and possibly also modifications in the ankle or foot that allow humans to walk on two legs… This study is the first to provide evidence of the existence of human-specific gene enhancers, which are switches near genes in the human genome." Evidence to date shows that of the 110,000 gene enhancer sequences identified in the human genome, HACNS1 has undergone the most change during the human evolution since the chimpanzee-human last common ancestor. HACNS1: Gene enhancer in evolution of human opposable thumb. (See, https://www.sciencecodex.com/gene_enhancer_in_evolution_of_human_opposable_thumb.)

glucose lit up as it accumulated in brain areas undergoing a cognitive process, i.e., "thinking."

Scientists observed the progression of PET images as subjects were put through a variety of mental activities, such as playing games, or solving problems. As early as 1988, studies began to reveal that the more active brain areas were associated with lower cognitive performance. This counterintuitive result suggested that higher intelligence required less brainwork.

In 1995, researchers used PET scans to see if men and women both showed equal brain efficiency in the same brain areas while they solved mathematical reasoning problems. Brains seem to work in different ways, and one example found that women and men, who have the same IQ, show different underlying brain architectures evidenced by different brain areas being engaged during problem-solving.[8]

In this study, male and female college students were selected from two groups, one having SAT-M scores over 700 and another in the 500 range. Men showed high math ability correlated with greater activity in the temporal lobes. Women, on the other hand, showed no such relationship between math ability and a specific activity anywhere in the brain. Other PET studies were conducted on students watching videos, which correlated lower IQ scores with greater activity occurring in the back of the brain.

Over the past 20 years, 40 brain studies were conducted using a variety of neuro-imaging machinery, where techniques have drilled down to identify brain mechanisms, structure, and function, which bear on intelligence. Still, large gaps in knowledge remain about the anatomical and physiological organization of the neocortex at the cellular and synaptic levels. Magnetic resonant imaging (MRI), as with PET scans, can produce information on brain function and structure such as gray matter, showing neurons at work. MRI can also reveal white matter as it links brains areas to carry information throughout the brain. Gray matter and white matter tissue have different water content, so they can be distinguished in these images. Other related innovations such as functional MRI (fMRI) show brain activity based on blood flow. Diffusion tensor

[8]In the neocortex forebrain, where perception, memory and language is found, men appear to have about 23 million neurons and women 19 billion neurons, but no differences have been detected in the average IQ between the two. See, Kock, C. (2019). "From Man to Mouse." *Scientific American*, Special Edition.

imaging (DTI) produces MRI sequences of images, such as white matter fibers, which when measured, provide input to mathematical algorithms and assess how well the fibers transmit signals.

Voxel-based morphometry (VBM) investigates focal differences in brain anatomy, using parametric statistical mapping. One of the first brain studies employed VBM to look at the hippocampus structure of London taxicab drivers, who have to memorize up to 25,000 routes in the city to carry a taxi license, and in the process develop brain areas, which are enlarged due to their brains' acquired specialization in navigating London. University College London scientists found that the back part of the posterior hippocampus in the taxi drivers, on average, was larger compared to control subjects, while the size differential in the anterior hippocampus was reversed. The scientists also noted that part of the hippocampus grew larger in taxi drivers who spent more time on the job.

We know that exercise, diet, and drugs can improve memory. But the condition of sub-structures within the hippocampus do as well, such as the condition of dentate gyrus (a site identified with memory loss in the aged), or the abundance of certain protein molecules, such as RbAp48. And, outside the hippocampus, the prefrontal cortex also affects the quality of both short and long-term memory. Because memory serves as a fundamental component of intelligence in humans, scientists are increasingly interested in whether gene engineering or enhancing protein availability will amplify intelligence, undoubtedly research that will find an application in the construction of *Homo futuro*.

It has been known for decades that the temporal lobes of the brain contain language centers, and sitting north of the temporal lobes, the parietal lobes process sensory inputs, controlling such things as our visual-spatial ability. The occipital lobe, to the rear of the parietal, primarily subserves vision. In 2007, R.E. Jung and R.J. Haier proposed a theory, referred to as the parieto-frontal integration theory (P-FIT) based on a review of 37 imaging studies of intelligence [15]. They theorized that clusters of neurons situated in the prefrontal and parietal lobes—the sides toward the back of the brain—correlate with performance on cognitive tasks. P-FIT areas, particularly the front and back of the brain, relate to memory, attention, and language. The theory holds that the brain integrates sensory information in the back areas, and then it further integrates higher-level processing as information flows to the frontal areas. Some of the parieto-frontal areas lie in the left hemisphere and others in the right, with one frontal area appearing deeper in the cortex. Information flow around these areas may contribute to individual differences in intelligence. And, to this end, a key white matter tract of fibers referred to as the arcuate fasciculus connects and thus integrates information flow between frontal and parietal areas of the brain. The arcuate fasciculus appears to be correlated with intelligence. In men, more gray matter generally correlates with higher IQ. In women, almost all of the areas, where gray matter correlated with IQ, were in the frontal lobe language areas, especially around a part of the brain related to language called Broca's area.

Creative innovation, defined as the ability to understand and express novel orderly relationships has been found correlated to a high level of general intelligence, domain-specific knowledge, and special skills. In 2010, Kenneth Heilman and his colleagues reported that specialized knowledge, which relates to creative innovation, is stored in specific portions of the temporal and parietal lobes [16]. They wrote:

> The frontal lobes have strong connections with the polymodal and supramodal regions of the temporal and parietal lobes, where concepts and knowledge are stored. These connections might selectively inhibit and activate portions of posterior neocortex and thus be important for developing alternative solutions. Although extensive knowledge and divergent thinking together are critical for creativity they alone are insufficient for allowing a person to find the thread that unites. Finding this thread might require the binding of different forms of knowledge, stored in separate cortical modules that have not been previously associated.

The consensus among scientists, e.g., R.E. Jung, R.J. Haier, K.M. Heilman, is that the frontal lobe contains the neurological correlates we associate with divergent thinking and thus innovation. We will return to this subject in a subsequent chapter.

More than one animal has a brain that exceeds the human sized brain in weight, volume, or number of neocortical neurons, but only the human brain appears isomorphic to other humans allowing communication on the level of thoughts, emotions, pheromones, and other forms of behavior.[9] But, as we venture into a world of transhumans, communicative sensitivities may be amplified, interpreted in different ways, perhaps

[9] The brain of the blue whale weighs 6.900 kg and it has a body weight of 100,000 kg, the Indian elephant brain weighs 6.0 kg, and has a body weight of 5000 kg.

reducing ambiguities or errors in what's intended. It's hard to imagine that humanoid robots will ever come to a full awareness of themselves, other robots, or the human condition of its biological counterpart. On the other hand, it certainly will exhibit extraordinary modes of pattern recognition, some of which we are just beginning to appreciate.[10] As to this latter point, succeeding chapters will explore what may come to fruition as the boundary between human and machine blurs and eventually disappears for particular kinds of cognitive functionality.

Notes

1. Hawks, J. (July 2013). "How Has the Human Brain Evolved Over the Years?" *Scientific American Mind* 24 (3): 76. https://doi.org/10.1038/scientificamericanmind0713-76b.
2. Spoor, F., et al. (2015). "Reconstructed Homo Habilis Type OH 7 Suggests Deep-Rooted Species Diversity in Early Homo." *Nature* 519 (7541): 83–86. https://doi.org/10.1038/nature14224. PMID 25739632.
3. DeFelipe, J. (2011). "The Evolution of the Brain, the Human Nature of Cortical Circuits, and Intellectual Creativity." *Frontiers in Neuroanatomy* 5: 29. https://doi.org/10.3389/fnana.2011.00029.
4. The Institute of Human Origins. http://www.becominghuman.org/node/homo-erectus-0 (Last visited 11/19/2018).
5. Holloway, R. (1996). "Evolution of the Human Brain." In: A. Lock and C.R. Peters, eds., *Handbook of Human Symbolic Evolution*. Oxford: Oxford University Press. pp. 74–125.
6. Herculano-Houzel, S. (2009). "The Human Brain in Numbers: A Linearly Scaled-Up Primate Brain." *Frontiers in Human Neuroscience* 3: 31. https://doi.org/10.3389/neuro.09.031.2009
7. Finlay, B.L., et al. (1995). "Linked Regularities in the Development and Evolution of Mammalian Rains." *Science* 268: 1578–1158
8. Kaskan, P.M., et al. "Encephalization and Its Developmental Structure: How Many Ways Can a Brain Get Big?" In: D. Falk and K.R. Gibson, eds., *Evolutionary Anatomy of the Primate Cerebral Cortex*. Cambridge University Press, pp. 14–29.

[10]Although somewhat controversial, systems are being developed and based on facial parameters used to determine a host of personal preferences. See, "Face-reading AI will be able to detect your politics and IQ." https://www.theguardian.com/technology/2017/sep/12/artificial-intelligence-face-recognition-michal-kosinski.

9. Prabhakar, S., et al. (2008). "Human-Specific Gain of Function in a Developmental Enhancer." *Science* (New York, NY) 321 (5894): 1346–1350.
10. Andersen, P. (1990). "Synaptic Integration in Hippocampal CA1 Pyramids." *Progress in Brain Research* 83: 215–222. https://doi.org/10.1016/s0079-6123(08)61251-0.
11. Lodato, S., and Arlotta, P. (November 13, 2015). "Generating Neuronal Diversity in the Mammalian Cerebral Cortex." *Annual Review of Cell and Developmental Biology* 31 (1): 699–720. https://doi.org/10.1146/annurev-cellbio-100814-125353. PMC 4778709. PMID 26359774. "The neocortex is the part of the brain responsible for execution of higher-order brain functions, including cognition, sensory perception, and sophisticated motor control."
12. Lui, J.H., Hansen, D.V., and Kriegstein, A.R. (July 2011). "Development and Evolution of the Human Neocortex." *Cell* 146 (1): 18–36. https://doi.org/10.1016/j.cell.2011.06.030. PMC 3610574. PMID 21729779.
13. Bogen, J.E., and Bogen, G.M. (1969). "The Other Side of the Brain III: The Corpus Callosum and Creativity." *Bulletin of the Los Angeles Neurological Society* 34: 191–220.
14. Aboitiz, F., Scheibel, A.B., Fisher, R.S., Zaidel, E. (1992). "Fiber Composition of the Human Corpus Callosum." *Brain Research* 598: 143–153.
15. Jung, R.E., and Haier, R.J. (2007). "The Parieto-Frontal Integration Theory (P-FIT) of Intelligence: Converging Neuroimaging Evidence." *Behavioral and Brain Sciences* 30: 135–187.
16. Heilman, M., et al. (2003). "Creative Innovation: Possible Brain Mechanisms, the Neural Basis of Cognition." *Neurocase* 9 (5).

CHAPTER 15

Crannies and Stacks

The highest activity a human being can attain is learning for understanding, because to understand is to be free.—Baruch Spinoza

If one were to ask, "Where does creativity come from?" we might start with, once upon a time a troop of genes fell headlong into a pool of neurochemicals, which ink-like filled an ancient author's pen, the one who composed the "Epic of Gilgamesh," and then set it upon the stacks at Ashurbanipal for posterity and poets to follow.[1] But, who really knows the origin. As with ancient cave paintings, music, or inventions, we have scant evidence when humans took to writing literature. If our metaphorical ink were ever discovered, it would likely be found running through the crannies and stacks of what we know is the cortex, the cerebral library where art, music, and literature await to satiate our constant craving and curiosity [1].[2]

The cortex remains plastic throughout most of our lives, allowing that vast network of neuronal membranes to shape and reshape the molecular analogs associated with creative expression: thought, emotions,

[1] The Epic of Gilgamesh is an epic poem from ancient Mesopotamia, regarded as the earliest surviving great work of literature, which was found in The Royal Library of Ashurbanipal, which also held a collection of 30,000 clay tablets and fragments containing texts from the seventh century BCE.

[2] Some scholars claim that humans have been shaped by evolution to be musical, while others maintain that musical abilities reflect an alternative use of more adaptive cognitive skills.

J. R. Carvalko Jr., *Conserving Humanity at the Dawn of Posthuman Technology*, https://doi.org/10.1007/978-3-030-26407-9_15

love, hate, fear, biases, prejudices, and preferences. And, trying to better understand how this materializes, scientists have been hard at work identifying the specific brain structure associated with what we consider creativity. For example, gray matter volume in motor, auditory, and visual-spatial brain regions, has been found to be greater for musicians relative to nonmusicians [2].

Analyzing musical performers from the perspective of the right and left cortical hemispheres suggests that both evolution and genetics have influenced preferences for creativity. As mentioned, generally our ability to integrate entire scenes, develop hunches, and to emote appears to come from the right side of the brain. Right-sided strengths include: intuition, holistic ideation, spontaneity, and nonverbal responses in activities such as art, performance, and music [3]. There are obvious emotional reactions to music, and it's been demonstrated that happy musical segments are associated with increases in left frontal EEG activity, whereas musical segments representing fear and sadness were associated with increased in right frontal EEG activity [4].

Studies indicate that the left hemisphere is called to task when we need precision in language and symbols. To sum up, to form a gestalt, we use the right hemisphere. The right hemisphere looks for differences that are not aligned with our expectations. It's the devil's advocate. The left hemisphere creates abstraction (equations, rules, bodies of discrete parts). Some liken the left as a closed system, and the right, an open system, where reason and imagination use both cortical hemispheres.

Creative people singled out for innovation tend to have several things in common: (a) high levels of specialized knowledge; (b) employ divergent thinking; and (c) and as mediated by the cortical frontal lobe, dispense neurotransmitter modulators, such as norepinephrine. In fact, the frontal lobe appears to be the part of the cortex that is most centrally related to creativity.

One area of ongoing research has identified a network of brain regions referred to as the default mode network [5].[3] These brain regions, located throughout the brain, include the dorsomedial prefrontal cortex, ventromedial prefrontal cortex, lateral temporal cortex,

[3]One working hypothesis is that the default network's primary function is to support internal mental simulations that are used adaptively. From this perspective, the network can be engaged in a directed manner, such as recalling the location of a parked car, and also when the mind wanders from the immediate task at hand.

posterior cingulate, and inferior parietal lobule [6]. The default mode interacts on a large scale and is activated, even when we are not paying attention, appearing to be divided into sub-networks, each dedicated to different brain functions [7, 8].

As mentioned previously, the cerebral cortex is split into two nearly symmetrical left and right hemispheres, on which are located the frontal, temporal, parietal, and occipital lobes.[4] The frontal, temporal and parietal lobes of the brain show a strong correlation between their respective volumes and intelligence [9, 10, 11]. The frontal lobes deal generally with the executive functions, e.g., self-control, and our ability to think critically: planning, reasoning, and abstract thought. The prefrontal lobe allows us to hold on to a piece of information temporarily when completing a task, while the frontal pole of the brain has been found important for analogical mapping [12]. Importantly, the frontal lobe has given rise to the disinhibition hypothesis, which as observed in EEG studies, posits that creative cognition increases under lower than normal levels of cortical activity [13]. Disinhibition and diffused attention has also been observed in studies of the prefrontal cortex that also have been associated with an increase in self-expression [14]. For example, increased brain activity during musical improvisation had been observed through fMRI [15].

Using functional MRI brain imaging, neuroscientists now routinely study regions of the brain responsive to the creation of music, such as a jazz pianist's improvisation [16].[5] Limb and Broun have reported that they used two paradigms that differ in musical complexity, and found that "improvisation (compared to production of over-learned musical sequences) was consistently characterized by a dissociated pattern of activity in the prefrontal cortex." The medial prefrontal lobe also plays an important role in limbic system, emotion, and cognitive control—so that makes sense because improvisation is kind of off-the-cuff and go-with-the-flow, so it perhaps involves more emotional brain processing regions. Lim and Landau have investigated the neural substrates of flow with respect to the mechanisms of musical improvisation [17]. The subject of

[4]The parietal lobe sits in the center of your head, above the temporal lobe, while the frontal sits in the fore and the occipital aft of the parietal.

[5]Neuroscience, psychology, and computer science are disciplines which now engage in the study of brain-based mechanisms underlying music, across a wide range of activities such as listening, performing, composing, reading, writing, musical aesthetics and musical emotion.

flow will be addressed below in some detail. Other similar studies compared improvised versus memorized rap performances, and also found significant activation in the medial prefrontal regions [18].

Studies have consistently shown that musicians' exhibit larger volumes of gray matter (neural cell bodies, dendrites, and axons). In the language center, such as Broca's area (temporal lobe), and depending on how many years they have been playing an instrument, the volume of gray matter is also increased. For singers, the cerebellum particularly, which resides beneath the cortex, has been shown to have an increased functional activation, as well as increased activation in the primary somatosensory cortex, basal ganglia, thalamus.[6]

Still, other researchers have reported that increased cortical surface area of the left planum temporale in musicians facilitates the categorization of phonetic and temporal speech sounds [19]. Similarly the corpus callosum which connects both hemispheres seems larger in musicians compared to nonmusicians [20]. Cortical gray matter is thicker in trained musicians, including primary auditory and motor regions and numerous prefrontal cortical regions [21]. Christian Gaser and Gottfried Schlaug, using a voxel-by-voxel morphometric techniques, found gray matter volume differences in motor, auditory, and visual-spatial brain regions, between professional keyboard players, and matched groups of amateur musicians and nonmusicians [22]. In yet another study, the Heschl's gyrus, part of the auditory cortex, was found to have 130% more gray matter in professional musicians compared to nonmusicians [23]. Clearly, science is making progress concerning where particular types of creative activities show distinct brain physiologies. How these kinds of discoveries can be translated into biological enhancement technology, such as through genetic engineering, is yet to be determined.

During the latter part of the twentieth century, psychologist Mihali Csíkszentmihályi considered the ways the brain can alter consciousness to improve performance [24]. Although evidence suggests that individuals report levels of satisfaction in intervals, during which they call on their abilities to meet a particular challenge, a strong science-based theory for how this manifests in the brain, has not yet materialized. Nonetheless there has been a wide-spread interest among psychologists and neuroscientists in the phenomenon he discovered, coined "Flow," which amounts

[6]Alex Doman, A. "Do Musicians Have Bigger Brains?" See, https://www.braintraining101.com/do-musicians-have-bigger-brains/ (Last visited 5/16/2019).

to a deep mental absorption observed when individuals work at their optimal capacity in tackling challenging problems with relative ease [25]. According to Csíkszentmihályi, this effect results from a match between skill and challenge, which occurs during intense and focused concentration in the present moment. In "flow," action and awareness merge into a loss of reflective self-consciousness. In explaining how this works, he introduces us to a two-dimensional plot. The x-axis marks, at one end, finding something extremely difficult, and at the other end, something extremely easy. The y-axis marks, at one end, an activity where one finds little pleasure, to the opposite end where extreme pleasure is found.[7] He discovered that individuals are happiest when a challenge is neither too easy nor too challenging, as it relates to one's ability, and neither too uninteresting nor too interesting. In other words, the most pleasurable activity is one that is challenging, provided it is not overwhelming.

Csíkszentmihályi hypothesized that people with several very specific personality traits may be better able to achieve flow more often than the average person. These personality traits include curiosity, persistence, low self-centeredness, and a high rate of performing activities for intrinsic reasons only. In other words, reasons that may have to do with meaningfulness. People with most of these personality traits are said to have an autotelic personality.[8]

Some researchers claim that during flow, large areas of the prefrontal cortex are deactivated, effectively turning off the "inner critic" [26]. As one enters the state of flow, there are neurological changes as well, affecting neurotransmitters such as norepinephrine and dopamine, endorphins and anandamide, serotonin, and oxytocin. For example, mesolimbic dopamine was found to influence novelty seeking and creative drives.[9] In one study, the researchers explored the neurochemistry and social flow of group singing, e.g., a vocal jazz ensemble performance, precomposed and improvised, while plasma oxytocin and

[7] See, EDxUChicago 2011, Mihaly Csíkszentmihályi, Rules of Engagement. https://www.youtube.com/watch?v=7e1xU0-h9Y8.

[8] An autotelic activity is one that involves doing something for the sake of it.

[9] Alice Flaherty proposes a three-factor model of creative drive, where she describes it as implicating the frontal lobes, the temporal lobes, and dopamine from the limbic system. The frontal lobes can be seen as responsible for idea generation, and the temporal lobes for idea editing and evaluation. See, Flaherty, A.W. (2005). "Frontotemporal and Dopaminergic Control of Idea Generation and Creative Drive." *The Journal of comparative Neurology* 493 (1): 147–153. https://doi.org/10.1002/cne.20768.

adrenocorticotropic hormone (ACTH) were measured to assess levels of social affiliation, engagement, arousal, and absorption in the task [27].

The physiological explanation for flow postulates that creativity is a function of one's level of consciousness. The hypothesis claims that we move from the normal awake state, exemplified by beta brain waves, where there is the production of cortical norepinephrine, to the alpha brainwave state (daydreaming), where nitric oxide is produced, to a middle state, as experience as one begins to doze off, of theta (dreaming), where dopamine, endorphins, and anandamides are released. The combination of theta/gamma borderline is the brain wave state associated with flow.

After a thorough review of the research surrounding flow, Cheruvu found that "The limited findings in existing literature have led to mixed and contradictory results on the attentional processes/networks related to flow, key brain hubs for flow, and the consistency/distinction of neurophysiological outcomes across multiple autotelic activities and environments" [28]. As such, the state of the science is not sufficiently well-understood to speculate as to whether the phenomena discussed under the rubric flow could be subjected to enhancement via genetic engineering. And perhaps the broader question is whether creative tasks can be reduced to particular traits, or even genetic patterns. If so, science and technology will certainly weigh in on the feasibility for genetic modification to improve the odds of enhancing one or another aptitude, talent or skill.

Notes

1. Patel, A.D. (September 2006). *Music Perception: An Interdisciplinary Journal* 24 (1): 99–104.
2. Gaser, C., et al. (2003). "Brain Structures Differ Between Musicians and Non-Musicians." *The Journal of Neuroscience* 23 (27): 9240–9245. PMID 14534258.
3. Nauert, R. (2015). "Imaging Finds Visual Creativity Uses Both Right, Left Brain." Psych Central. Retrieved 9 August 2018, from https://psychcentral.com/news/2012/03/06/imaging-finds-visual-creativity-uses-both-right-left-brain/35612.html.
4. Schmidt, L.A., and Trainor, L.J. (2001). "Frontal Brain Electrical Activity (EEG) Distinguishes Valence and Intensity of Musical Emotions." *Cognition & Emotion* 15 (4): 487–500. https://doi.org/10.1080/02699930126048.

5. Buckner, R.L. (2013). "The Brain's Default Network: Origins and Implications for the Study of Psychosis." *Dialogues in Clinical Neuroscience* 15 (3): 351–358.
6. Bashwiner, D.M., et al. (2016). "Musical Creativity "Revealed" in Brain Structure: Interplay Between Motor, Default Mode, and Limbic Networks." *Scientific Reports* 6: 20482. https://doi.org/10.1038/srep20482.
7. Jung, R.E., Mead, B.S., Carrasco, J., and Flores, R.A. (2013). The Structure of Creative Cognition in the Human Brain. *Frontiers in Human Neuroscience* 7: 330. Also see, Flaherty, A.W. (2005). "Frontotemporal and Dopaminergic Control of Idea Generation and Creative Drive." *The Journal of comparative Neurology* 493 (1): 147–153. https://doi.org/10.1002/cne.20768.
8. Beaty, R.E., et al. (2014). "Creativity and the Default Network: A Functional Connectivity Analysis of the Creative Brain at Rest." *Neuropsychologia* 64C: 92–98.
9. Andreasen, N.C., Flaum, M., Victor Swayze, I.I., O'Leary, D.S., Alliger, R., and Cohen, G. (1993). "Intelligence and Brain Structure in Normal Individuals." *American Journal of Psychiatry* 1: 50. https://doi.org/10.1176/ajp.150.1.130.
10. Flashman, L.A., Andreasen, N.C., Flaum, M., and Swayze, V.W. (1997). "Intelligence and Regional Brain Volumes in Normal Controls." *Intelligence* 25 (3): 149–160. https://doi.org/10.1016/s0160-2896(97)90039-8.
11. MacLullich, A.M.J., Ferguson, K.J., Deary, I.J., Seckl, J.R., Starr, J.M., and Wardlaw, J.M. (2002). "Intracranial Capacity and Brain Volumes Are Associated with Cognition in Healthy Elderly Men." *Neurology* 59 (2): 169–174. https://doi.org/10.1212/wnl.59.2.169.
12. Green, A.E., Kraemer, D.J.M., Fugelsang, J.A., Gray, J.R., and Dunbar, K.N. (2011). "Neural Correlates of Creativity in Analogical Reasoning." *Journal of Experimental Psychology Learning Memory and Cognition.* https://doi.org/10.1037/a0025764.
13. Martindale, C. (1977). "Creativity, Consciousness and Cortical Arousal. *Journal of Altered States of Consciousness* 3 (1), 69–87.
14. Daffner, K.R., Ryan, K.K., Williams, D.M., Budson, A.E., Rentz, D.M., Wolk, D.A., et al. (2006). Increased Responsiveness to Novelty Is Associated with Successful Cognitive Aging. *Journal of Cognative Neuroscience* 18: 1759–1773.
15. Limb, C.J., and Braun, A.R. (2008). "Neural Substrates of Spontaneous Musical Performance: An fMRI Study of Jazz Improvisation." *PLoS ONE* 3 (2): e1679. https://doi.org/10.1371/journal.pone.0001679.
16. Ibid.

17. Landau, A.T., and Limb, C.J. (2017). "The Neuroscience of Improvisation." *Music Educators Journal* 103 (3), 27–33. https://doi.org/10.1177/0027432116687373.
18. Liu, S., et al. (2012). "Neural Correlates of Lyrical Improvisation: An fMRI Study of Freestyle Rap." *Scientific Reports* 2: 834.
19. Elmer, S., Hanggi, J., Meyer, M., and Jancke, L. "Increased Cortical Surface Area of the Left Planum Temporale in Musicians Facilitates the Categorization of Phonetic and Temporal Speech Sounds." *Cortex* 49: 2812–2821 (2013). 30. Halwani, G.F., Loui, P., Rüber, T., and Schlaug, G. Effects of Practice and Experience.
20. Steele, C.J., Bailey, J.A., Zatorre, R.J., and Penhune, V.B. (2013). "Early Musical Training and White-Matter Plasticity in the Corpus Callosum: Evidence for a Sensitive Period." *Journal of Neuroscience* 33: 1282–1290.
21. Hyde, K.L., et al. (2009). "Musical Training Shapes Structural Brain Development." *Journal of Neuroscience* 29: 3019–3025.
22. Schlaug, G., et al. (2003). "Brain Structures Differ Between Musicians and Non-musicians." *Journal of Neuroscience* 23 (27): 9240–9245. https://doi.org/10.1523/JNEUROSCI.23-27-09240.
23. Schneider, P., et al. (2002). "Morphology of Heschl's Gyrus Reflects Enhanced Activation in the Auditory Cortex of Musicians." *Nature Neuroscience.*
24. Csíkszentmihályi, M. (1990). *Flow—The Psychology of Optimal Experience.* HarperCollins. Also see, Csikszentmihalyi, M. (1999). "Implications of a Systems Perspective for the Study of Creativity." In R.J. Sternberg, ed., *Handbook of Creativity.* New York: Cambridge University Press.
25. Csíkszentmihályi, M. (1996). *Creativity: Flow and the Psychology of Discovery and Invention.* New York: HarperCollins.
26. Dietrich, A. (2004). "Neurocognitive Mechanisms Underlying the Experience of Flow." *Consciousness and Cognition.* https://doi.org/10.1016/j.concog.2004.07.002.
27. Keeler, J.R., Roth, E.A., Neuser, B.L., Spitsbergen, J.M., Waters, D.J., and Vianney, J. (2015). "The Neurochemistry and Social Flow of Singing: Bonding and Oxytocin." *Frontiers in Human Neuroscience* 9. https://doi.org/10.3389/fnhum.2015.00518.
28. Cheruvu, R. (2018). The Neuroscience of Flow, Harvard University Extension School, PSYCE-1609, The Neuroscience of Learning: An Introduction to Mind, Brain, Health and Education.

PART III

Technology of Creativity

CHAPTER 16

Different Outcomes

Sarouk: Were the technological changes installed just to alleviate disease or sickness?

Mensa: Yes, at first. Then little by little other technologies were available to reduce emotional barriers, so we could see things for what they were, impartially. Introversion, introspection, forms of self-awareness, neuroticism and narcissism, were suppressed. The technology favored the cognitive over feeling and emotion, and in time produced personalities with lesser degrees of openness, extraversion, and great levels of agreeableness.

Sarouk: Were there attempts to deal with attitudes, say empathy or indifference.

Mensa: They came at this a little differently, developing technology to engender curiosity for science and math, reducing it for things like poetry or literature.

Sarouk: What do you mean, exactly?

Mensa: The biocomputers were integrated into the sympathetic nervous system to slow down the production of adrenaline, and thus channel the affects of arousal. It reduced my anxieties. But, I began to get less and less excited about things.

J. R. Carvalko Jr., *Conserving Humanity at the Dawn of Posthuman Technology*, https://doi.org/10.1007/978-3-030-26407-9_16

Sarouk: But, how did that change people? Did it alter their perceptions, their dispositions?

Mensa: I can only say that I'm not the person today that I was a hundred or even fifty years, ago. My focus, and even interest in reading a good fiction, or getting excited about a new idea, waned, unless the idea had to do with something objective or real. I began to notice, a changing attraction to things that once interested me, each time a new operating system was downloaded into my brain-related peripherals. I suspect that they were causing frontal lobe deficits in order to decrease the generation of ideas that had no practical application. I found myself becoming more and more judgmental about new ideas, unless of course they offered some immediate utility.

Sarouk: How did it affect your quality of life?

Mensa: Somehow the technology was de-socializing, partitioning me and my friends into electronic spaces where we could no longer communicate on the same level. Protocols were not standard, inputs produced different outcomes. The world was becoming more vacant as what was once considered art and literature slowly vanished. And, what remained was considered a relic of an era when people wasted time. And, more troubling, especially for those who persisted in practicing these forms of expression…, they were considered eccentric or even mentally ill.

CHAPTER 17

Machines to Molecules

It's commonly accepted among anthropologists that the crafting of primitive tools dates back to the time when *Homo sapiens* arrived on the scene. Over time simple tools were combined into what today we refer to as a "machine." And, for thousands of years these implements were used to move stones, logs and, with rare exception, animals and people. They carried out causal, physical activities, where the effect of something like moving materials from point A to point B was immediately observed.[1] But, other than simple machines, ones based on levers to lift or wheels to transport, it was only 2000 years ago, in ancient Greece, 87 BCE, when a mechanism called "The Antikythera," employed a complex assembly of 30 meshing bronze gears to compute astronomical positions. What made the gears turn in this machine, as it does in all such constructions, depended on causal relationships between opposing forces. However, these rotations represented something more than physical movement, they represented an abstraction.[2] Unlike machines of physical

[1] Causality operates independent of the mind, as it's a consequence of dynamically working Universe resulting from fields of force, gravitational, electronic, atomic, and so forth. Biological metabolisms likewise operate subject to causality, and in this regard are no different than inanimate chemical processes ever-present throughout nature.

[2] In an early calculation system, a board covered with sand was divided into lines, each line representing a different numerical position. Numbers and quantities were calculated by means of various signs drawn along the lines. The abacus was thought to have been

J. R. Carvalko Jr., *Conserving Humanity at the Dawn of Posthuman Technology*, https://doi.org/10.1007/978-3-030-26407-9_17

force and motion that move objects, "The Antikythera" computed numbers, representing the movements of planets and the occurrences of eclipses, the abstract sort of things that only a human mind grasps.[3] And, thus over time, the idea of machine evolved from its physical form, to forms of abstract computation run on computers.

In AI theory, "machine" does not mean a physical device, but a class of mathematical functions that perform calculations or logical operations signifying various situations or states. Computation theory in particular defines "machine," as a scheme for encoding and decoding structures and processes. In the way we are using the word "machine," it may apply to how we explain the order underlying the evolution of a species, ecosystem or culture. In fact, we are following this line of thought to prepare us to consider how even small increases in a population IQ, could potentially amplify a collective intelligence—leading to an accelerated reorganization of knowledge domains and culture.

Artificial intelligence in the modern era refers to computations, which as indicated have no physical substance. John Barden points out that AI has at least three separate, though interrelated, aims:

> An "engineering" aim: ... [to] provide computational principles and methods for engineering, useful artefacts that are arguably intelligent, without necessarily having any mechanistic similarity to human or animal minds/brains... A "psychological" aim: ... [to] devise computational principles, computationally-detailed theories, or running computational systems that provide a basis for possible testable accounts of cognition in human or animal minds/brains....A "general/philosophical" aim: ... [to] devise computational principles, ... that serve as or suggest possible accounts of cognition in general, whether it be in human-made artefacts, in naturally-occurring organisms, or in cognizing organisms yet to be discovered, or

invented in the Sumer region between 2700–2300 BCE, and was representational of an abstract entity, i.e., a number, dependent on a table of successive columns, which delimited the successive orders of magnitude of the Sumerian sexagesimal number system. However, the article comprised beads, which were moved manually. See, https://www.abacus-maths.com/abacus-history1.html.

[3] I take the view that although distances represent physical reality, the quantification of such is a human construct, which if all life were to become extinct, the quantifier, i.e., the numbers representing the distance would disappear as well.

that illuminate philosophical issues such as the nature of mind, thought, intelligence, consciousness, perception, language, representation, learning, rationality, society, etc. ... [1]

An additional premise of this book is that AI instantiated in biocomputers, will come to pass. These devices will find a multiplicity of applications in the human anatomy. In some instances they will allow humans to think quicker, or have access to greater amounts of information. In other instances, whether installed in the anatomy or as separate systems accessible by the anatomy, e.g., the brain, they may be programmed to assist in the production of music, art, and literature.

AI employs numerous technologies, dependent on the application: machine learning (ML), machines that process natural languages, or others that mimic cognitive or creative processes (e.g., composing music), control systems as for autonomous vehicles). In some quarters the scope of AI is circumscribed. For example, AI may refer to ML on the one hand, and on the other, computational processes akin to the production and application of fractals or cellular automata [2]. Machine learning also may be modeled within the construct of a biological computer, and one which could someday be embedded into the anatomy to serve as an adjunct to human intelligence. Cellular automata on the other hand is applicable in modeling complex physical behaviors, involving self-organization and emergent properties in particular cases.[4] These are types of behaviors often associated with computer art and music, as well as modeling the evolutionary effects of technology on society.

Artificial intelligence, as applied to ML, discovers patterns for developing inferences. Some of these are supervised learning systems, which build mathematical models based on sampled data, referred to as "training data." Training data helps develop what patterns to look for, and helps establish parameters for the subsequent classification of objects. These paradigms are used in pattern recognition systems, as for example, bio-specimen identification or facial recognition. In other cases a system is not supervised, but essentially teaches itself, using constructed networks—as one envisions a neural network that feeds a multiplicity of lower layers of neural networks,

[4]In philosophy, systems theory, science, and art, an emergence occurs when an entity is observed to have properties that its parts do not manifest, but that properties or behaviors emerge only when the parts interact in a wider whole—Wikipedia.

which overtime, conditions itself to parse differences in classes of objects or data that it has been exposed to.

Beyond ML, AI takes on many other forms, such as those directed toward the kinds of creative activities to be discussed in subsequent chapters:

1. Symbolic AI: knowledge-based, rule-based regimens that use "knowledge-about" the subject in addition to the construction of grammars, applicable to the particular art or music form. Perhaps among the oldest approaches, it is based on the "physical symbol systems hypothesis," which holds that many aspects of intelligence can be achieved by the manipulation of symbols and assumptions. [3]
2. Optimization: techniques that use population-based methods, such as evolutionary algorithms. Classic evolutionary algorithms include genetic algorithms, gene expression programming, and genetic programming.
3. Computational methods: Complex systems involving self-similarity and cellular automata.

What associates a real-world phenomenon to a representative model is a concept referred to as isomorphism, which is based on a one-to-one correspondence that associates a hypothetical machine output, typically in the form of a computer simulation, to a real-world behavior.[5,6] To understand how these machines and processes work, let's take a brief

[5] A, B, C have a conventional order, namely an alphabetical order, and similarly 1, 2, 3 have an order of the noted integers, and thus one particular isomorphism is "natural," namely $A \mapsto 1, B \mapsto 2, C \mapsto 3$.

[6] One of the precepts of AI theory is that the problems to which it is applied must be finitely describable. Human intelligence, although not provably so, is nonetheless believed to be not finitely describable, which leads to the conclusion that AI can never attain the level of potential human intelligence. It can however simulate aspects of intelligence, which are finitely describable. I add that computability theory holds that some problems have been determined unsolvable either by a human or a machine. These assumptions play significantly in consideration of whether a computational process could rise to the level of human creativity, especially as it concerns the production of something as complicated as a stage play.

look at models in the abstract, ones based on analogies and metaphors that stand-in for our realities, social and physical.[7]

Engineers conceptualize physical dynamic utilitarian systems by considering: (1) an input structure or process, (2) the underlying transforming rules representing causal associations, and (3) a new structure or process, as determined by the first two factors. This can be thought of as (1) Input, (2) Transformation, and (3) Output, or ITO for short.

ITO elements are present in all dynamic systems: for example, the operation of a spring with an attached weight under the forces of gravity and an applied force. If one were designing a simple amplifier, ITO might be, a microphone that converts sound to an electronic signal (I), a circuit that transforms or amplifies the signal (T), and an output (O), such as a audible sound. And for more complex processes the concept of ITO would have to be expanded. Illustrative of complex processes would be evolution, climate change, weather, the behavior of the Internet, the growth of bacterial colonies; or the way a culture or society expands in complexity under the influences of population growth, technological innovation and levels of education.

A type of model called a formal system, as illustrated in Fig. 17.1, may be hypothesized that corresponds to a sliver of the real world, but a sliver where we don't actually deal with masses, forces and frictions, but instead grapple with a world of mathematical axioms, symbols and theorems. As indicated above, AI represents a form of computation. As such, axioms serve as the starting point for constructing a theorem and producing an end state, which would not have been self-evident but for the application of a rule applied to a set of symbols—numbers for instance. In the background of every software language axioms, symbols and theorems are present, although hidden, but dealt with quite matter of fact at a much higher level. The architectures for complex AI systems, which compose music, art, and literature, start here.

In spite of sounding abstruse, axioms, symbols, and theorems, are concepts we all dealt with when we learned $1+1=2$. This simple expression uses symbols 2, 1, +,=, which are applied via a rule that states: when the symbols are assembled to look like "$1+1=$" the answer

[7] Bear in mind that obviously not all problems in the world can be analyzed by, solved by or reduced to an algorithm or mathematical formula.

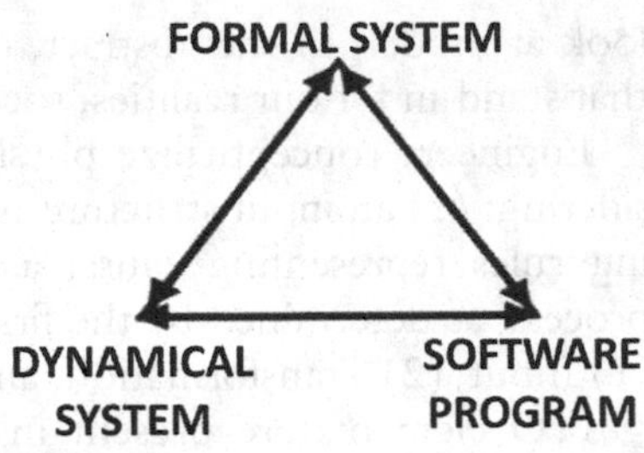

Fig. 17.1 Relationship among real and virtual systems

is 2.[8] It happens that this assembly of rules and symbols can relate to the real world if we give it interpretation. Thus meaning or semantics is not intrinsic to the symbols or the theorem; instead, it comes from a mind that reflects on the meaning of the mathematical statement, brought about by definition and the syntax. So, a computer program that produces code that, in turn, generates the sounds like those heard in numerous Beethoven compositions, has no sense for what the sounds do to a human amygdala reveling in the "beauty" of the music, any more than a fly does. This idea extends to languages having syntax and words in a context that provides content or intelligence.[9] The point is that simple axioms or starting points are capable of generating complex signals that we perceive as music. And, this same idea can theoretically become the starting point for AI constructions that someday may be installed into biocomputers, which find applications in the human anatomy, perhaps the brain. More on this in a later chapter.

Rules programmed and designed into electronic hardware and software operate on substance or content, which we refer to as "data." Computer data exists as a series of symbols, usually designated by a "1" or a "0," and importantly can be viewed as corresponding with behaviors in the real world, which permits us to impart an "assigned meaning." According to Hofstadter, animate molecules can emanate from inanimate matter, as

[8] More formally, a theorem is defined to be a mathematical statement that is proven to be true. The statement $1+1=2$ has definitely been proven, however, whether it's a theorem depends on how you define 1 and 2, and + (or even just +1). If you use Peano axioms then $1=S(0)$ and $2=S(S(0))$. If you further say that $m+1=S(m)$ for a natural number m, then $1+1=2$ can be considered as a theorem in such system of axioms. See, https://math.stackexchange.com/questions/348889/is-11-2-a-theorem.

[9] We are in the territory of the philosophical theory of signs and symbols that deals with their function in both artificially constructed and natural languages, comprising syntactics, semantics, and pragmatics.

inherent in abstract "symbols" [4]. By animate, he means self-referenced, or what we regard as the "self," the "*I*" alluded to, earlier. He theorizes how the properties of self-referential systems, most famously in Gödel's incompleteness theorems, can be used to analogize the unique properties, more particularly patterns that course through the mind.

So, moving between a synthesized world and a real world depends on the model and the real world being isomorphic, that is in a one-to-one relationship. As will be discussed later, some music creation systems offer a good match between what we imagine a real-world composer and her music should sound like, and what we hear as generated by computer. In some instances, programmers/engineers succeed by encoding the formal system with data extracted from the natural system, and conversely by decoding data residing in the formal system into the real world.

Things like music and art are abstract and intangible, but the idea of moving from a model to the physical world is made explicit in the case of genetic alteration. Turning a model into a physical manifestation is the heart of genetic engineering, i.e., when new genes are spliced into a genome. Therafter, according to biological rules instantiated by the genes itself, the organism organizes to direct gene expression and express proteins essential for its continued survival. The encoding/decoding operation amounts to using a dictionary or translator to associate the one-to-one relationships. It also illustrates how a model takes on substance in the form of a biological embodiment—a human brain, for instance.

Where these symbols sit in the Universe, and what we call them, are of no significance in and of themselves. The symbols can be represented by ink on a piece of paper that register in the mind, marking which photons bounce off and hit our retinas. Symbols, which prompt neurons to excitedly fire, and even fire in a manner so that they take on meaning. Meaning that, in fact, causes me to move my arm, or my mouth to form and express words. These symbols also do not need to be embodied in print, but may be represented by an electrical charge (voltage = 2 or 3 volts) or even the chemical state of molecules of the kind associated with living organisms: Oxygen, Sulphur, Hydrogen, Nitrogen, Carbon, Phosphorus.

If we imagine atoms as representing symbols in a chemical language, molecules would represent strings, analogous to letters or words used in constructing sentences, and their assembly would be subject to a set of rules, analogous to a syntax, i.e., the set of rules, principles, and processes that govern the structure of sentences. Oxygen and sulphur each have two sites to bind other molecules, hydrogen has one, nitrogen three, and carbon at least 4. Phosphorous can combine with three different atoms and sometimes a fourth—usually oxygen. The order of these combinations are analogous to the order of words in a sentence, each depending on a kind of syntax to carry out their function.

In real-world chemistry, genes act recursively, iterating upon building blocks called nucleotides to form molecules of larger and more complex statements and, corresponding functions. DNA code directs cell growth and the production of enzymes and other substances in the body, and importantly, reproduction through RNA replication. As previously mentioned, chemical sentences, such as mRNA, taken in groups of three nucleotides at a time, called codons, assembles proteins. In this instance the proteins are used to grow neurons, which transmit intelligence, that is, communicate, in the strictest sense, as illustrated in Fig. 6.2, in the chapter entitled "The Platform."

Let's turn to how combinations of symbols or strings, e.g., A, +, B, =, C, begin as meaningless statements in a formal system, but progress into things that are significant. As mentioned, we refer to starting points as axioms, and others as transformation rules, or rules of inference, which point toward how one or more strings are permitted to evolve or can be converted to another string. In a mathematical or logical system, the conversion of an axiom or string via a transformation is either to another axiom or string, which becomes what's called a theorem, terminal state or terminal string. Remember in a strictly mathematical sense strings and rules have no assigned meaning.

Hofstadter's Godel, Escher, Bach (GEB) introduced us to his MIU system, which brings into focus how even meaninglessness, in the mathematical sense, can have causal implications, e.g., reading these words cause neurons to fire. Deepak Chopra makes a similar point referring to the Eastern spiritual tradition: "Vedanta says that mind is innate in creation. … Mind creates matter every time we have thoughts that generate unique electrochemical activity in the brain. But no one has credibly shown how molecules learned to think."[10]

[10]See, Reply to seven commentaries on "Consciousness in the Universe: Review of the 'Orch OR' Theory" Stuart Hameroff, Roger Penrose. See, https://ac.els-cdn.com/S1571064513001905/1-s2.0-S1571064513001905-main.pdf?_tid=679b0520-2b1d-4454-b200-258e662760b3&acdnat=1552411380_a6abe06781cdff2f214b36d14bf43e53.

Earlier I'd indicated that models can be used as representations of the real world, but what if, at least a subset of these were more than a representational model; and that this subset was in fact, actually part of the real world. Suppose, as illustrated in Fig. 17.1, that the Formal system were instantiated in a Program, which could carry out a series of actions, such as those that scientists engaged in showing that life might form from disparate chemicals. In 1952 Stanley Miller, with assistance from Harold Urey, carried out at the University of Chicago, what has been considered the classic experiment investigating abiogenesis, a theory that conditions on primitive Earth favored chemical reactions that synthesized organic compounds from inorganic precursors. Simulating what he thought were those conditions, Miller passed an electric current through a flask filled with water, hydrogen, methane, and ammonia. Within the week he discovered that the experiment produced 20 amino acids—the kind found in DNA.

In the Miller-Urey experiment, the transformation serves as an example of the self-replication of meaningless symbols (molecules) into biological entities, which we know that in turn create social institutions, cultures and its artifacts. There seems to be a bridge here, although scientists line up on two sides of a divide. In GEB, Hofstadter, after an exhaustive argument related to Gödel's theorem, suggests that they can. Scientist, Sir Roger Penrose differs, arguing that the mental quality of "understanding" cannot be summarized by any computational system and must derive from some "non-computable" consideration because a "neurocomputational approach to volition, where algorithmic computation completely determines all thought processes, appears to preclude any possibility for independent causal agency, or free will. Something else is needed. What non-computable factor may occur in the brain?" [5, 6, 7].

One cannot dispute that living organisms assemble through replication. As James Thomson, the biologist who patented the method to produce stem cell lines in the laboratory observed, "A stem cell replaces itself through proliferation for prolonged periods (self-renewal) and gives rise to one or more differentiated cell types [8]." At the molecular level, elements from a finite group of atoms, join, based upon affinities of electrons in outer atomic orbits around the nucleus. These are the embodiment of the axioms, symbols and transformative rules we have been discussing. These affinities attract other elements as they assemble into larger molecules. On a microscale, we can consider that these molecules comprise the "letters" or strings in what become words—macromolecules, which form the elements, essential for a lifeform. The manner of biological communication varies; it may be deemed chemical or electrical, but in reality, we refer to one or the other depending on our objective or point of view.

It seems safe to suggest that AI technology, following the outlines of biological replication, someday may be used to direct the progressive modification of genomes. Certainly, one can imagine an assembly line for artificial cells or synthetic biological computers. It may not be a stretch to envision a future where error-checking, or more to the point, spell-checking that makes corrections to the genome to eradicate a birth defect or susceptibilities to certain diseases [9].[11] Direct in vivo correction of genetic or epigenetic defects in somatic tissue would offer a permanent solution by addressing root causes of a genetically encoded disorder. The same idea applies to germlines, as for instance embryos, which would also address root causes of genetically encoded disorders, but in this case the correction would be passed on to subsequent generations.

The obvious implication is that employing our tiny machines for purposes of genetic modification to express traits for intelligence will come to pass. This will likely be the case in an iterative embryo selection (IES) process or through combinations of CRISPR/Cas9 technologies, which would accelerate enhanced intelligence. As of yet, technology has not ventured into this biological territory in quite this way, but few would argue that it's not an impossibility, and perhaps only a matter of time. That having been said, there are mathematical systems that can potentially help actualize the concept of how to move from the abstraction of a machine or formal system of symbols to a computer program, to something having a biophysical form, fit, and function. Or, at the other end of the spectrum, our formal system may explain how it's possible that society as we know it radically changes under the influence of a widespread increase in IQ, coupled with an exponential rate of growth in the productivity of technology, generally.

[11]In 2016, Canadian researchers working at the Program in Genetics and Genome Biology, Research Institute, involved in eradicating muscular dystrophy, used CRISPR/Cas9.

Notes

1. Barnden, J. (2014). "Metaphor and Artificial Intelligence: Why They Matter to Each Other." https://www.researchgate.net/publication/228374445_Metaphor_and_artificial_intelligence_Why_they_matter_to_each_other.
2. Samuel, Arthur. (1959). "Some Studies in Machine Learning Using the Game of Checkers." *IBM Journal of Research and Development* 3 (3): 210–229. CiteSeerX 10.1.1.368.2254. https://citeseerx.ist.psu.edu/viewdoc/summary?doi=10.1.1.368.2254, https://doi.org/10.1147/rd.33.0210.
3. Newell, A., et al. (1976). "Computer Science as Empirical Inquiry: Symbols and Search." *Communications of the ACM* 19 (3): 113–126. https://doi.org/10.1145/360018.360022.
4. Hofstadter, Douglas. (2007). *I Am a Strange Loop*. Basic Books.
5. Penrose, R., and Hameroff, S. (2014). "Review Consciousness in the Universe: A Review of the 'Orch OR' Theory." *Physics of Life Reviews* 11: 39–78.
6. Penrose, R. (1989). *The Emperor's New Mind: Concerning Computers, Minds, and the Laws of Physics*. Oxford: Oxford University Press.
7. Penrose, R. (1994). *Shadows of the Mind: An Approach to the Missing Science of Consciousness*. Oxford: Oxford University Press.
8. See, *The Human Embryonic Stem Cell Debate, Science, Ethics, and Public Policy* (ed. S. Holland, K. Lebacqz, and L. Zoloth), MIT Press (2001); Human Embryonic Stem Cells, James Thomson.
9. Wojtal, D., et al. (January 7, 2016). "Spell Checking Nature: Versatility of CRISPR/Cas9 for Developing Treatments for Inherited Disorders." *The American Journal of Human Genetics* 98: 90–101.

CHAPTER 18

Machines, Computers, Software

The best way to predict the future is to create it.—Abraham Lincoln

In less than two generations, new lexicons appeared: metadata, fail whale, Google bomb, shock site, troll, electronic medical prescription, facial recognition, crime-seeking drone. So-called friends, real, virtual, social, and political now post façades on Facebook, Instagram, or Twitter bringing an illusion of friendliness, tribalism, open connectedness, while satellites surreptitiously circle the planet compiling dossiers on the masses. Multinationals launch an array of robotized medicine, banking, and education, Apps that drive cars. We measure progress in bits, bytes, and dollars, but at what cost? Have we become dehumanized embodiments, a necessary cog, in all manner of genetic and electronic computation and control? And, is technology so parasitic that it has trained us to surrender our identity, our essence, to the faceless forces behind the organization and its intelligent machines?

Machines have traveled alongside humankind throughout history beginning modestly as a simple wedge, which defines the operative principle behind the power of the spear, and has been constantly redefined as civilization progressed, e.g., early clockwork mechanisms that predicted time, or calculated the movement of the heavens. Traditionally evolution has been seen as separate from the world of mechanisms, the former part of nature, the latter a product of human nature, but the ever-tightening interrelationship between machines and human biology cannot be ignored.

J. R. Carvalko Jr., *Conserving Humanity at the Dawn of Posthuman Technology*, https://doi.org/10.1007/978-3-030-26407-9_18

We tend to lose sight of the technology that surrounds us. Smartphones and cars represent cognitive and functional extensions of ourselves [1]. But, whatever the forces that moved the *hominid* to become facile with the idea of machine, has not likely changed, and the intellectual process that steadfastly assures the continuation of invention has not changed either. For, what has been the long trek of innovation over the course of millions of years has at bottom been either a machine or process.

Two words dominate the technological lexicon: machine and process. As technology combines and sparks new fields of intellectual thought these two words have had to stretch to wrap their underlying principals around new ways of explaining how the simple machine and conceptual analogical modeling has led to nanoparticles, microelectromechanical (MEM) devices, manufactured viruses and even theories of how the mind works. Let's sketch out the path we have charted from the ancient concept of what a machine or process originally applied to, to where we are today, an entire industry engaged in understanding the genomic architecture of human disease, one that requires the collection of massive datasets (big data), and computational modeling (machine learning) that leads directly to embryo genetic testing and combining dense, genome-wide genotyping methods to advance human physiological potential.

In engineering the word "machine" refers to a device that transforms energy into work, casting off the idea that it has something to do with being automatic. The most widely used idea relates to a physical, tangible device that performs work, comprising parts, that come together to serve a function, as does a pump. Under this classification, we could include naturally occurring machines, such as a heart that performs functions as a pump. And, in this last use of the word machine, we see how it has morphed from mechanical devices of springs and levers to natural things like human organs, perhaps a brain, and eventually takes us into metaphorical territory, when it becomes synonymous with words like computer, and mathematical constructs as mentioned earlier.

In the early nineteenth century, George Boole created the concept of Boolean logic and binary systems—counting systems that only used "ones" and "zeros." During this era, a British inventor Charles Babbage and his protégé Augusta Ada Byron Lovelace, the daughter of the English poet Lord Byron, developed a calculating engine incorporating Boole's logic, the kind of logic that serves as an underpinning for today's computer.[1] In another industry, Joseph Jacquard, a French weaver developed the earliest programmable loom, using thin, perforated wooden boards to control the weaving of complicated designs [2]. This invention served to branch early computing technology in two directions, one which used a machine to create artistic designs, and in this sense, was a forerunner to the artificial intelligence that today creates forms of art, music and literature.

And, in another direction the Jaquard invention led Herman Hollerith to conceive the idea of using similar perforated cards in a system that passed punched cards over electrical contacts in a device to gather statistics during a national census. In 1889, Hollerith was granted patent protection related to the automation of tabulating and compiling statistical information. In 1911, a company utilizing the Hollerith invention formed out of several smaller companies, under the name Computing-Tabulating-Recording Company. In 1924, Thomas J. Watson, Sr. changed the company name to International Business Machine Corporation (IBM). It of course went on to become the world's largest manufacturer of business machines. IBM would develop the peripherals, such as card readers, printers and electric typewriters that later would attach to the large mainframe computers of the 1950s.

In 1936, mathematician Alan Turing wrote "On Computable Numbers with an Application to the Entscheidungs problem" [3]. Computable numbers are the set of real numbers that, using a suitable, finite, and terminal algorithm, can be calculated to any degree of precision. This idea served as a computational paradigm for the modern computer that read instructions and initiated an action to alter its next step. The hypothetical Turing machine was imagined as an electrical device, and as the machines were doing the work of computation, the name computer stuck, but also did its synonym machine. About the same time, John von Neumann, another mathematician, invented a method for storing instructions within a memory freeing the requirement to feed the instructions from paper cards or perforated tapes. The computer as we know it was coming together.

[1]The Analytical Engine was a proposed mechanical general-purpose computer designed by English mathematician and computer pioneer Charles Babbage, as assisted by Ada Lovelace. It was first described in 1837 as the successor to Babbage's difference engine, a design for a simpler mechanical computer.—Wikipedia

Unlike the computers of today, the computers of yesteryear used electronic vacuum tubes, then discrete solid-state transistors, which were followed by early small-scale integrated devices. An integrated device may have contained from ten to thirty transistors. Today most semiconductors are large-scale integrated circuits with millions of transistors. In the early days, memory was hardwired, meaning that its program or its data could not easily be altered. But, new companies, Fairchild, Intel, and Texas Instrument were to develop microcomputers that would revolutionize computation with ever-accelerating process performance and semiconductor memory.

Future computers will continue to use the standard silicon-based materials of today, but we are also headed toward the molecular computer and quantum computer of tomorrow. And, eventually new computational models will emerge, based computational paradigms tailored to DNA coding, cellular automata, fractals, the entangled qubits in quantum superposition. Other computers, will have as their objective replication, that is reproducing computers like itself, or manufacturing drugs, or mixing artificial biological products within the body. Imagine that a future computer will employ sensors to enable it to chemically analyze a molecule, and then duplicate the structural twin of that molecule. In this way, such a computer may operate as a self-replicating machine, or constructor of proteins or enzymes. Obviously such a device is not inorganic hard silicon, but conceivably an organic soft molecular computer designed to use analogs of CRISPR technology, which can interface with biological platforms, perhaps a DNA machine that transcribes and assembles DNA base pairs to make protein. It's clear at least a significant share of this technology it headed in the direction of—well our brains.

I have discussed the concept of machine in some detail, but in fact a machine qua computer functions as a process. And, it would serve us well to briefly turn to "process," to understand conceptually how a series of actions or steps achieve a particular end.

Process stands in relation to the causal implication, which is lacking in the description of a machine. Unlike a machine composed of matter, which implies tangibility, a process does not have tangibility or substance. As a word used in reference to technology, it means a continuing development involving many changes. Process applies to the states that a physical thing undergoes in a dynamic system. Take for example a matchstick, which is composed of a chemical comprised of phosphor. If I strike the surface of the match against a rough surface, it ignites, burns the phosphor and the underlying matchstick, finally extinguishing itself when the supply of combustible material has been exhausted. We describe this process in terms of various steps having physical states. The steps in the process consist of (1) applying a frictional force to raise a temperature, (2) igniting the phosphor, (3) burning the phosphor at a certain rate, (4) producing a flame of certain temperature, (5) changing temperature as the stick burns, and finally the gradual, and (6) decreasing of the flame as it extinguishes.

Although the foregoing example applies to the process of lighting a matchstick, the same conceptual idea surrounds the process of computer, which may involve the steps of reading data, storing data, accessing the data, performing a logical operation on the data, and taking an action, such as printing the result. And, as we have been discussing, communication, control or cybernetic processes can be programmed to simulate the behavior of natural systems or expressions found in the arts and music. Let's turn to how scientists are using these ideas about machines and processes to develop technology to advance brain prosthetics.

Notes

1. Verbeek, P.P. (2005). "Beyond the Human Eye, Mediated Vision and Posthumanity." In: P.J.H. Kockelkoren and P.J. Kockelkoren, eds., *Proceedings of AIAS Conference 'Mediated Vision.'* Ensched: Veenman Publishers en ARTez Press.
2. Delve, J. (October–December 2007). "Joseph Marie Jacquard: Inventor of the Jacquard Loom." *IEEE Annals of the History of Computing* 29 (4): 98–102; see, p. 98.
3. "On Computable Numbers with an Application to the Entscheidungs Problem." *Proceedings London Mathematical Society* 42: 230–265 (1936), pp. 544–546, vol. 93, England.

CHAPTER 19

Pathways to the Brain

I am not what happened to me, I am what I choose to become.—Carl Jung

In reengineering our biology to improve the odds of survival, technologists focus on a machine of a different sort—one that is biological, but offers strikingly appropriate analogies to the apparatuses and processes just discussed. It's through the transformation of information pervading the ether "out there" into signals that penetrate the deeper anatomical sensorial world, to shed light on an otherwise blind cortex, to reveal the manifold dimensions of the world in which we live.

As we engage in messaging others, even in the plain old fashioned face-to-face way, it necessitates a mind to world connection through senses: sight, hearing, touch, smell, and taste. Beyond these obvious thresholds to the outer world, we also have those which we pay less attention to, like our vestibular sense of balance, of movement and proprioception, i.e., orientation, space, and time. We employ over 260 million visual cells, and tens of thousands of auditory and receptor cells for touching, smelling, and tasting. Some sensory subsystems (e.g., retinal cells in the eye, hair cells in the ear's cochlea, taste buds in the gustatory system) preprocess electrochemical signals. Other sensory receptors take a circuitous path through the brain first processed by the thalamus, an important sensory waystation. Receptors analyze, separate, and integrate signals and noise, the raw data of the Universe, even before it is interpreted by the brain's sensory-specific cortices, such as the motor cortex for movement, occipital cortex

J. R. Carvalko Jr., *Conserving Humanity at the Dawn of Posthuman Technology*, https://doi.org/10.1007/978-3-030-26407-9_19

for vision, or superior temporal gyrus for sound. It is these areas of the anatomy, where the scientist endeavors to develop artificial analogs to support and augment human potentiality, via computer-like technology or genetic engineering, which someday will be bring an awareness to events and things, that are at present beyond the scope and capability of our current physiology. These will be embedded technologies, referred to as unawareables, able to control important physiological function.

Biological sensors, specifically neuro-chips, have been implanted into the neuroanatomy for well over five decades. The first application had to do with helping individuals with profound hearing loss due to a sensorineural disability. The cochlea is the sensory organ of the auditory system that turns sound vibrations into neural firing due to its mechanosensory "hair cells." A cochlear implant, containing a single-channel electrode, was surgically embedded in a patient's cochlea, at Stanford University in 1964 [1]. The modern multichannel cochlear implant came to fruition in 1977 [2]. Cochlear implant systems typically consist of two parts: one is worn behind the ear and consists of a microphone/processor; the second part, a stimulator/receiver, is implanted within a recess of the patient's temporal bone.

The implanted unit typically includes an antenna receiver coil that receives a coded signal and outputs a stimulation signal to an intracochlear electrode assembly, which applies the electrical stimulation to the auditory nerve to produce a hearing sensation corresponding to the original sound.

Neurons located in the cortex respond to stimuli from the thalamus as it relays signals from the visual field, i.e., the 125 million visual receptors in each eye. The signals are a series of electrical spikes. These spikes run down axons (or neural cell body), in response to such color, movement, and size of objects that are sensed by the eyes.[1] As of now, we do not know the gene complex that delivers the molecular code for creating the structure of the cells involved in this activity. We also do not know whether the genes at issue are of the sort that adds to the structure of a biological cell or serve an epigenetic function.

[1]David Hubel and Torsten Wiesel received the Nobel Prize in 1981 for their discovery that nerve cells in the visual cortex of an anesthetized cat showed a series of responses when light was shone on the cat's eye.

In 2010, researchers at the Salk Institute found that synchronous cortical input from the thalamus occurs upon detecting a salient event in the sensory environs, such as an object that moves into the receptive field. The scientists also determined the number of synchronous thalamic spikes needed to reliably report a sensory event to the cortical neurons [3]. When excited by an event, the spike rate increases to typically 50–100 Hertz, and for short periods may reach as high as 500 Hertz [4]. And, when excitation is removed, the resting frequencies of the spikes decrease to between 1 and 5 Hertz. The timing of spikes relates to an object's color or orientation. For example, a spike corresponding to the perception of the color "pinkish red" fires in synchrony with a spike for "round contour," enabling the visual cortex to merge these signals into the image of a flower pot.

Other studies also report that synchronization of neuronal activity in the brain can affect memory formation [5]. Nothing here seems to indicate what role genes play, although more on this will be addressed further on. One could speculate that by understanding data representations, scientists will eventually develop the hardware in the form of semiconductor technology or synthetic genes to interface with the brain, and thus alter the functionality of perception, cognition, processing speed, and memory [6].

In 2001, scientists implanted a microchip into a chimpanzee to control its neuromotor functionality, thereby allowing it to control a computer cursor using its thoughts. Then four years later, in 2005, a tetraplegic became the first person to control an artificial hand using a brain–computer interface. In 2012, *Nature* reported that two tetraplegic individuals, who were unable to move their limbs as a result of damaged brain stems, had nearly 100 electrodes implanted in their motor cortex allowing them to guide a robotic arm with recorded neuronal signals, ones associated with their intention to move.

In January 2016 the U.S. Defense Advanced Research Projects Agency announced that they would be developing an implantable neural system to enable communication between the brain and the digital world. The device would interface electrochemical signaling, used by neurons, into digital code.[2] This of course would involve an army of disciplines representing neuroscience, low-power electronics, photonics,

[2]Towards a High-Resolution, Implantable Neural Interface (2017). See, https://www.darpa.mil/news-events/2017-07-10.

medical device packaging, systems engineering, mathematics, computer science, and wireless communications. If this can be achieved, how far are we from construction of an artificial brain? And, will such a prosthetic be capable of expressing something as confounding as a sense about itself, a self-consciousness—, perhaps?

Notes

1. Mudry, A., and Mills, M. (May 2013). "The Early History of the Cochlear Implant: A Retrospective." *JAMA Otolaryngology–Head & Neck Surgery* 139 (5): 446–453. https://doi.org/10.1001/jamaoto.2013.293. PMID 23681026.
2. "2013 Lasker-DeBakey Clinical Medical Research Award: Modern Cochlear Implant." The Lasker Foundation. Retrieved 14 July 2017.
3. Wang, H.-P., et al. (2010). "Synchrony of Thalamocortical Inputs Maximizes Cortical Reliability." *Science* 328: 106. https://doi.org/10.1126/science.1183108.
4. Crick, F. *The Astonishing Hypothesis* (Scribner, 1994).
5. Jutras, M.J., and Buffalo, E.A. (2010). Synchronous Neural Activity and Memory Formation. *Current Opinion in Neurobiology* 20 (2): 150–155.
6. Sejnowski, T., and Delbruck, T. (October 2012). "The Language of the Brain." *Scientific American* 307 (4).

CHAPTER 20

Imaginative Construction

Imagine a super-sized alien, from another galaxy, approaches earth, but because of its immense size sees our planet in the same proportion to its body, as we see a soccer ball. It hovers over the world and with an advanced optical instrument, observes through a troposphere of clear skies and clouds, oceans showing little but the undulating waves of winds and tides. Finally, it spots an urban city, which to the alien's eye appears as an orderly array of colorless, odorless, static geometric blueprints that upon closer inspection reveals a pulsation, as people and their machines travel roads and byways, and tend to the business of everyday life. It focusses from a slightly different angle and sees a range of activities stretching from a quiet assumption of civility to the chaos of war. It zooms-in closer still and detects zones of intimacy, people sharing caverns of secrets, of nakedness, of sexuality, primordial obsessions, sweat, odors, privacy, and love.

The alien's field of view widens and narrows, until it internalizes the lives of several billion humans and the commercial routes that crisscross the planet. But, even the alien can't explain all that it perceives, or to recreate the multifaceted network that accounts for the separate and compounding effects of this dynamic system. Could it be that each of us has a mind, or more accurately a process, which we call consciousness with a complexity comparable to worlds in stochastic flux? And could it be that this immense complexity accounts for our inability to explain how our mind works?

J. R. Carvalko Jr., *Conserving Humanity at the Dawn of Posthuman Technology*, https://doi.org/10.1007/978-3-030-26407-9_20

The neural correlates of consciousness consist of a set of neuronal events and mechanisms sufficient for an explicit memory or conscious percept [1]. Neuroscientists discover these neural correlates of subjective phenomena experimentally by manipulating neurons using molecular biology in combination with imaging to model organisms that can be used in large-scale genomic analysis and manipulation. What seems surprising is that understanding the neural correlates of consciousness, has not led to any widely accepted theory of consciousness.

Physicist David Deutsch explains why this may be the case:

> … Consider the nerve signals reaching our brains from our sense organs. Far from providing direct or untainted access to reality, even they themselves are never experienced for what they really are—namely crackles of electrical activity. Nor, for the most part, do we experience them as being where they really are—inside our brains. Instead, we place them in the reality beyond. We do not just see blue: we see a blue sky up there, far away. We do not just feel pain: we experience a headache, or a stomach ache. The brain attaches those interpretations—'head', 'stomach' and 'up there'—to events that are in fact within our brain itself. Our sense organs themselves, and all the interpretations that we consciously and unconsciously attach to their outputs, are notoriously fallible… [2]

Yes, the mind works more like a planet than a machine, making it impossible to capture the whole of what goes on at any given moment, and thereby ably describe or explain it deterministically. Reductionism assumes determinism. Models based on reduction to a materiality assume that the brain and its mental states comprise machine-like operations. Material embodiments of the brain, which function according to physics and chemistry, might be modeled, if and when we solve the mind–body problem, the theory of mind, and thereby functions to produce emotions and feelings. As neuro-philosopher, Patricia Churchland suggests, "…when we make decisions, when we go to sleep, when we get angry, when we're fearful, these are just functions of the physical brain" [3].

John Searle believes that "… in the philosophy of mind, obvious facts about the mental, such as that we all really do have subjective conscious mental states and that these are not eliminable in favor of anything else, are routinely denied by many, perhaps most of the advanced thinkers in the subject" [4]. But, he believes that once we understand how the brain works, as we have for "phenomenon like digestion or photosynthesis," consciousness will be reduced to its material substance [5].

What makes consciousness impossible to explain has to do with it being a collection of practically innumerable random processes internal to each of us. Unlike how you and I can point to and describe entities like neurons, mountains, and boiling water, both qualitatively and quantitatively, only I can describe the pain within me, because it's internal to me. Such is my consciousness. Only I have access to it.

We've come a long way in actualizing artificial organs, such as hearts and kidneys. These are constructions that function in ways that are physically describable, as we discussed earlier with reference to machines. But, consciousness may not be physical, at least not in a way we understand the concept of physicality in terms of physics and chemistry. And, the lack an understanding of human consciousness, and how it works, may in the long run prevent us from fully synthesizing, through the application of a computer, what passes for human imagination. Nonetheless, even if we do not fully understand how the mind works, mechanistically, scientists are making progress in interfacing electronic devices to such organs as the hippocampus, which serve as memory prosthetics. If memory, like consciousness is not a well understood physical phenomenon, but nevertheless can be enhanced, does this not suggests the possibility that consciousness also may be enhanced, despite not being understood, as for example something mechanistic or physical.[1] The essence of the entity may be alterable without knowing precisely how it works.

The notion that even if we did not fully understand how the brain worked, we could nonetheless manipulate it, is not a new idea. The seventeenth-century philosopher René Descartes first raised the specter of an evil demon that the philosopher Hilary Putnam later resurrected [6]. Putnam imagined a person whose brain was separated from his body and then kept alive in a vat. When the brain was connected to a computer it revealed that it had the illusion that things were normal [7]. So, query, in a world where brains may be interfaced to a computer, will the hybrid entity, either the brain or the computer, exercise intelligence,

[1] Michael LaTorra, makes the point that different opinions on the question of consciousness, subjective experience, between materialists such as Patricia Churchland and dualists such as David Chalmers. Bottom line: As of now we do not have the technology for human-level AI, nor do we have technology for scanning human brains to help create software brain emulations.

and yet feel autonomous? Or, even more bazaar, perhaps manifest in a newfound expression or embodiment of the soul, one that emerges within the crevices of a creative organism, self-awareness or an "*I*," which might emulate or exceed the discoveries of our most prominent scientists or the artistry of a Beethoven, Monk, da Vinci, or Picasso? [8].

Without question, state-of-the-art software and the computational engines that run it, easily exceed humans in certain kinds of brain-power. The grandmaster at the game Go was sorely beaten by DeepMind Technologies' AlphaGo, a system consisting of 48 standard-faire, central processing units (CPUs) running neural network algorithms for machine-learning and logic.[2] But, as Jim Kozubek writes: "intelligence is not a simple input-output system, as much as it is a developed ability to hold in mind and toggle between two or more opposing thoughts, as much as it is a capacity for memory" [9]. As the subject relates to creativity, it raises the question of the relationship between the mind and how it assembles thoughts through representations and combinations of concepts. For example, humans arrange thoughts as associative memories, which fundamentally operate differently from computer memories, which are arranged more like books in a library. But as we turn more and more to computers as extensions of ourselves, we see that their architectures are beginning to coalesce around and conjoin biology and computer science, a subject we now explore in the next few chapters.

Notes

1. Koch, C. (2004). *The Quest for Consciousness: A Neurobiological Approach.* Englewood, CO: Roberts & Company Publishers. ISBN 0-9747077-0-8.
2. Deutsch, D. (2011). *The Beginning of Infinity, Explanations That Transform the World.* Viking.
3. The Benefits of Realizing You're Just a Brain, New Scientist, Interview Patricia Churchland, by Graham Lawton, November 27, 2013. https://www.newscientist.com/article/mg22029450-200-the-benefits-of-realising-youre-just-a-brain/ (Last visited 08/11/2018).
4. Searle, J.R. (1997). *The Mystery of Consciousness.* New York: New York Review Books.
5. Searle, J.R. (1998). *Minds, Language and Society.* Basic Books.

[2]At a match held in Seoul in South Korea, on 12 March 2016, the world Go champion Lee Sedol was defeated.

6. Descartes, R. (1641). *Meditations on First Philosophy.*
7. Putnam, H. (1981). *Reason, Truth and History.* Cambridge University Press.
8. Hofstadter, D. (2007). *I Am a Strange Loop.* Basic Books.
9. Can CRISPR–Cas9 Boost Intelligence? Jim Kozubek, September 23, 2016. https://blogs.scientificamerican.com/guest-blog/can-crispr-cas9-boost-intelligence/ (Last visited 8/7/2018).

CHAPTER 21

The Techno Mind

Artificial neurons were first proposed in 1943 by two giants in the pantheon of artificial intelligence, neurophysiologist, Warren McCulloch, and logician, Walter Pitts. The McCulloch-Pitts model was rooted in mathematics. Conceptually, it appears as a summation of inputs, as supplied by signals, for example voltages, which are summed according to their salience or weight. At some point a threshold is reached, which fires a neuron (See, Fig. 21.1). Following this model, the study of neural networks began to split into two fields: neurobiological processes and artificial intelligence, setting afoot projects since that era to model versions of thought processes.

Building a brain is not a new ambition; it dates back decades. W. Ross Ashby's 1952 book, Design for a Brain, focused on cerebral mechanisms that produce adaptive behaviors. Hundreds of papers and book have been written on the subject since then. More recently, Futurist Ray Kurzweil offered his thoughts on the creation of an artificial brain in his 2012 book How to Create a Mind: The Secret of Human Thought Revealed. Discussing his pattern recognition theory of mind, Kurzweil posits that the neocortex is a hierarchical system of pattern recognizers, which if emulated in a computer could lead to an artificial superintelligence.

Apropos to the aspirations of Ashby and Kurzweil (and others unmentioned), scientists are building prototypes of cortical prostheses to mimic the brain's natural patterns in the information encoding phase of learning and memory. The cortical prostheses can (1) monitor input patterns

J. R. Carvalko Jr., *Conserving Humanity at the Dawn of Posthuman Technology*, https://doi.org/10.1007/978-3-030-26407-9_21

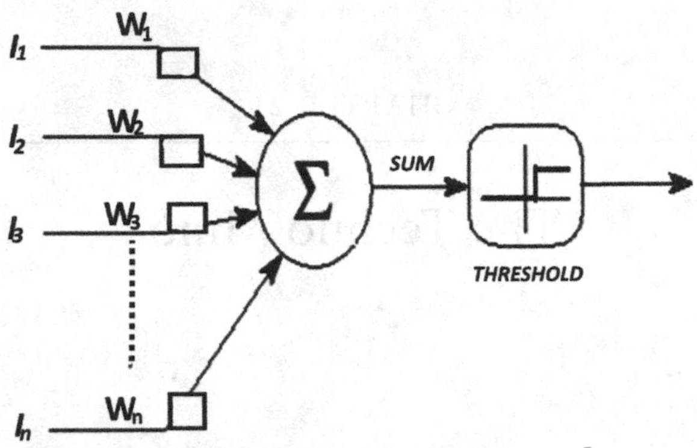

Fig. 21.1 Model of a neuron

to hippocampus, during the information encoding phase, (2) predict associated hippocampal output patterns, and (3) deliver stimulated electrical pulses, during the same phase of the task in a pattern that conforms to the normal firing of the hippocampal output region. The utility of cortical prosthetics is established by their capacity to substitute encoding stimulation, when for example, a hippocampal ensemble cannot generate the necessary codes to successfully perform memory functionality [1].

Beyond assisting in memory recall, other technologies are being developed to affect our psychological outlook. In order to observe mood swings, University of California researchers inserted tiny electrodes onto the surfaces and within the deeper structures of patients' brains, such as the amygdala and hippocampus. Known as intracranial electroencephalography (iEEG) recording, the procedure measured neural activity to locate seizure-triggering areas of the brain and is frequently used to gain insight into brain activity by conducting in vivo cranial recordings [2]. Using computers with machine learning algorithms, the researchers examined the complex relationships between patterns of neural activity indicative of one's mood. Part of their findings revealed that a common subnetwork fluctuated at a specific frequency between the amygdala and hippocampus, evidence that the two parts of the brain were likely communicating.

If we back up twenty years, we see a steady stream of progress toward developing brain prosthetics. In August 1999, a conference was held, entitled *Toward Replacement Parts for the Brain:*

Intracranial Implantation of Hardware Models of Neural Circuitry. The sponsors of the event were the National Institute of Mental Health (NIMH), the University of Southern California (USC), Alfred E. Mann Institute for Biomedical Engineering, and the USC Center for Neural Engineering. The assembly of neuroscientists, engineers, and medical researchers came together to discuss and report on (1) mathematically modeling the functional properties of different regions of the brain, (2) the design and fabrication of microchips for incorporating these models, and (3) the engineering of neuron/silicon interfaces to integrate microchips and brain functions [3].

In 2001, as part of the U.S. government Defense Advanced Research Projects Agency a team of interdisciplinary researchers headed by Theodore Berger reported the results of performing an implantable biomimetic electronic device in a brain. Berger, and the team developed a prosthetic, which comprised a microcomputer processor, programs and an interface. The system was then installed it into the hippocampus of a rat, where they were able to duplicate the pattern of interaction between two subregions of the hippocampus, CA1 and CA3. These subregions interact to create long-term memory, and by monitoring the neural spikes in these cells, using an electrode array, they could then playback the same pattern, within the same array. The result was a prosthetic that when implanted in rats with a normal, functioning hippocampus, demonstrated the ability to improve memory.

The experiment provided for a better understanding of the interface and functional requirements for a brain prosthetic, whether employed for enhancement or therapeutic effect. As part of their research, the Berger team outlined the requirements for the system as follows: (1) it must be biomimetic, i.e., neuron models incorporated in the prosthetic must have properties of real biological neurons; (2) a physiological or cognitive function must be detectably impaired (according to neurological or psychiatric criteria); (3) the neuron and neural network models must be miniaturized sufficiently to be implantable; (4) the microchip system must communicate with living neural tissue in a bidirectional manner; (5) the variability in phenotypic and developmental expression of both structural and functional characteristics of the brain necessitates adaptation of each prosthetic to the individual patient; (6) there should be a consideration included in the neuron/network design for "personalizing" the prosthetic according to the user; and (7) and the power required for the prosthetic device must minimize the generation of heat in brain subregions to remain biocompatible [4].

In the near future, computer processing rates of the order of magnitude of 10 trillion flops per second (FPS) will exceed the processing rate of the human brain, estimated at roughly 10^{15} FPS. This kind of clout will permit adjuncts to human memory, running at the same processing rate as the human brain, and will coexist within our bodies, to provide IBM Watson-like artificial intelligence and directly communicate with services now offered via the Internet, much like a Smartphone, directly.

Within the next few years, computers will perform 10^{16} calculations per second and have storage capacities exceeding 10^{15} bytes, which come close to the processing power of the human brain, but the appropriate "brain" model and software will be required if the brain itself is to be faithfully simulated. The brain has functionality that controls not only conscious thought, but directs and controls sensory inputs and outputs, as well as metabolic systems, which regulate the entire anatomy.

IBM Watson-like artificial intelligence allows computers to do theoretically what an efficient mind, tasked with solving problems using logical associations does, and which can be represented by sets of rules. Alan Turing predicted that a computer would eventually pass for another human, when an interviewer would not be able to distinguish the computer's response from that of a human. The computer seems well past that point.

Henry Markram runs IBM's Blue Brain Project, which uses a supercomputer to simulate neural micro-circuits. His aim has been to construct models of neuromodulators, analyze different forms of plasticity, gap-junctions, and the neuro-vascular glia system, and to couple them to neuro-robotic systems, which in the long run will simulate perception, cognition and behavior. To this end, scientists have been collecting data on neuron morphology and electrical behavior of the juvenile rat in the laboratory for years. Part of the project seeks to simulate a rat's brain, which contains over 100 million neurons and one trillion synapses [5].

Blue Brain Project scientists are also attempting to engineer neocortical column functionality through a greater understanding of its composition, density and the distribution of the numerous cortical cell types.[1]

[1]The columnar hypothesis, framed by Vernon Mountcastle in 1957, holds that the cortex is composed of discrete, modular columns of neurons characterized by a consistent connectivity profile. And although the columnar hypothesis has not yielded a microcircuit that corresponds to the cortical column, or even any genetic basis for its existence, or how to

The density of each cell type and the volume of the occupied space provides clues for cell positioning and constructing the cortical circuits.

Among others in the field of reverse engineering the human brain, Joe Z. Tsien, one of the scientists that developed the transgenic mouse Doggie, referred to earlier, studies the relationship of neural circuits and behavior that underlie brain computation and intelligence [6]. According to Tsien's Theory of Connectivity, a "power-of-two" based computation serves as the design logic for constructing cell assemblies into a microcircuit-level building block termed a functional connectivity motif (FCM) [7]. But, for all the research into constructing an analog to the anatomical brain, predictions as to when this will occur vary considerably, with a preponderant support base putting it in the 2050 timeframe. But, progress on parts of the brain that can enhance intelligence are already within our grasp. This progress primarily has to do with prosthetics that improve performance in selective ways.

The reason constructing brain-like supercomputers is important in our consideration of transhumans deals with the difference between science and engineering. Science is a bottom-up process, whereas engineering is top-down. On their way to achieving the overarching objective like constructing drilling rigs, circus tents, or tunnels, engineers have to join smaller elements, upon which to build upon. These are the foundations, forms, beams, and trusses, which inevitably undergo inventive transformation, becoming the inventory of technology for such things as insulin pumps, catheters, and heart stents. As we engineer a synthetic brain, the smaller elements will become the prior art in a succession of inventions that take on new functionality, somewhere with the biological brain. But, because it is difficult to imagine a breakthrough technology, such as the invention of the transistor or the laser, a successful brain synthesis, by today's lights, will depend on the continued miniaturization of logic circuits.

Intel's main digital processor workhorse, introduced in 2014, packs nearly 40 million, 14-nanometer-sized chips into one square millimeter. A fingernail-sized substrate can contain over one billion transistors.

construct a column, it remains highly influential in explaining the cortical processing of information. Mountcastle's description of the columnar organization has been cited over 1462 times! For a summary of its history, see, The neocortical column, DeFelipe, J., et al. *Frontiers in Neuroanatomy*, 26 June 2012.

In a few short years, a 12-nanometer chip will reduce the processor size threefold, to about 5 nm. By comparison, one DNA molecule base pair measures 0.34 nm. About 20 base pairs roughly, a nano-sized transistor, making 5 nm sized circuits well-suited for applications at the cytological level.

Closer to the objective of integrating computers into the brain, scientists involved with the SyNAPSE project at IBM Research created the most advanced neuromorphic (brain-like) computer chip to date. Called TrueNorth, it consists of 1 million programmable neurons and 256 million programmable synapses, across 4096 individual neurosynaptic cores. Built on Samsung's 28 nm process, TrueNorth has 5.4 billion transistor count, making it the largest and most advanced computer chip ever made. Perhaps more important, though, TrueNorth is incredibly efficient: The chip consumes just 72 milliwatts power at loads that equate to 400 billion synaptic operations per second, per watt—that makes it about 176,000 times more efficient than a modern CPU running the same brain-like capacity, or 769 times more efficient than other state-of-the-art neuromorphic approaches" [8].

More than a few projects, such as, Markam's, Tsien's and SyNAPSE are in the exploratory stage, although there does not appear to be coordinated effort to join or assimilate them into a single technological mainstream that moves in one direction. But, over time we may see how innovation actually does converge. The parallel here is that between 1940 and 1980, many ideas were floated and modeled as to a wide range of solutions related to computers, personal and commercial, but only through years of winnowing and absorption did the most efficacious ideas survive. And, somewhat unpredictably advancements such as the Internet connectivity and telecommunications capacities, e.g., bandwidths and transmission speeds, were hardly considered. And unlike brain technology these days, then a significant commercial direction played into the development of large mainframe computers. It took years before a paradigm shift brought the feasibility of computer technology into the home, literally—supplying nearly every individual throughout the world with a computer in the form of their personal smartphone. Much of this due to the increasing miniaturization of semiconductors.

Similar to the developmental comparison between mainframe computers and personal computers, today great interest abounds in in-the-body computers, which will use microelectromechanical systems (MEMS), such as nano-sized motors. These devices will range in size from 10^{-9}

(a few hundred atoms across) to 10^{-5}—the diameter of a white blood cell. These traveling nanobots will course through arteries and cellular membranes to deliver drugs and destroy pathogens. It is conceivable and likely that these machines will be equipped to communicate with the outside their anatomical boundary. And, to the extent they do, to what degree can we anticipate that such technologies will improve our virtual intelligence, that is, information within the anatomy that can access search engines, such a Google, using simple thoughts? And to this end, development of the molecular computer will eventually move into the anatomy to improve various elements within the psychological g-factor complex: improved visuospatial processing (Gv), working memory (Gwm), and fluid intelligence (Gf). Let's next turn our attention there.

Notes

1. Berger, T.W., et al. (2011). A Cortical Neural Prosthesis for Restoring and Enhancing Memory. *Journal of Neural Engineering* 8 (4): 046017.
2. Kirkby, L.A., Luongo, F.J., Lee, M.B., Nahum, M., Van Vleet, T.M., Rao, V.R., Dawes, H.E., Chang, E.F., and Sohal, V.S. (2018, October 29). An Amygdala-Hippocampus Subnetwork that Encodes Variation in Human Mood. *Cell* pii: S0092–8674(18)31313–31318.
3. Berger, T., et al. (2005). "Brain-Implantable Biomimetic Electronics as a Neural Prosthesis for Hippocampal Memory Function." In: Berger and Glanzman, eds., *Toward Replacement Parts for the Brain*. MIT Press.
4. Berger, T.W., et al. (2001, July). "Brain-Implantable Biomimetic Electronics as the Next Era in Neural Prosthetics." *Proceedings of the IEEE* 89 (7).
5. Markram, et al. (2015, October 8). "Reconstruction and Simulation of Neocortical Microcircuitry." *Cell* 163: 456–492.
6. Tsien, J.Z. *Frontiers in Systems Neuroscience*, 3 February 2016 | https://doi.org/10.3389/fnsys.2015.00186.
7. Xie et al. "Brain Computation Is Organized via Power-of-Two-Based Permutation Logic." *Frontiers in Systems Neuroscience* 10. https://doi.org/10.3389/fnsys.2016.00095.
8. See, https://www.extremetech.com/extreme/187612-ibm-cracks-open-a-new-era-of-computing-with-brain-like-chip-4096-cores-1-million-neurons-5-4-billion-transistors.

CHAPTER 22

Bioengineered Computers

Earlier we discussed bacterial viruses as sources of DNA for recombinant modification of genes that were in turn used to create new life forms. Beginning in the early 1990s, scientists started using combinations of organic chemistry and molecular biology to develop synthetic forms of DNA to achieve similar aims. Referred to as synthetic DNA, this technology has the potential for producing a wide variety of products, such as drugs, new life forms, and even computers [1].[1]

In 2003, Nobel Laureate, Hamilton Smith was part of a group that synthetically assembled the genome of the virus, Phi X 174 bacteriophage. His efforts, at the J. Craig Venter Institute (JCVI), were to partially synthesize a species of bacterium derived from the genome of *Mycoplasma genitalium*.[2] Then in 2008, synthetic biology was brought into prominence when Dr. Venter impressed a TED forum on his work

[1]The term synthetic biology was used as early as 1974 by Waclaw Szybalski who saw molecular biology evolving from description to manipulation of genetic systems. See, Kohn, A., and Shatkay, A. (eds.). (1973). "In Vivo and In Vitro Initiation of Transcription." In: *Control of Gene Expression*. New York: Plenum Press.

[2]The J. Craig Venter Institute is a nonprofit genomics research institute founded by J. Craig Venter, in October 2006. The Institute consolidated four organizations: the Center for the Advancement of Genomics, The Institute for Genomic Research, the Institute for Biological Energy Alternatives, and the J. Craig Venter Science Foundation Joint Technology Center.

J. R. Carvalko Jr., *Conserving Humanity at the Dawn of Posthuman Technology*, https://doi.org/10.1007/978-3-030-26407-9_22

in creating living bacteria from synthesized molecules.[3] Two years later, JCVI announced creation of the world's first self-replicating synthetic genome in a bacterial cell of a different species [2, 3]. In 2010, Venter's team announced they had assembled a complete genome of millions of base pairs, inserted it into a cell, and which caused the cell to start replicating [4].

JCVI's success for self-replicating synthetic genomes, led President Obama to convene the President's Commission for the Study of Bioethical Issues, in 2010, to identify the risks and ethical concerns. The Commission held a series of public meetings to assess the science, ethics, and public policy and then issued a report entitled *New Directions: The Ethics of Synthetic Biology and Emerging Technologies.* While recognizing that synthetic biology could likely lead to cures for diseases, such as muscular dystrophy, Parkinson's disease, cystic fibrosis and cancer, the commission concluded that it would be "imprudent either to declare a moratorium on synthetic biology until all risks can be determined and mitigated, or to simply 'let science rip,' regardless of the likely risks" [5, 6]. Successful synthetic biological systems now include multicellular applications, e.g. bacteria that sense and destroy tumors and organisms that produce drug precursors.

Researchers are experimenting with various sources of DNA, such as the *E. coli* genome, and have developed synthetic biological circuits that mimic toggle switches measured in micrometers. Installed into the human anatomy, these devices, potentially as complex as microprocessors, will enhance well-being, improve intelligence, and promote longevity. In the future, this technology may also serve as a component in supercomputers that may not run conventional programs, which store and fetch data, and execute programs serially, but rather act as parallel processors, where data and programs are intertwined. Bio-circuits may be constructed that function as neural networks, and perform chemical computations making obsolete the centuries-old matrix forms of algebra, substituting instead, in vivo analysis of bioelectronic molecules. And, although much is written about future AI-driven silicon-based robotics,

[3] Craig Venter a biologist and entrepreneur instrumental in sequencing the human genome, has worked with a team of researchers at JCVI to create the largest man-made DNA structure by synthesizing and assembling the 582,970 base pair genome of a bacterium, Mycoplasma genitalium JCVI-1.0.

in time, seamlessly interfaced digital-like biological platforms, installed into transhumans, may lead to the dominant form of *Homo futuro*—a *Homo sapiens* platform augmented via one or more biocomputer processors.

Let's turn our attention to the power of the genome's natural structure to solve certain mathematical problems, which surprisingly, modern computers either cannot feasibly achieve or, at least, not achieve without expending great time, energy, and expense. When carrying out the process of DNA transcription, a DNA-based molecular computer generally follows the same principles as a Turing machine, a theoretical computer that moves along an "input" tape reading and interpreting the data stored on the tape [7]. A simple explanation describes the Turing machine as a device having a "read head" that reads a tape with squares arranged in a line. With the head placed over one square at a time, the machine can perform three basic operations: (1) read the symbol on the square under the head; (2) edit the symbol by writing a new symbol or erasing it; (3) move the tape left or right by one square, so that the machine can read and edit the symbol on a neighboring square; and (4) optionally repeat operations 1–3 or halt.

Turing's mind experiment was deemed a universal computer—as it could be programmed to compute anything that was computable. And likewise, a related corollary, offered by mathematicians Alonzo Church, Alan Turing, and Kurt Gödel, held that: any human computation can be done by a Turing machine [8, 9]. This is theoretically true, but some computations could take hundreds of years. And, even with the aid of conventional computers, considerable time. For particular problems, parallel computation could theoretically reduce computation time. But, the question remained for some time as to how to a construct parallel computer to solve a class of problems. One idea looked at how DNA transcribes as a possible solution to a type of combinatorial problem.

The idea that DNA works like a Turing machine was driven home in 1994, when physicist Leonard Adleman used a DNA molecule as a computational element to solve a classical mathematics' problem referred to as the "Traveling Salesman" or Hamiltonian path problem [10].[4]

[4]The traveling salesman problem deals with the question: if we have a list of cities and the distances between each pair of cities, what is the shortest possible route that visits each city and returns to the origin city? In mathematical jargon it's called an NP-hard problem in combinatorial optimization, which finds applications in operations research and theoretical computer science. The mathematician Leonard Euler posed a similar problem referred to as The Seven Bridges of Konigsberg.

Centuries ago, mathematician Leonard Euler posed The Seven Bridges of Konigsberg conundrum, which set up an island where two rivers branches passed on each side, respectively. Seven bridges crossed the two river branches. The objective was to plan a walk that crossed each of seven bridges just once. In other forms, a map is presented and the challenge is to find the shortest path for a traveler that starts at city A, visiting each of the other cities only once, and returning to A.

Adleman saw the Seven Bridge problem as one where an airline had to fly through several cities without crossing over the same path twice.[5] The experiment involved finding a route through a network of "cities" (labeled "1" to "7") connected by one-way "routes." The problem specifies that the route must start and end at specific cities and visit each one only once. His approach to solving the problem can be summarized as follows: (1) using the genome's nucleotides molecules A, T, C, and G in combinations to spell, in a code-word, each city by name and possible flight paths to the respective city. Think of an airline ticket with a city's name and a flight number. (2) the molecules representing each of these cities and corresponding flight numbers were mixed in a test tube, where some of the DNA strands stuck together, based on the way A, T, C, and G form combinations, i.e., A and T stick together in pairwise fashion, as do G and C, thus the sequence AGCT would stick perfectly to TCGA. Some of the nucleotides that stuck together represented possible answers. (3) a few seconds later, all of the possible combinations of DNA strands were formed in the test tube. (4) Adleman eliminated any wrong combinations through chemical reactions, which left behind only the flight paths that connected to the cities, and more important only through one continuous flight path from start to finish.

Adleman's experiment was the first to show proof of concept that biology, specifically its DNA apparatus, could be used to solve a real-world mathematics' problem. In 2007, a patent issued on an invention using the Adleman-type computer for carrying out more general computations [11]. Since Adleman, the technology for developing computer-like devices and circuits using synthetic DNA has developed at a rate not unlike that witnessed in the 1960s relative to semiconductors [12]. Molecular computers have dense information storage capacities (1 bit per cubic nanometer), and exhibit a rich computational parallelism (as shown by

[5] Adleman wrote: "[A] path exists that will commence at the start city (Atlanta), finish at the end city (Detroit).

Adleman), which adds to its speed. DNA molecules can also operate at effective speeds of as much as 10^{15} operations per second. As such, these devices are capable of performing about 1 billion operations per second, needing no active power sources, operating at the equivalent of about one watt radiated or dissipated over one second. One gram of DNA (approximately one cubic centimeter) can store one trillion CD's worth of information.

An early example of an early biocomputer was based on protein development. Its inventor Robert Birge, in his 1995 Scientific American article, "Protein-Based Computers," claimed a computer having compact size and faster data storage, which could be used in applications amenable to parallel-processing, three-dimensional memories and neural networks [13]. It should be recognized that inorganic computers, those based silicon-based microprocessors, are unlike computers based on naturally occurring organic matter. As such, it's likely that the types of problems, which these computes will address have the potential to solve problems, as Adleman demonstrates, that a conventional computer would find difficult or impossible to solve.

Let's back up a quarter-century and revisit the early years of synthetic biology development, when a team at ETH Zurich created DNA consisting of two artificial nucleotide genetic base pairs designed to work with natural AGTC base pairs. Then, in 2000, scientists reported constructing two devices that worked inside *E. coli* cells, by altering its DNA sequence, causing the cells to blink predictably—that is, turn on and turn off [14]. These and other similar acting modules have the potential for regulating gene expression, protein development, and cell-to-cell communication. Sixteen years ago scientists at Stanford University first constructed a bio-transistor from genetic material that worked inside of living bacteria [15].

In 2007, researchers demonstrated a universal logic evaluator that operates in mammalian cells, which, in 2011, was proof-of-concept for therapy that used biological digital computation to detect and kill human cancer cells [16].

Enzyme logic gates provide Boolean operations such as combining the presence of two inputs into one output. These are referred to as "AND" and "NOT AND" or "NAND" functions, which according to mathematical theory, if properly arranged, can carry out any imaginable, realizable mathematical function. This type logic, of course, underpins the digital computer. Another similar device contained a feedback loop allowing it to flip-flop, that is, toggle between two states, much like the electronic circuit used in computer memory, pulse generators, time-delay circuits, oscillators, and logic formulas [17].

Chris Voigt, a synthetic biologist at the University of California—San Francisco and his team engineered a bacterial system to regulate gene expression in response to a red light and another to sense its environment and conditionally invade cancer cells.[6] As reported in Nature: "[T] he system consists of a synthetic sensor kinase that allows … bacteria to function as a biological film, such that the projection of a pattern of light on to the bacteria produces a high-definition two-dimensional chemical image (about 100 megapixels per square inch). The spatial control of bacterial gene expression could be used to 'print' complex biological materials and to investigate signaling pathways" [18].

Voigt and his colleagues also successfully demonstrated a logical AND gate inside bacteria. This latter achievement ranks on a par with the invention of the electronic AND gate invention, used by electrical engineers in developing control circuits, which eventually led to the creation of the modern computer.

With advances spurred on by CRISPR-type developments, Voight and his colleagues reported that they are employing dCas9, which uses small guide RNAs, (sgRNAs) to repress genetic loci via the programmability of RNA:DNA base pairing, for building transcriptional logic gates to perform computation in living cells [19]. In 2017, other researchers also demonstrated a Boolean logic and arithmetic system to engineer digital computation in human cells [20].

In a similar vein, synthetic sensors and logical switches have been used to control the assembly and function of such things as secretion needles that export spider silk proteins or a photosynthetic apparatus for converting light into chemical energy. But, as Purnick and Weiss point out: "Whereas traditional engineering practices typically rely on the standardization of parts, the uncertain and intricate nature of biology makes standardization in the synthetic biology field difficult. Beyond typical circuit design issues, synthetic biologists must also account for cell death, crosstalk, mutations, intracellular, intercellular and extracellular

[6]Autopoiesis is the property of a bacterial cell or a multicellular organism that allows it to renew itself by regulating its composition and conserving its boundaries, and has called attention to biological mechanisms of self-production. See, Varela, F.J., in *Self-Organizing Systems: An Interdisciplinary Approach*, 1981.

conditions, noise and other biological phenomena" [21]. No one disputes the power of this technology to improve our health, but only recently has proof of concept appeared. The question is how soon before we see products that actually employ this technology.

Until recently, synthetic cells were products of the natural nucleotide bases comprised of ATGC, but in 2014, a semisynthetic organism, with an expanded genetic alphabet was created, employing artificial nucleotides not found in Nature [22]. This further demonstrates the rapid confluence occurring between natural biological systems and synthetic systems, between hybrid living and completely synthetic cells. Devices of this sort may be combinations of nonbiological materials and synthesized materials designed for biological applications. They are built in several different ways: (1) top-down by either by knocking out nonessential genes or by a completely replacing the genes with synthetic ones, and (2) bottom-up by assembling the cell using nonliving materials.

Artificial cells, are not computational devices per se, or synthetic sensors or logical switches, but are tiny hollow spheres having innards that function as a biological cell. Construction is usually part biological and part synthetic, employing polymeric materials, having properties of permeability that can enclose liposomes, nanoparticles, with polymersomes made from amphiphilic synthetic block copolymer spheres. These spheres can have radii ranging from 50 nm to 5 μm or more. Like natural cells, they can be made to self-assemble and self-reproduce to interlock chemical properties and contain electrical components of the variety that can be assembled to compute.

In one embodiment, the artificial cell mimics properties of biological cells in both morphology and function. Today scientists can construct blood cells in the form of Hemoglobin-based nano-sized oxygen carriers as a type of red blood cell substitute. Artificial cells for hepatocyte transplantation has proved feasible in providing liver function in animal liver disease and bio-artificial liver devices. Encapsulated bacterial cells have been proposed for the modulation of intestinal microflora, prevention of diarrheal diseases, treatment of *H. Pylori* infections, atopic inflammations, lactose intolerance, and immune modulation. Encapsulated artificial hepatocytes cells now play role in alternative therapies for liver failure.

One bioengineering goal is to build devices as adjuncts to living cells, where cells supply the energy and materials to ecological systems that decode DNA into messenger RNA and finally manufacture protein. A recent example of this was reported by ETH Zurich, where engineers infused hamster cells with networks of genes whereby adding antibiotics turned the output of the synthetic genes to low, medium or high, which could eventually lead to gene therapies, synthesizing drugs and the manufacture of proteins.

The concept of adding electronics to artificial cells has been around since about 2004, when the European Commission sponsored a project to develop the Programmable Artificial Cell Evolution, with a specific aim of embedding Information Technology (IT) using programmable chemical systems that approach artificial cells having properties of self-repair, self-assembly, self-reproduction, and the capability to evolve. In 2008, the project reported that it had achieved three core functions of artificial cells (a genetic subsystem, a containment system, and a metabolic system), and generated novel spatially resolved programmable microfluidic environments for the integration of containment and genetic amplification. As part of the mission, the team studied vesicles as a container for one family of artificial cells, which would self-assemble into higher order building blocks using processes like computer-aided manufacturing systems.

To fully develop computer systems on biological platforms, engineers have to invent computational components, as well as circuits that transform the real world of continuous variables, such as temperature, motion, pressure, and electronic analog signals.[7] One such example of this technology resulted in U.S. Patent 9,697,460, ('460) entitled "Biological analog-to-digital and digital-to-analog converters," issued (2017) to James Collins and Timothy Kuan-Ta Lu.

It will be instructive to explain this invention in some detail to illustrate the state of the art as it presently exists. The disclosure describes an artificial cell that performs a standard conversion between the analog and the digital world, a function essential in the operation of many complex electronic systems today. A binary numeral system is utilized, which therefore directly corresponds to its electronic counterpart, the digital computer.[8] It is remarkable that this device may be employed in

[7]According to one developer of these devices, "The ability of DNA-based tracers to store information makes them attractive for performing distributed measurements and delivering localized information upon recollection… We have demonstrated that smart DNA-based tracers can measure temperature, oxidative stress, and light intensity or duration." See, http://www.fml.ethz.ch/research/fosslab.html#CED4.

[8]For example, a combination of three genetic toggle switches produce outputs 00000000, when all genetic toggle's are "off," and 00000001 00000010, and 00000011, when 1, 2 and 3 genetic toggle switch, respectively are "on."

engineering complex behavioral phenotypes in cellular systems, such as prokaryotic, eukaryotic, or synthetic cells, by combining the power of nucleic acid-based methods with biological approaches to effectuate targeted responses in both cellular and noncellular systems. This device and other devices such as this one are what the future holds in store for embedding processors into the anatomy to realize sophisticated computer-mediated applications.

What follows is drawn from the '460 patent specification. In overview its biological switches convert analog inputs into digital outputs and digital inputs into analog outputs. An analog electrical signal is fed into a bank of converter switches, different voltage thresholds produce a digital output, similar to its counterpart semiconductor analog to digital converter. Thus, depending on the strength of the analog signal, a different number of switches are flipped, thus yielding a digital output, which is represented by the combination of switches that are toggled—precisely providing the kind of functionality used in conventional electronic circuitry.

To provide for digital-to-analog conversion, a digital input is represented by a bank of switches in binary format, where each switch drives a transcriptional promoter of differing strengths, with the switch representing the least significant bit, driving a promoter of least strength, and the switch representing the most significant bit driving a promoter of the greatest strength. These promoters express identical outputs, such as proteins, including fluorescent reporters, transcriptional activators, and transcriptional repressors, or RNA molecules, such as iRNA molecules, which inhibit gene expression or translation, by neutralizing targeted mRNA molecules. Thus, the digital input represented in the bank of switches is converted to an analog output based on the additive activity of the different promoters that are activated based on the specific digital combination of switches.

The '460 invention also provides for genetic toggle switches, made up of nucleic acid and protein components, such as promoters, transcriptional activators, transcriptional repressors, and recombinases. The ability to manipulate and combine these type of devices is what promises to turn this technology into a platform for a fully fledged digital computer. A "genetic toggle switch," as defined in the patent is an addressable cellular memory unit or module that can be constructed from any two repressible promoters arranged in a mutually inhibitory network. The one described in the patent exhibits bistable behavior over a wide range of parameters, where the two states are tolerant of fluctuations inherent in gene expression, i.e., a genetic toggle switching does not flip randomly between states [23].

Because a DNA-like computer uses principles and materials from organic chemistry they should better interface with human physiology than silicon-based computers. DNA may supply analogous programming information, but mechanisms are required to affect tissue in computationally meaningful ways. Scientists at UCLA are advancing protein engineering targeted to better deliver drugs and artificial vaccines, and building nanostructures by assembling computer-designed protein domains in rigid configurations [24]. In the case of the brain, for example, the biochemical alterations would involve protein production and epigenetic expression—this last a mechanism, as mentioned previously, controls gene expression.

Before molecular computers solve practical problems, scientists still need to: (1) develop an artificial cell that communicates and networks with its environment and with other cells; (2) improve replication, division and evolution of these kinds of cells; and (3) manipulate artificial cells to ingest nutrients and fully integrate into living organisms [25].

These kinds of hurdles have to be dealt with each time a paradigm shift occurs. Our Stone Age ancestors must have worked long and hard before they joined a rock and a stick able to club anything worthwhile. In that case, a vine likely served as the tie between two differently shaped and functional things so that they worked as one unitary whole. As the matter applies to biological computation within the sphere of an artificial cell, it may involve forming common boundaries between two complex molecules, spaces, or phases, or in the case of the brain, the blood–brain barrier.[9] Interfacing for specific inventions always presents novel challenges, especially where independent and isolated systems join for a common purpose. The Cas9 protein utilized as a gene engineering tool is an example of an interface that induces site-directed double-strand breaks in DNA. These breaks allow the introduction of heterologous genes through nonhomologous end joining, and homologous recombination in organisms.

[9]The blood–brain barrier is formed by brain's endothelial cells which line the cerebral microvasculature. It serves to protect the brain from fluctuations in plasma composition, and from circulating agents, such as neurotransmitters and xenobiotics, which have the potential for disturbing neural function.

Today the revolution in synthetic biology has already changed the landscape in the domains of computer and biological science: (1) creating new life forms; (2) engineering control devices that work like micromachines to replace errant gene sequences; and (3) developing new forms of computational devices that have astonishingly close analogs to the conventional computer. Each one of these developments are significant in disease, therapeutics, and for pushing forward enhancements to the human anatomy. Without question in the near future synthetic microcomputers will course through our bodies, first for improving well-being and then potentially enhancing the human experience.

Notes

1. Specter, M. (2009). "A Life of Its Own, Where Will Synthetic Biology Lead Us?" *New Yorker*, Annals of Science. http://www.newyorker.com/reporting/2009/09/28/090928fa_fact_specter?printable=true#ixzz26xAwyxUh.
2. Craig Venter, J., et al. (2010) "Creation of a Bacterial Cell Controlled by a Chemically Synthesized Genome." *Science*.
3. For a full discussion of synthetic biology read: New Directions: The Ethics of Synthetic Biology and Emerging Technologies, Presidential Commission for the Study of Bioethical Issues, 2010.
4. Cheng, A., and Lu, T.K. (2012). "Synthetic Biology: An Emerging Engineering Discipline." *Annual Review of Biomedical Engineering* 14 (1): 155–178. https://doi.org/10.1146/annurev-bioeng-071811-150118. PMID 22577777.
5. Purnick, P.E.M., and Weiss, R. (2009). "The Second Wave of Synthetic Biology: From Modules to Systems." *Nature Reviews Molecular Cell Biology* 10 (6): 410–422. Health Reference Center Academic, http://link.galegroup.com/apps/doc/A201086861/HRCA?u=a13qu&sid=HRCA&xid=cabdf1d3.
6. Kaebnick, G.E., Gusmano, M.K., and Murray, T.H. (2014). "The Ethics of Synthetic Biology: Next Steps and Prior Questions." *Synthetic Future: Can We Create What We Want Out of Synthetic Biology?* Special report. *Hastings Center Report* 44 (6): S4–S26. https://doi.org/10.1002/hast.392.
7. Frisco, P., et al. (1998). "Simulating Turing Machines by Extended mH Systems." *Computing with Bio-Molecules* (Springer), 221–238.
8. Copeland, B.J. (2017). "The Church-Turing Thesis." In: E.N. Zalta, ed., *The Stanford Encyclopedia of Philosophy* (Winter 2017 edition). https://plato.stanford.edu/archives/win2017/entries/church-turing/.

9. Rabin, M.O. (2012, June). Turing, Church, Gödel, Computability, Complexity and Randomization: A Personal View.
10. Adleman, L.M. (1994, November 11). "Molecular Computation of Solutions to Combinatorial Problems." *Science* 266: 1021–1024.
11. U.S. patent, 7,167,847, Sasagawa, et al. (2007). DNA Computer and a Computation Method Using the Same.
12. Kahan, M., et al. (2008). "Towards Molecular Computers That Operate in a Biological Environment." *Physica D: Nonlinear Phenomena* 237 (9): 1165–1172.
13. Birge, R. (1995, March). Protein-Based Computers. *Scientific American*.
14. Gardner, T.S., et al. (2000, January). "Construction of a Genetic Toggle Switch in Escherichia Coli." *Nature* 403 (6767).
15. Bonnet, J., et al. (2013, May 3). "Amplifying Genetic Logic Gates." *Science* 340 (6132), 599–603. http://www.sciencemag.org/content/early/2013/03/27/science.1232758.abstract?sid=27de36f9-9333-4a17-b891-cdb5c2567577 (Last visited 1/02/2014).
16. Rinaudo, K., Bleris, L., Maddamsettim, R., Subramanian, S., Weiss, R., and Benenson, Y. (2007, July). "A Universal RNAi-Based Logic Evaluator That Operates in Mammalian Cells." *Nature Biotechnology* 25 (7): 795–801. https://doi.org/10.1038/nbt1307. PMID 17515909.
17. Singh, V. (2014, December). "Recent Advances and Opportunities in Synthetic Logic Gates Engineering in Living Cells." *Systems and Synthetic Biology* 8 (4): 271–282. https://doi.org/10.1007/s11693-014-9154-6. PMC 4571725. PMID 26396651.
18. Levskaya, A. (2005). "Synthetic Biology: Engineering Escherichia Coli to See Light." *Nature* 438 (7067).
19. Nielsen, A.A.K., and Voigt, C.A. (2014). "Multi-input CRISPR/Cas Genetic Circuits That Interface Host Regulatory Networks." *Molecular Systems Biology* 10 (11): 763.
20. Weinberg, B.H., et al. (May 2017). "Large-Scale Design of Robust Genetic Circuits with Multiple Inputs and Outputs for Mammalian Cells." *Nature Biotechnology* 35 (5): 453–462. https://doi.org/10.1038/nbt.3805. PMC 5423837. PMID 28346402.
21. Ibid., see note 5 above.
22. Malyshev, D.A., et al. (2014). "A Semi-synthetic Organism with an Expanded Genetic Alphabet." *Nature* 509 (7500): 385–388. Available from: http://www.ncbi.nlm.nih.gov/pubmed/24805238.
23. Ibid., see note 14 above.

24. King, N.P., et al. (2012, June). "Computational Design of Self-Assembling Protein Nanomaterials with Atomic Level Accuracy." *Science* 336 (6085): 1171–1174. Also, see, "New Method to Identify Intermediates in Protein Folding, Advancing Nanotechnology with Protein Building Blocks." *Foresight Institute at Space Frontier Conference*. http://www.foresight.org/nanodot/?p=5196 (Last visited 10/01/2012).
25. Xu, C., Hu, S., and Chen, X. (2016, November). "Artificial Cells: From Basic Science to Applications." *Materials Today* 19 (9).

CHAPTER 23

Fantastic Voyage

Anything one man can imagine, other men can make real.—Jules Verne

Fantastic Voyage, a 1966 science fiction film follows a submarine crew shrunken to a microscopic size as they venture into the body of a scientist to repair his damaged brain. We are approaching the time when, not a human crew, but a robotic crew moves fiction into reality. How small can we make the robot?[1] To answer the question we need to step into the world of new carbon discoveries and the older silicon materials.

Nanomedicine is being applied to a host on initiatives from nanoelectronic biosensors to molecular computers and micro-mechanical devices. Anatomically implanted devices are being employed for drug delivery, therapeutics, and the identification of disease [1]. Projects now underway are developing prototype nano-robots to perform repairs at

[1]The blood–brain barrier prevents materials from crossing from the blood supply into the brain, proper. The smallest vessels or capillaries are lined with endothelial cells, which have small spaces between each cell so substances can move readily between the inside and the outside of the vessel, but in the brain, these cells fit tightly and substances, with some exceptions like glucose, cannot pass out of the bloodstream. The ribosome, which is of course of a component of every cell in the body, is therefore something of a biological machine already found in the brain, functioning in protein synthesis (translation). Ribosomes link amino acids specified by mRNA molecules, where small subunits read RNA and the large subunits then join amino acids to form a polypeptide chain, a linear organic polymer of a number of amino acid residues bonded together in a chain, which form part or the whole of the protein molecule.

J. R. Carvalko Jr., *Conserving Humanity at the Dawn of Posthuman Technology*, https://doi.org/10.1007/978-3-030-26407-9_23

the cellular level. Just how small technology can ultimately go seems, for now, unbounded, giving much space to AI technology that will do more than deliver a drug, but perhaps analyze, diagnose, prescribe, and provide medical treatment in ways and places never imagined.

Examples of AI, too numerous to list, have existed since the early days of computers, certainly well-established by the 1970s. An example of a typical AI system, developed in that era illustrates how a pattern recognition system of comparable complexity plausibly can be downsized to satisfy an in-the-body AI application. The system had been developed for automating the white blood cell differential count, performed by lab technicians. It used a computer (Varian 620 I) and a co-processor that employed discrete transistors and small-scale semiconductor integration in a conventional central processor comprised of an arithmetic/logic unit, data and program registers, input/output circuit interfaces, and a memory.[2] Physically the system measured about the size of kitchen microwave oven. It contained fewer than 500 transistors in its central processor and had a 32,768 16-bit word memory. The computer software system was comprised of an operating system, an AI pattern recognition application's program, and input/output drivers for acquisition of data from sensors attached to a scanning microscope. In contrast IBM announced, June 2017, that it is commercializing a 5-nanometer chip, packing 30 billion transistors into a space, the size of a fingernail. By scaling the size of the AI pattern recognition system described, viz. 32,500 transistors, using the new IBM technology, i.e., each 5-nanometers/transistor in size, we can estimate a modern version of an array of 32,500 transistors, arranged as 179×179 transistors on a square substrate, results in 640 nm length per side, the diagonal of which is 1.26 micrometers, which is about 16% the diameter of the nucleus of a white blood cell. Therefore, the cell would be larger than the new computer by a factor of 6, and easily contain a computer system that actualizes a sophisticated pattern recognition system, of the kind used to recognize the 5 major white blood cell types, in the late 1960s. Without question, computer processes can exist within the framework of the internal human anatomy. A quick summary of where computer technology sits today will more than make the point clear.

[2]The pattern recognition described here was reported in Scientific American, November 1970, entitled Analysis of Blood Cells (K. Preston, M. Ingram). The 620 I was a general-purpose digital computer, with a parallel, binary 16-bit word for instruction and data, with magnetic-core memory. A basic machine cycle took 1.8 microseconds.

According to Moore's law, the number of transistors incorporated into integrated circuits double approximately every 24 months, which implies that computing power doubles as well. Technologists estimate that at the current state-of-the-art, processors, having billions of transistors, will continue to shrink to the size of a bacterium, 2 micrometers (μm) in length and 0.5 μm in diameter, with a cell volume between 0.6 and 0.7 μm^3. Before the end of 2018, transistors, but a few atoms wide, will populate state-of-the art processors with upwards of 13 billion transistors [2]. Related to these changes computer processing speeds will approach 10^{16} (10 quadrillion) operations per second, referred to as flops. This microscopic size will permit computers to reside both within and alongside the human cell taking us a step closer to when genetics, nanotechnology, and robotics converge, to realize, as Ray Kurzweil sees it, the modification of the brain molecule by molecule [3].

Let's speculate on what we see if we set no limits on the material embodiment of an AI computer, i.e., whether constructed from Voigt-like biological bag of assembled parts or nanotechnology technology, installed into a artificial cell, which could be infused into an anatomical organ. Let's also add a subminiature CRISPR processor located in the same anatomical framework, which can summon Cas9 enzymes to perform editing on command. Now, we need a subroutine that executes an algorithm for installing self-replicated genes at sites dictated by a diagnosis and prescription.

Our speculation is closer to reality than some fantastic voyage of a previous era, when we consider progress being made into the realization of the electronic artificial cell, as discussed earlier. We should broaden our thinking and consider that the artificial and synthetic gene models suggested here may not be directly effected at the nucleotide (AGCT) level, but at the epigenetic level where AI influences the tags, which act as cellular memories reacting to signals from inside hormones, enzymes, or other chemicals that someday may be released via an anatomically installed nano-lab to turn genes on or off. In most of these technologies, we are effectively building machines and the plants in which they will be installed. And, it is precisely this "machine" that our not so hypothetical AI is intended to power and in turn, find a space within which it will operate.

Notes

1. Ranganathan, R., Madanmohan, S., Kesavan, A., Baskar, G., Krishnamoorthy, Y.R., Santosham, R., Ponraju, D., Rayala, S.K., and Venkatraman, G. (2012). "Nanomedicine: Towards Development of Patient-Friendly Drug-Delivery Systems for Oncological Applications." *International Journal of Nanomedicine* 7: 1043–1060. https://doi.org/10.2147/ijn.s25182. PMC 3292417. PMID 22403487.
2. GlobalFoundries, Samsung, and TSMC have been more forthcoming, and they seem to be following the same playbook. They are each introducing EUV in a second iteration of a 7-nanometer manufacturing process—the 7-nm node… Silicon wafers have to make many stops along the way in their transformation from smooth blanks to iridescent platters jam-packed with 13-billion-transistor microprocessors. https://spectrum.ieee.org/semiconductors/nanotechnology/euv-lithography-finally-ready-for-chip-manufacturing (Last visited 1/29/2018).
3. Kurzweil, R. (2012). *How to Create a Mind: The Secret of Human Thought Revealed*. New York, NY: Viking.

CHAPTER 24

Automata Artificial and Otherwise

That brain of mine is something more than merely mortal; as time will show.
—Ada Lovelace

Automata theory refers to the study of abstract machines and automata, which involves both mathematics and computer science [1].[1] To better understand this concept, take, for example, a tessellation, which is a flat surface of a repeating pattern or tiling, using one or more geometric shapes, usually with no overlaps or gaps. Historically, tessellations were used in Ancient Rome and in Islamic art, such as the tiles at the Alhambra palace. In the twentieth century, M. C. Escher used tessellations for artistic effect.[2] In mathematics, cellular automata creates tessellations of the plane, and it is becoming apparent that these not only produce fascinating visual patterns, but more and more these patterns show up in computer-generated music and theories of consciousness [2, 3]. Let's turn our attention to tessellations appearing in computational systems, specifically, what is referred to as *cellular automata*, which will help us better understand how it contributes to machine-generated transformational creativity.

[1] Automata theory serves as the foundation for AI research.

[2] M.C. Escher Tessellations Gallery, see, https://www.google.com/search?q=m.c.+escher+tessellations+gallery&tbm=isch&source=univ&client=firefox-b-1-d&sa=X&ved=2ahUKEwie2frj-pPiAhURna0KHaj4CCIQ7Al6BAgLEBE&biw=1093&bih=477.

J. R. Carvalko Jr., *Conserving Humanity at the Dawn of Posthuman Technology*, https://doi.org/10.1007/978-3-030-26407-9_24

Initially automata theory was viewed as a model for the logical and mathematical properties involved in the nervous system or more accurately the state and behavior of neural networks, which were regarded as the mechanistic substrate of intelligence [4]. As mentioned earlier, in the early 1940s, McCulloch and Pitts considered the possibility of an artificial neuron, which led to the idea that these networks could be modeled mathematically.[3] Their contributions helped formalize the notion of finite automata, and in significant ways addressed the first modern computational theory of mind and brain [5]. During the late 1940s, others were similarly thinking through computational schemes and computer architectures related to automata. To this end, John von Neumann and Stanislow Ulam, the former working on self-replicating systems and the latter studying crystal growth, conceived of a machine where the data itself determined the next state [6].

The von Neumann/Stanislaw manner of computation was later dubbed cellular automata, and although initially thought of in abstract mathematical terms, they recognized that theses forms had strong analogs to dynamic physical systems [7]. A physical analogy that works here is a water molecule, where at a given temperature, it forms a small hexagonal plate that forms six smaller plates that sprout six arms that grow in synchrony, yielding a complex, yet symmetrical shape we call a snowflake. The snowflake encompasses the substance water or H_2O, which in and of itself is "programmed" to grow a repeated pattern, and then halt further development. Not so, for crystal growth generally, which differs from the growth of a liquid droplet, in that during growth the molecules or ions must fall into the precise lattice positions to grow into orderly repeating pattern, in three dimensions.

The DNA sequence, albeit biological, represents a semantic processor, or as analogous to a machine with formal machine properties, might be viewed as having a symbology and syntax for interpreting molecules that transmute properties from one substance into another, resulting in a third, consisting of none the original properties nor potentialities of the first two materials. The result is transformative. John Casti, wrote: "While human languages are unimaginably complex, there is another type of language employed by nature that offers many possibilities for analysis using the automata-theoretic ideas … as a first approximation we can consider

[3] McCulloch and Pitt's 1943 paper, "A Logical Calculus of the Ideas Immanent in Nervous Activity," is regarded as the seminal paper on neural networks.

the DNA molecule as a one-dimensional cellular automaton having four possible values per cell, which we shall label A, C, G, and T, …." [8].[4]

As with any field of study, whether basic science or pure mathematics, cellular automata, as a field of study, does not represent anything, unless and until we give meaning to its rules and states, and provide a statement of functionality, as represented by the symbols we choose (e.g., points, lines, zeros, ones, or even pixels or lights). Until we assign meaning anyone of the several examples of automata amounts to nothing, except to demonstrate that we can create novel patterns, strings or sequences [9].

Without digressing into a treatise on cellular automata, these patterns imitate a variety of natural phenomena and human activity from the emergence of forms of life, music, and language. Princeton University scientist Stephen Wolfram, who has extensively studied cellular automata wrote: "there are examples all over physics and biology of systems that look like that, that grow in exactly that way: crystal growth, for example, cell growth in embryos, the organization of cells in the brain, and so on. The important thing is that the mathematical features of cellular automata are the same mathematical features that are giving rise to complexity in a lot of the world's physical systems" [10].[5] To Wolfram's point, the field cellular automata has demonstrated utility: for example, exploring the theoretical effects of social policy, or modeling forest fires, rodent populations, pandemics, immune response, and the racial integration of urban housing [11].[6]

Regarding other self-replicating paradigms, in 2011, news broke that a team of scientists had created a "bent triple helix" structure, based around three double helix molecules, each made from a short strand of DNA. By treating each group of three double helices as a code letter, they demonstrated that, in principle, one can construct self-replicating structures, plausibly to be used as a molecular computer. It seems that

[4] Physical DNA replication is an example, where the data (the amino acids) are arranged as codons that produce proteins.

[5] Two-dimensional parallel data processing falls into the class of machines that work on the principles of cellular automata, and as illustrated by Unger's U.S. Pat. No. 3,106,698 and Golay's U.S. Pat. No. 4,060,713.

[6] I'd worked with M.J.E. Golay and Kendall Preston, in the 1960s, building a computer for cellular automata processing, and then researching algorithms for two-dimensional pattern recognition.

one could imagine that more complex organisms might be created, depending on encoded and decoded information [12].

In God and Golem, Inc. (1964), cybernetics pioneer Norbert Wiener predicted that self-learning systems would be capable not only of un-programmed learning, but of self-replicating, i.e., reproducing and evolving. Self-replication has the potential for constructing interface-able systems to carry out autonomous processes within the human anatomy, processes that would undertake special tasks related to sensing, computation, and evaluation, which humans would be incapable of performing. Along similar lines, more than one mathematician, computer scientist or physicist has pondered the amount of information one would need to auto-construct an android, that is one capable of replicating itself [13].

Automata data (also known as automata elements) can be mapped in any number of dimensions, beginning with dimension one, referred to as an elementary cellular automata, prototypically a simple Turing machine. In a two-dimensional space, we can picture the automata as resembling a tessellation, tiles or even a checkerboard or piece of graph paper, with marks or pieces placed upon the squares to be rearranged, that is transformed according to a set of rules.[7]

In Hofstadter's MIU system, the rules dealt with involved the creation of one-dimensional strings, or vectors, whereas in a two-dimensional or even three-dimensional constructs, the effect of the rule and transformation could produce symbolic strings that grew in the dimension of the respective array, either 2-D or 3-D.

One way to imagine a two-dimensional cellular automaton is to consider a Cartesian grid or checkerboard arrangement (Fig. 24.1) and a set of rules, which are applied to any single square having an element. Other arrays, for example hexagonal could also be used develop these patterns [14]. In this case, a rule would apply in two directions, representing what to do with the state of the square (i.e., turning it black or white) depending on its proximity to its nearest-neighboring elements, i.e., in the immediate vicinity [15, 16]. The rule as applied to each element determines whether the element's next state or future, is to remain in play—represented as a black square—or to be summarily removed, which

[7]Gardner, M. (1970, October). "Mathematical Games: The Fantastic Combinations of John Conway's New Solitaire Game Life." *Scientific American* 223: 120–123. Also see, Gardner, M. (1983). *Wheels, Life and Other Mathematical Amusements.* San Francisco: Freeman.

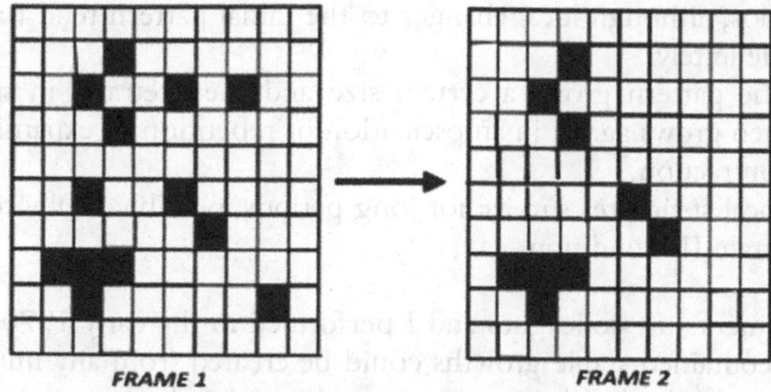

Fig. 24.1 Trajectory of a 2-dimensional cellular automata

is represented by a white square. For example, assume that you have two checkerboards side by side: Frame 1, having an initial complement of checkers on some spaces and other spaces vacant, and Frame 2 initially empty. A rule might be that, if the checker in Frame 1, upon which you place your finger, has a neighboring checker in an adjoining square, then place a checker on Frame 2, the second checkerboard, in the corresponding location. If no adjacent neighbor exists, then do not place a checker in the corresponding location in Frame 2. If we repeat this process at each square on each of the checkerboards, until we have traveled each of the 64 squares to the end, we'd likely see that the second checkerboard has a checker pattern, in positions they were originally placed in the first checkerboard. The first move is shown in Fig. 24.1. Now if we wipe the first Frame 1, checkerboard clean, and repeat the procedure on the second checkerboard, moving the pieces in the reverse direction, Frame 2 to Frame 1, we would again likely see a change in the structure of the pieces and the checkerboard as seen as a whole.

Going back and forth in this manner and examining the patterns that emerge we notice that one or more things eventually happen [17]:

I. A first pattern where one observes that the states evolve into a stable state or gradually disappear, where any hint of a pattern vanishes.

II. As a next possibility, initial patterns evolve chaotically, and stable states, which appear, are quickly destroyed by the surrounding chaos, although local changes to the initial pattern tend to spread indefinitely.

III. The pattern grows a certain size and then retracts in size and then grows again, in an oscillation or repetition of expansion and contraction.

IV. Local structures survive for long periods, possibly finally reaching a state III condition.

Experiments my colleagues and I performed in the early 1970s found that self-contained stable growths could be created from any number of arrangements of points in one, two or *n*-dimensional space, and depending on the rule in play, some of these resulted in independent gliders that mutated, evolved, and went off on their own [18].

Biologists and AI programmers have advanced systems that emulate the evolution of species, individual behaviors, and cognitive processes [19]. Similar to cellular automata, evolutionary computation use algorithms developed from simple rules [20]. In an illustration of the last point, ethologist and evolutionary biologist, Richard Dawkins created a Biomorph Seed instilling it with very simple reproductive properties [21]. Essentially the process was used to analogize a complex embryological development. Typically, a process encodes a theme in what would be comparable to a genome with pattern and regularity. Thereafter, the entire population evolves, from the genome, according to a regime referred to as evo-devo dynamics.[8] In the Dawkins' process, the properties included hypothetical "genes" that were directed to the growth of a basic initial vertical line, to a branch resembling a twig, having subbranches at certain allowed angles. The first few generations looked much like a forked twig, followed by forked twigs, having forked twig appendages. By further imposing rules that would mimic genetic variability, he created a society of biomorphs, each basically differing in that a mutation in one gene occurred. In Dawkins' Biomorph Land, brave new worlds were created, fortunately only in the information-theoretic memory of a computer.

Francisco Vico and his colleagues at the University of Malaga, in Spain, utilize a form of evolutionary programming, i.e., music pieces

[8]Evolutionary developmental biology (evo-devo) compares the developmental processes of different organisms to infer the ancestral relationships between them and how developmental processes evolves.

obtained by simulated evolution, which composes music *ex nihilo*, or out of nothing. More will be said about this in later chapters.

From the perspective of cellular automata and evolutionary programming, we know that small changes in rules, or starting conditions (mathematicians refer to these as axioms) produce large changes in the evolution of games, population growths, genes, and life forms generally. Query: whether life and society have any resemblance to these paradigms, an if so, what might we anticipate? Do these phenomena remain static, vanish, expand to outstrip all boundaries, or simply oscillate in some state-space, forever?

To keep the matter of these artificial constructions and the reality of biological world in perspective, let's stipulate that a cell and its inherent pattern has objective, observer independent reality, unlike the abstract codes we design into and which cause a computer to behave in an assigned manner. A cell contains structure and organization that serves to carry out its intended mechanical, electrochemical, and ultimately what we generally refer to as biological functioning. However, our essential being manifests in a chain leading from DNA to the complex societies in which we live, the social reality we have constructed, a continuous link, representing fundamentally to how molecules organize and thereby take on functionality, but in these different forms [22]. And, further, that this metamorphosis, from DNA to society, reveals emergent properties that extend to the manner in which society itself, as evolutionary computation suggests, propagates in dimensions dependent on technological expansion, cultural transformation, environmental change, and population growth. We will address this point in some detail below, but before that let's explore AI and some of its limitations in emulating human creativity.

Notes

1. von Neumann, J. (1956). "The General and Logical Theory of Automata." In: A.H. Taub, ed., *Collected Works* (vol. 5). New York: Pergamon, pp. 288–328.
2. Miranda, E.R. (2007). "Cellular Automata Music: From Sound Synthesis to Musical Forms." In: E.R. Miranda and J.A. Biles, eds., *Evolutionary Computer Music.* London: Springer.
3. Penrose, R., and Hameroff, S. "Consciousness in the Universe: A Review of the 'Orch OR' Theory." *Physics of Life Reviews* 11 (2014): 39–78.

4. Jordan, M., et al. (2004). Neural Networks. In: A.B. Tucker, ed., *Computer Science Handbook* (2nd edition). Section VII: Intelligent Systems. Boca Raton, FL: Chapman & Hall/CRC Press LLC. ISBN 978-1-58488-360-9.
5. Piccinini, G. (2004). "The First Computational Theory of Mind and Brain: A Close Look at Mcculloch and Pitts's 'Logical Calculus of Ideas Immanent in Nervous Activity'." *Synthese* 141: 175–215, Kluwer Academic Publisher.
6. Beyer, W.A., et al. (1985). "Stanislaw M. Ulam's Contributions to Theoretical Biology (PDF)." *Letters in Mathematical Physics* 10 (2–3): 231–242. Bibcode:1985LMaPh..10..231B. CiteSeerX 10.1.1.78.4790. https://doi.org/10.1007/bf00398163. Archived from the original (PDF) on 27 September 2011. Retrieved 5 December 2011.
7. von Neumann, J. (1951). "The General and Logical Theory of Automata." In: L.A. Jeffress, ed., *Cerebral Mechanisms in Behavior—The Hixon Symposium*. New York: Wiley, pp. 1–31.
8. Casti, *Reality Rules, Picturing the World in Mathematics, the Fundamentals.*
9. M.J.E. Golay devised a scheme utilizing an hexagonal array of points in a method that provided a topological metric used for two-dimensional objects. See, Golay, M.J.E. (1969, August). "Hexagonal Parallel Pattern Transformations." *IEEE Transactions on Computers*, C-18, #8: 733–740.
10. Regis, E. (1987). *Who's Got Einstein's Office?* Addison-Wesley. See, http://www.stephenwolfram.com/about-sw/interviews/87-einstein/text.html. Also see, Wolfram, S. (1983). "Statistical Mechanics of Cellular Automata." *Reviews of Modern Physics* 55 (3): 601–644. Bibcode:1983RvMP...55..601W. https://doi.org/10.1103/revmodphys.55.601.
11. Batty, M., Couclelis, H., and Eichen, M. (1997). "Editorial: Urban Systems as Cellular Automata." *Environment and Planning B: Planning and Design* 24: 159–164; White, S.H., et al. (2006); White, S.H., Martin del Rey, A., and Rodríguez Sánchez, G. (2006). "Modeling Epidemics Using Cellular Automata." *Applied Mathematics and Computation* 186: 193–202; Soares-Filho, B.S., et al. (2002); Soares-Filho, B.S., Cerqueira, G.C., and Pennachin, C.S. (2002). "DINAMICA—A Stochastic Cellular Automata Model Designed to Simulate the Landscape Dynamics in an Amazonian Colonization Frontier." *Ecological Modelling* 154: 217–235.
12. "Self-Replication of Information-Bearing Nanoscale Patterns." *Nature*. http://www.nature.com/nature/journal/v478/n7368/full/nature10500.html. Retrieved 14 September 2012. Also see, "Self-Replication Process Holds Promise for Production of New Materials." ScienceDaily. http://www.sciencedaily.com/releases/2011/10/111012132651.htm. Retrieved 14 September 2012.

13. Golay, M.J.E. (1961, June). "Reflections of a Communications Engineer." *Analytical Chemistry* 33 (7).
14. Bays, C. (2005). "A Note on the Game of Life in Hexagonal and Pentagonal Tessellations." Department of Computer Science and Engineering, Complex Systems, 15, University of South Carolina, Columbia, SC.
15. Kier, L.B., Seybold, P.G., and Cheng, C.-K. (2005). *Modeling Chemical Systems Using Cellular Automata*. Springer. ISBN 9781402036576.
16. Golay, M.J.E. Analysis of Images U.S. Patent 4,060,713, 1977.
17. Ilachinski, A. (2001). *Cellular Automata: A Discrete Universe*. World Scientific. ISBN 9789812381835.
18. Golay, M., Carvalko, J., and Preston, K., investigated initial configurations that produced a spontaneous cancerous growth utilizing an hexagonal matrix of points. Research Image, Perkin-Elmer Inc. Vol. 6, No. 1, January 1971.
19. Holland, J.K. (1975). *Adaptation in Natural and Artificial Systems*. Ann Arbor: University of Michigan Press.
20. Fogel, L.J., et al. (1966). *Artificial Intelligence Through Simulated Evolution*. New York: Wiley; Hofstadter, D.R. (1999). *Gödel, Escher, Bach*. Basic Books, pp. P–2 (Twentieth-anniversary Preface). ISBN 0-465-02656-7.
21. See, Langton, C., et al., eds. (1992). *Artificial Life-II*. Redwood City, CA: Addison-Wesley. Also see, Dawkins, R. (1986). *The Blind Watchmaker*. London: Longman.
22. Hofstadter, D. (2007). *I Am a Strange Loop*. Basic Books.

CHAPTER 25

What's the Matter with Hal

Nature gives to every time and season some beauties of its own.—Charles Dickens

Lady Ada Lovelace, known for her contributions to Charles Babbage's Analytical Engine, recognized that although computers had applications beyond pure calculation, in her words, "they cannot originate anything" [1].[1] Contrary to Lovelace's opinion, will computers become so powerful that someday they will create art, music and literature, without human intervention? After all, by 2021 Cray Computing will have launched a supercomputer that can handle 10^{24} or 1 quintillion (1 billion times 1 billion) processes per second. Having storage capacities in excess of 10^{15} bytes, or 1.0 petabyte, brings them in line with the processing power of the human brain.[2] The brain has functionality, which obviously not only controls conscious thought, but directs and controls sensory inputs and outputs as well as metabolic systems that regulate the anatomy. As such, brute processing power a brain does not make.

[1] Early in the nineteenth century, Charles Babbage exhibited a small working model of a digital computing machine for the automatic production of logarithm tables, tide tables, and astronomical tables incorporating the logic of George Boole, the kind of logic that serves as a foundation for today's computer. His subsequent more ambitious proposed Analytical Engine was to have had a memory and a central processing unit and would have been able to handle conditional branching.

[2] There are varying opinions about the processing power of a average brain. Part of differences in opinion, especially when comparing the brain to a computer, stems from defining what is meant by a computer cycle or "operation." Computer programmers deal in

J. R. Carvalko Jr., *Conserving Humanity at the Dawn of Posthuman Technology*, https://doi.org/10.1007/978-3-030-26407-9_25

But, on some level, IBM Watson-like artificial intelligence can do what any intelligent individual can, or at least appear to do. Watson runs artificial intelligence programs that search answers to unstructured questions. In 2011, before millions of viewers, it outscored the all-time money winning contestant on the T.V. program Jeopardy. The machine is capable of processing 80 trillion operations (teraflops) per second. It runs about 2800 processor cores and has 16 terabytes of working memory. Watson researchers collected 200 million pages of content, both structured and unstructured, across 4 terabytes of disks. It searches for matches and then uses about 6 million logic rules to determine the best answers. Watson runs fast and has an incredible source of knowledge. But, can it emote?[3] Does it have self-awareness?[4] [2] It would seem both these characteristics are necessary to carry forward any semblance of creativity, as least as it relates to the humanities.

Machines, whether they flop one or 80 trillion calculations per second, merely follow instructions, whether explicit, calculable or embedded in the "data" as we learned is the case with cellular automata. At any rate, computing power does not bestow self-awareness, which as I argue is a necessary condition for creativity, relative to the human experience, essentially characterized by what we call the humanities. And although other creatures, my dog for instance, has limited self-awareness, it suggests that while it's necessary for purposive creativity, it is not sufficient.

Although powerful number-crunching machines can solve incredibly complex problems involving physics, chemistry, and biology, computers cannot invent on the level of Edison, nor paint on the level of

the term FLOP, which is a measure of floating point operations per second. Another estimate may rely on computer instructions per second. As such there can be a widely differing number that estimates a brain's power of computation versus a computer's from about 10^8 or 1 times 10 to the power 8 computer instructions per second, to 10^{18} or 1 times 10 to the power 18 calculations per second. Ray Kurzweil, a leading figure in the matter of computers, AI and engineered brains, estimates that brain capacity is 100 billion neurons times an average 1000 connections per neuron (calculations taking place primarily in the connections) times 200 calculations per second. This would amount to $2*10^{16}$ or 2 times 10 to the power 16, or 2 followed by 16 zeros. See, https://www.kurzweilai.net/the-law-of-accelerating-returns (last visited 10/3/2019).

[3] Engineers at the AI research lab DeepMind, a subsidiary of Google, created AlphaGo, a machine, designed to play Go that works on deeply layered neural nets to mimic the brain and nervous system, say they do not understand how the machine's intuition works.

[4] Self-awareness is not to be confused with self-evaluation, which has been found to negatively affect creativity.

Michelangelo, nor answer how the Universe works. Computers are capable of supplying design alternatives to a wide range of engineering and biological development problems. Humans alone are the thought-species. We have the capacity to ask, answer and conjecture, beyond offering alternatives. According to physicist David Deutsch:

> The real source of our theories is conjecture, and the real source of our knowledge is conjecture alternating with criticism. We create theories by rearranging, combining, altering and adding to existing ideas with the intention of improving upon them. [3]

We also use analogies and metaphors for explaining, describing, and hypothesizing how science works, whether quantum particles, electromagnetic waves, atoms, DNA molecular helices, electrochemical ion channels, protein folds, genetic editing, and on and on into the compendium of science [4]. Professor of cognitive science and literature, Douglas Hofstadter and psychologist Emmanuel Sander argue that analogy sits at the core of thinking [5]. But analogies and metaphors can mislead us, as to when the thing being referred to may itself not be well understood. One example that quickly comes to mind is consciousness, where theories are advanced and models are proposed based on the slimmist facts. At times, the difficulty of trying to explain how something works has to do with boundaries, e.g., where nature ends and the artificial entity begins. Often the matter has more to do with drawing the proper constructs, examples, and isomorphisms that reasonable exist between artifacts of nature and those that are invented.

As metaphor relates to AI based creativity, attempts are underway to improve human-language processing, particularly to develop automated understanding and production of emotional expression, which of course, whether explicitly or implicitly, represents a primary objective in the use of metaphor [6]. It should be noted that metaphor applied to human-language processing is very much removed from metaphor as applied to developing a scientific theory or explaining how one or another technology functions.

We discover nature, but inventions are not discovered, they come into existance as products of thought, conception and innovation. Except for this difference, a natural phenomena and a human invention are often indistinguishable in the abstract. For example, the renowned Chakrabarty patent, which first held that lifeforms were patentable, claimed: "A bacterium from the genus Pseudomonas containing therein at least two stable energy-generating plasmids, each of said plasmids providing a separate hydrocarbon degradative pathway."[5] From the plain language it cannot be determined, if the claim reads on an invention or a product of nature. Let's turn to the world of models, ones that highlight the similarities, rather than the differences between things natural and things artificial.

A mathematical description or model represents a logical process as expressed in a formal system of rules, generally logic and symbols. This will remain true of software programs generally. If the logic and symbols are supposed to represent physical forms as found in the natural sciences then as applied to scientific laws and observation, models construct slices of reality; and typically, the reality describes a natural system.[6] A model portrays a larger system to the extent salient quantifiable aspects of the system are identified and properly associated, as well as imparting information about the system's causal effects on the environment. The description and linkages endeavor to specify the feature of the natural system in terms of its materiality and modality. The material takes into account both aspects of its physical nature and the modes of behavior we observe. I am referring here to the scientific description of a natural system.

Let's draw a few distinctions between intelligence and creativity, where the former includes the aptitude for logic, desire for facts, objective self-awareness, ratiocination, the process of exact thinking or a reasoned train of thought, planning, quantitative and logical problem-solving (deductive and inductive reasoning processes), motivated toward closure.[7] The latter, creativity includes conjecture, diverse thinking, abductive reasoning processes, resistance to premature closure, imagining possibilities, teaching away from the obvious, and improvisation [7].[8] It

[5] US Pat. 4,259,444A, for Microorganisms having multiple compatible degradative energy-generating plasmids and preparation thereof (1981).

[6] When an isomorphism between two objects exists, they are in some way the "same thing," but described in different ways. However when a homomorphism exits between two objects, only a part of the structure in one is found the other.

[7] Deductive reasoning reaches a conclusion based on a general premise followed by a universal premise. Inductive reasoning infers a conclusion from particular observations. Abductive reasoning begins with an observation and then seeks to find the simplest and most likely explanation for it.

[8] The Torrance Tests of Creative Thinking are used to assess creative potential. The TTCT comprises a series of figural exercises (thinking with pictures) and verbal activities (thinking with words) that are indicative of one's creative abilities. The list here derives in part from the TTCT.

Table 25.1 Comparison of characteristics between creativity and intelligence

Creativity	*Intelligence*
Openness to novelty	Evaluation and assessment*
Idea generation*	Categorization and classification*
Curiosity*	Reasoning via logic and quantification*
Inventiveness	Identification*
Imaginative	Interpretation and linkage, cause/effect*
Holistic	Analytical*
Use of metaphor and analogy	Use of metaphor and analogy
Abductive reasoning*	Deductive/inductive reasoning*
Divergent thinking	Convergent thinking*
Blind variation and improvisation	Selective retention*
Implicitness	Explicitness*
Neuronal disinhibition	Neuronal fidelity
Use of default mode network	Use of cognitive control network
Playfulness	Perseverance/rigidity*
In rare cases Schizotypal/Psychosis	In rare cases Autism

may be that computers are suited to process forms of intelligence, where computation and logic play a significant role. On the other hand, creativity cannot be easily reduced to forms of computation and logic, except in rare instances where computational constructs, such as automata may lead to forms of art and music.

Table 25.1 lists behaviors associated with human intelligence and creativity. As to the intelligence column, some characteristics, as starred "*," can be deemed within the realm of an AI emulation via searchable databases, natural language programs, the quantification of human psychological, biological and physiological reactions, facial and behavioral recognition algorithms, learning machines and neural networks. On the other hand the list, which includes creativity, suggests as starred "*," computer processing paradigms that may lead to outcomes considered inventive or innovation, notably, where progress is being made in the composition of music, abstract art, and some forms of simple authorship related to jokes, flash fiction, and poetry.[9]

Anthropologist Agustin Fuentes writes: "the essence of creativity is to look at the world around us, see how it is and imagine other

[9] For an impressive array of haiku and short poems generated by computer, see, Trentini, Y.A. (2017). *Computer Generated Poetry Will Knock Your Socks Off.* https://medium.com/@Yisela/computer-generated-poetry-will-knock-your-socks-off-763c815a1b52.

possibilities that are not immediately present or based on our immediate personal experience" [8]. This of course requires a measure of self-awareness. Artificial intelligence, regardless how sophisticated, can never emulate human imagination, as it lacks insight, self-awareness, subjectivity, and cannot form the diverse abstractions of the kind by which humans fathom concrete instances. It cannot know or express human wants, feel needs and desires, or to look out into the world and wonder.[10] Thus it's a closed system, which in the words of Marvin Minsky, consists of a collection of "senseless loops and sequences of trivial operations," which prevents it from forming extensible concepts, the one's needed to initiate acts of creation, things that heretofore never existed, the heart of human existence at almost every level in life [9]. E.O. Wilson writes:

> Creativity is the unique and defining trait of our species; and its ultimate goal, self-understanding: what we are, how we came to be, and what destiny, if any, will determine our trajectory. What then is creativity? It is the innate quest for originality. The driving force is humanity's instinctive love of novelty-the discovery of new entities and processes, challenges and disclosure of new ones, the aesthetic surprise of unanticipated facts and theories, the pleasure of new faces, the thrill of new worlds. [10]

As mentioned earlier, in 1950, Alan Turing proposed a test to ascertain whether a machine could think [11]. It required an interrogator to depose an individual and a machine. If, through a series of questions, the interrogator was unable to distinguish the answers given by the individual and machine, then the machine was deemed to have passed the test, which implied that the machine could think. Turing wrote:

> The new form of the problem can be described in terms of a game which we call the 'imitation game'. It is played with three people, a man (A), a woman (B), and an interrogator (C) who may be of either sex. The interrogator stays in a room apart from the other two. The object of the game for the interrogator is to determine which of the other two is the man and

[10] Lady Lovelace wrote in her notes to Memabrea work below[1]: "the analytic engine has no pretensions to originate anything. It can do ***whatever we know how to order it*** to perform. It can follow analysis; but it has no power of anticipating any analytical relations or truths" (her italics). 1. Menabrea, L.F. (1843). *Sketch of the Analytical Engine Invented by Charles Babbage*. Scientific Memoirs, 3, archived from the original on 15 September 2008. Retrieved 11 March 2018, with Lovelace notes.

which is the woman. He knows them by labels X and Y, and at the end of the game he says either 'X is A and Y is B' or 'X is B and Y is A'. The interrogator is allowed to put questions to A and B thus:

> C: Will X please tell me the length of his or her hair?
>
> Now suppose X is actually A, then A must answer. It is A's object in the game to try and cause C to make the wrong identification.
>
> His answer might therefore be
>
> 'My hair is shingled, and the longest strands are about nine inches long.'

The Turing Test for many decades has stood for the idea that one could establish that a computer were thinking, if a suitably programmed computer could emulate a human response, without detection. It may well be that IBM's Watson-like Big Blue passes that test, although philosopher John Searle and other critics counter on grounds that a computer knows nothing about what it is responding to [12]. Searle, raised the Chinese Room Argument, which disputes that a computer understands or appreciates for the actual content of the question.

> Computational models of consciousness are not sufficient by themselves for consciousness. The computational model for consciousness stands to consciousness in the same way the computational model of anything stands to the domain being modelled. Nobody supposes that the computational model of rainstorms in London will leave us all wet. But they make the mistake of supposing that the computational model of consciousness is somehow conscious. It is the same mistake in both cases. [13]

Searle questions, "how," the computer knows, i.e., the epistemic relationship between the computer, the query, and the answer it delivers [14]. Even if Searle's conclusion isn't accepted, and one takes the position that the Turing Test shows a machine can exhibit a form of intelligence, the argument doesn't stop there. We need to further inquire into what it means to think. Does thinking invoke notions such that an "idea" constitutes an existing physical process, like pumping, or are there different categories of what it means to think.

To paraphrase Turing's question, "Can machines think, in a way indistinguishable from humans?" Or, for our purposes we should frame it as

"Can machines create, in a way indistinguishable from humans?" [15].[11] Human have the capacity to engage in a creative process that includes the ability to diverge from the norm, the known, and improvise, discover or invent.

One might be persuaded to accept the proposition, "machines may emulate the human creative process, which includes improvisation," if a reasonably complete understanding of the mind exists, which first requires a satisfactory theory of consciousness. With such understanding, hypothetically, a supercomputer might be constructed to express fear, intention, desire, jealousy and rage—features of the human condition inaccessible to scientific measurement or expressed by natural law.

The idea of digital computers exhibiting creativity had been considered almost as early as Turing's 1950 essay on the Imitation Game, which as indicated focused on whether a computer was capable of thinking in the traditional sense [16, 17]. Those that study computer-generated creativity attempt to build programs that produce products across a wide domain of interests, such as science, art, literature, and gaming, which plausibly can be considered creative.[12] Courses taught at various colleges and universities throughout the world offer studies in computational creativity, information-processing theories related to human creativity, and how to employ AI for augmenting/amplifying autonomous systems to produce computer-generated music and art.

Although the creative act may involve thinking, it is not the same as thinking. In Turing's famous essay, Computing Machinery and Intelligence, he avoids the verb "to think" and calls attention to the capacities of a machine to imitate human behavior, not creativity. As Professor Jefferson wrote in his 1949 Lister Oration:

> … the schism arises … over the use of words and lies above all in the machines' lack of opinions, of creative thinking in verbal concepts. …

[11] Alan Turing wrote: We now ask the question, "What will happen when a machine takes the part of A in this game?" Will the interrogator decide wrongly as often when the game is played like this as he does when the game is played between a man and a woman? These questions replace our original, "Can machines think?" See, Turing, A. (1950). "Computing Machinery and Intelligence." *Mind* LIX (236): 433–460, doi: https://doi.org/10.1093/mind/LIX.236.433.

[12] The International Conference on Computational Creativity (ICCC) is an open, international, academic conference. See, http://computationalcreativity.net/home/.

> Not until a machine can write a sonnet or compose a concerto because of thoughts and emotions felt, and not by the chance fall of symbols, could we agree that machine equals brain-that is, not only write it but know that it had written it. No mechanism could feel (and not merely artificially signal, an easy contrivance) pleasure at its successes, grief when its valves fuse, be warmed by flattery, be made miserable by its mistakes, be charmed by sex, be angry or depressed when it cannot get what it wants. [18]

Turing also recognized that the computer, beyond dialogue, was incapable of an act of independent creation:

> Q: Please write me a sonnet on the subject of the Forth Bridge.
> A: Count me out on this one. I never could write poetry.
> Q: Add 34957 to 76764.
> A: (pause about 30 seconds and then give as an answer) 105621.
> A: Yes.
> Q: I have K at my K1, and no other pieces. You have only K at K6 and R at R1. It is your move. What do you play?
> A: (After a pause of 15 seconds) R-R8 mate. [19]

In going forward, we examine the sense conveyed by Jefferson, exploring whether transhumans or artificial entities can faithfully create or imitate music, storytelling and other art forms, indistinguishable from that produced by humans.[13]

Notes

1. Fuegi, J., and Francis, J. (2003, October–December). "Lovelace & Babbage and the Creation of the 1843 'Notes'." *Annals of the History of Computing, IEEE* 25 (4): 16–26. https://doi.org/10.1109/mahc.2003.1253887. Also see, Turing, A.M. (1964). "Computing Machinery and Intelligence." In: A.R. Anderson, ed., *Minds and Machines*. Englewood Cliffs, NJ: Prentice Hall.
2. Silvia, P.J., et al. (2004). "Self-Awareness, Self-Evaluation, and Creativity." *Personality and Social Psychology Bulletin* 30: 1009–1017.

[13]Art is broadly defined as: authorship, including among other things literary, dramatic, musical, architectural, cartographic, choreographic, pantomimic, painting, pictorial, graphic, sculptural, and audiovisual creations.

3. Deutsch, D. (2011). *The Beginning of Infinity: Explanations That Transform the World*. New York: Viking.
4. Brown, T.L. (2003). *Making Truth: Metaphor in Science*. Urbana: University of Illinois Press.
5. Hofstadter, D.R., and Sander, E. (2013). *Surfaces and Essences*. Basic Books.
6. Delfino, M., et al. (2005, April). "Figurative Language Expressing Emotion and Motivation in a Web Based Learning Environment." In *Proceedings of the Symposium on Agents That Want and Like: Motivational and Emotional Roots of Cognition and Action* (pp. 37–40), at AISB'05 Convention, University of Hertfordshire, UK. Brighton, UK: Society for the Study of Artificial Intelligence and Simulation of Behaviour.
7. Cramond, B., et al. (2005). "A Report on the 40-Year Follow-Up of the Torrance Tests of Creative Thinking: Alive and Well in the New Millennium." *Gifted Child Quarterly* 49 (4): 283–291. http://dx.doi.org/10.1177/001698620504900402.
8. Fuentes, A. (2017). *The Creative Spark: How Imagination Made Humans Exceptional*. Penguin Random House.
9. Minsky, M. "Steps Toward Artificial Intelligence." In: E. Feigenbaum and J. Feldman, eds. *Computers and Thought*, p. 447.
10. Wilson, E.O. (2017). *The Origins of Creativity*. Liveright Publishing Corporation.
11. Turing, A.M. (1950). "Computing Machinery and Intelligence." *Mind* 49: 433–460.
12. Searle, J. (1980). "Minds, Brains, and Programs." *Behavioral and Brain Sciences* 3 (3): 417–424. https://doi.org/10.1017/s0140525x00005756.
13.. Searle, J. (2002). *Consciousness and Language*. Cambridge University Press, p. 16. ISBN 0521597447.
14. Bello, P., Bringsjord, S., and Ferrucci, D. (2001, February). "Creativity, the Turing Test, and the (Better) Lovelace Test." *Journal Minds and Machines* 11 (1): 3–27. https://dl.acm.org/citation.cfm?id=596904.
15. Harnad, S. (2008). "The Annotation Game: On Turing (1950) on Computing, Machinery, and Intelligence." In: R. Epstein and G. Peters, eds., *The Turing Test Sourcebook: Philosophical and Methodological Issues in the Quest for the Thinking Computer*. Kluwer.
16. Turing, A. (1950). "Computing Machinery and Intelligence." *Mind* 59 (236): 433–460.
17. Oppy, G., and Dowe, D. "The Turing Test." In: E.N. Zalta, ed., *The Stanford Encyclopedia of Philosophy* (Spring 2018 edition). https://plato.stanford.edu/archives/spr2018/entries/turing-test/.
18. Jefferson, G. (1949). "The Mind of Mechanical Man." *British Medical Journal* 1 (4616): 1105–1110.
19. Ibid., see note 16 above.

PART IV

Creative Psychology & Essence

CHAPTER 26

The Struggle for Perfection

Sarouk: Mensa, tell me again how we came to be. What exactly was your configuration when you were born?

Mensa: We only had one model—aboriginal. The model was a couple of million years old. Nothing artificial, only what nature supplied—and not connected to a network.

Sarouk: And, physical appearance?

Mensa: Outwardly, not altogether different from today, a body, two arms, two legs, a head, with a brain for abstract thinking and feeling. My height, nearly two meters. A tall girl in my day.

Sarouk: I understand the genetic part, but what did the bio computers add?

Mensa: We embedded AI bio processors to increase our visuo-spatial responses, direct digital access to the brain to store memes, and real-world facts, encyclopedia-like, for searching and recall.

Sarouk: All sounds wonderful to me, like going from a prehistoric being to a fully fledged person.

Mensa: Yes, I suppose, if you were to liken it to how our ancestor Homo sapiens were different from their earlier Homo erectus relatives. In a way, we Homo futuro are quite different from Homo sapiens.

J. R. Carvalko Jr., *Conserving Humanity at the Dawn of Posthuman Technology*, https://doi.org/10.1007/978-3-030-26407-9_26

Sarouk: So, if we keep the analogy going, the sapiens were less constrained than their predecessor. They after all were the ones that broke free from mental constraints that from the beginning of time had prevented the species from having abstract thoughts. From forming mental representations of things.

Mensa: Yes, I guess. But, now we are heavily AI bound, and it has its own constraints. We can think super-abstractly. But, we are not free to think our own raw thoughts, to go off the rails, as my parents used to say. We are bounded by looking at problem solving more strictly, in terms of the syntactical and semantic capabilities inherent in a computer language.

Sarouk: Meaning what? Only solutions found in the application of AI?

Mensa: In some ways yes. For instance our artistry only simulates what was once human artistic expression. In other cases, we do not bother to explore the limits, the theoretical limitations. We know in advance the problems that can never be solved, so we don't or rather can't go there.

CHAPTER 27

Power of Imagination

We may stand alone in the Universe, the only creature that self-reflects, that sees beauty, truth, and comes to know ambiguity, the strange side of its existence, the twilight zone, imagined parallel worlds, all in a space exclusive to and within ourselves. Then suddenly, Wow! And, one among us changes forever that which follows—and where for each of us life moves in another direction. In ancient times it might have been an invention, like the wheel, or an idea like one god, or an ideal like freedom. A moment, apparently, but rare moments don't just arrive from anywhere, unannounced. They have origins, and Plato counsels that true inspirations often need to be drawn out. His proposition applies to humankind, not other species or as we can safely assume, AI-driven devices. Unlike computers, humans are agents of both input and output creativity, agents that form mental constructs that lead to Wow moments of artistic undertaking, productions or products, the latter often in the utilitarian sense.[1]

We associate input creativity with artistic creativity, that which employs perception and the analysis of incoming information, as one's stimulus or interest in expressing a landscape (regardless of the medium: poetry, painting, music). Output creativity deals with the production of

[1]The distinction between artistic and utilitarian is not sharp, and bear in mind that either term can be applied to ideas, products, processes, and expressions, using every kind medium of manifestation, such as artistic, musical, linguistic, scientific, technological and mathematical.

J. R. Carvalko Jr., *Conserving Humanity at the Dawn of Posthuman Technology*, https://doi.org/10.1007/978-3-030-26407-9_27

scientific theories, or inventions, dealing with more formal constraints, as opposed to input creativity—although in some cases the two kinds of creativity merge [1]. For example, *chiaroscuro*, Leonardo da Vinci's idea that light falling against an object produces a solidity of form, was used in the production of his paintings. In most cases, what seeps into our awareness is recursively exchanged between input and output creativity. We hear evidence of this when a musician listens to her ensemble, its theme, and albeit constrained by the range of the instrument and the mood of the piece, improvises a novel riff, scale or arpeggio. Among a multitude of other creative tactics, humans use their intuition, take risks, and capitalize on chance opportunity or serendipity.[2] Artificial intelligence as it stands is poorly equipped to deal with these notions, which no one fully understands, either in their origin or their influence on the creative process. There are no known ways for an algorithm, as Plato suggests, to draw out the Wow moment.

An AI constraint implies boundaries and rules. Related to software it applies to solving a problem within the syntactical and semantic capabilities inherent in a computer language, and only simulated in the applications of AI that attempt to mimic human artistic expression. In other cases, theoretical computational limitations exist, which manifests as the impossibility of creating an algorithm to achieve a particular result. We will discuss these in other chapters. In some instances AI works well, as it does in facial recognition, or medical diagnosis and certain kinds of musical composition. In this latter sphere, neither language nor algorithmic constraints impose restrictions on composed music of a kind. In fact, some AI generated music can hardly be distinguished from compositions authored by the likes of Mozart or Bach. Other AI constructs, as in the game Go or chess, often appear unexpectedly to outsmart the world's grandmasters. So, the constraints insofar as these types of outcomes are not problematic. But, as a general rule, constraints to emulate human activity, especially emotions, limit what AI can effectively achieve, according to conventional notions of what is deem aesthetic or creative.

But, at bottom, AI software doesn't possess intentionality, that is beliefs, wants, needs, desires, fears, or instantiate semantics, that is ways of interpreting and thinking about ideas, before, during or after their accumulation. The formative process, the one that needs to congeal,

[2] Thousands of inventions were a result of accident, e.g., penicillin, Teflon, slinky, silly putty, post-it notes, saccharin, popsicles and safety glass.

needs to be drawn out. AI is reserved to specific problem-solving, accessing finite libraries of possible combinations, employing decision trees, feedback loops, Markov chains, restricting its transformational or paradigmatic potential to emulated human creativity.[3]

We are not just aware of our surroundings, we are aware in the context of a Universe that contains historical import as well as cultural influence, the impacts of one or another artistic embellishment, the established norms, products and productions. The parallel here follows philosopher, Thomas Nagel's 1974 article, "What is it like to be a bat?" In regards to our resourcefulness, there exists a "subjective character of experience"—which higher order mammals possess, but which cannot be computer generated, as of yet. Computers, as products of human invention, which whether sitting on my desk or hardwired to my brain, do not have the faculty of perception. They remain subjectively ontological artifacts, programmed to perform computations, subject to human abstract interpretation, noncausal, limited by the scope of a hardware and software specification.

As compared to a software program, regardless how sophisticated, human experience derives from a well-spring of perceptual associative combinations, which may be drawn upon as may be relevant to meaningfulness, which itself serves to escort creative intentionality. This boundless well-spring of perception puts AI-driven artifacts at a disadvantage, and thus as mentioned into a distinct ontological category. AI machines that seemingly create are not creative, at least not in the way humans are [2].[4]

Further to what constrains intelligent machines from ascending into the realm of creative machines, let's return to the idea we focused on earlier, as to the constraint brought about by available categories of creative output. Consider two perspectives: the first, what is specific and

[3]A Markov chain is a stochastic model describing a sequence of possible events in which the probability of each event depends only on the state attained in the previous event.—Wikipedia

[4]Building perception into machines is something considered at the embryonic stage of AI. In 1958 the perceptron algorithm was invented at the Cornell Aeronautical Laboratory. Designed for image recognition, it contained 400 photocells, randomly connected to electrical "neurons." Weights were encoded in potentiometers, and updated during a learning process. *The New York Times* reported the perceptron to be "the embryo of an electronic computer that [the Navy] expects will be able to walk, talk, see, write, reproduce itself and be conscious of its existence."

unique about a creative product from the viewpoint of the potential creator, and the second, what is specific and unique to a particular class of products. The first deals with a form of representation, or the sensory states having distinctive phenomenological properties, for instance, the size, shape of a flute and its harmonic and tonal qualities. But, we know that some things about the flute only exist subjectively, like the flute's mirror-like finish or its timbre [3]. The reflections and colors we see are a consequence of neurological processes that convert photon wavelengths into conscious representations. As to timbre, it's a result of neurological processes that convert frequency of vibrations of the surrounding air, to a conscious representation of sound.[5] In each instance, these phenomena cannot be experienced by anything other than organisms, which, like humans, convert energy into psychological perception. Inanimate objects, like electronic or chemical-based sensors, do not possess perceptions, but are mere constructs of technology, as for example a thermometer that pushes a column of mercury dependent on temperature.

If a subject is restricted in what it can perceive, because it does not have the requisite neurological fitness, it therefore has a categorical limitation. Locke's "veil of perception," serves as an analogy, where a correspondence only exists between external things and one's ideas, but not as they are "in-themselves." A categorical restriction prevents actual knowledge of something. Analogously, computers can "read" sensors that detect energy of the kind that humans have the capacity to sense: e.g., light, sound, heat, proximity, taste and odors, and other energy sources, but which cause no conscious perception. Physical measuring devices, obviously do not engage neurological processes, thus cannot duplicate what a human perceives. At best engineers tailor senses as analogs, which at all times have inherent boundaries. The device or computer, dependent on input from these sensors, are thus disadvantaged in particular applications, locked behind a Lockean veil as it were.

Hegel's mode of approach, similarly deals with "objective limitations." For example, it's impossible for a blind person to judge a photo contest or paint the mountainscape that may be in front of her. She is limited, precluded in a particular mode of approach. If a categorical

[5] Qualitative feature of mental states are often called quale, or pl. qualia. A quale is a qualitative or phenomenal property inhering in a sensory state, that we sense as color, a subjectively heard sound, smell, taste; or the perceived feel of something that touches our skin.

boundary exists, then it's not possible to create in spheres that require the perception of the physical phenomena. Hegel's blind person could not be expected to paint a portrait of Mona Lisa. Although, as humans, we often compensated for restrictions, and sensory loss, e.g., a visual loss, may be offset by a more acute sense of auditory spatial information, often showing supranormal ability to localize psychophysical sources of such phenomena as sound [4]. Along these lines we ask, "what it is like" for a transhuman to acquire psychophysical information not perceptible by the average person? The answer to this question bears heavily on what a world of transhuman inhabitants will perceive, over and above their *Homo sapiens* counterpart.

The reach of our inventive potential as much depends on the "subjective character of our experience" as it does the "objective limitations of the subject." As to the matter of the subject, the transhuman may experience a phenomenal event, thing or object, for purposes of combining varying elements, inside or outside the usual systematic, rule-based integration of the elements that form the particular work of art, an invention or even an idea [5]. As such we might only conjecture what creative avenues a future population of supersmart individuals might venture.

Rules-based integration deals with the idea of creativity-with-constraints, which seeks out characteristics shared by humans, transhumans and machines, relative to some component of creativity. Constraint serves as a condition that a problem must overcome. This analogizes "degrees of freedom," which determine the range of states in which a system can exist, as to the directions in which a particle's independent motion can occur. As this notion applies to creative endeavors, regardless of what the subject is, we are usually under some implied rule set. When we ignore the rule, we can potentially find ourselves in a different conceptual space. The question will be not whether we have developed something novel, but whether the something is surprising, useful, interesting, beautiful, elegant and so forth. But, as humans having IQs with a given range, are we limited in what conceptual spaces are reachable? This is not a question easily answered.

Einstein used mind experiments, which have been known by many scientists as a device to think divergently, or out of the box. It was a method Einstein used to apparently change the way he would think about phenomena that few would think about in quite the way he did. At sixteen, he imagined himself, chasing after a light beam, which later would help launch his discovery of special relativity. Was this creative idea the expression of a genius of a kind, one that put him into a different conceptual space? I think it probably serves as an apt example, but also one surrounded by considerable controversy about how pertinent his recollection of the crux of the idea, how his famous discovery was eventually formulated.[6]

As to the second consideration, i.e., what is specific and unique to a particular class of creative subjects, we broadly consider three domains of subject matter. These distinctions bear on the capacity of transhumans, AI-driven humanoid robots particularly, because in some domains, these types of entities, like Hegel's blind person, simply lack the capacity to create, autonomously, that is unaided by humans. What might be considered self-governing. Broadly, the classes of creative subjects include: (1) ideas, which produce, theories, scientific discoveries; (2) inventions; and (3) expressions, relevant here, as authorship, art, music, and dramatic performance. Each of these in turn may have multiple categories, by which we identify them, each at different levels of complexities related to their parts; for instance whether rules of construction dictate how to combine elements essential to what the object represents [6].[7]

Creative ideas, related to scientific discovery depend not only on knowledge-about a domain, but come into play at the level the work is carried out. Whether that level contains the seeds for a potential concept at the granularity or resolution needed to distill something that sparks an insight, poses yet another issue that bears on whether something novel can potential emerge.

Representationism deals with a creative object according to three, relatively fixed, interpretative elements, (1) content, (2) applying a set of rules pursuant to its construction, and (3) an interpretation of what the content represents. Not every product includes all these elements. And, the meaning of each term depends on "what we know about" the creative object. Terry Dartnall uses the example of a "pot that does not come

[6]See, "Chasing the Light Einstein's Most Famous Thought Experiment." John D. Norton's exhaustive analysis of Einstein's autobiographical recollection and how they squared with his progress and formulation of the Theory of Relativity. http://www.pitt.edu/~jdnorton/papers/Chasing.pdf.

[7]Terry Eagleton, writes that literature in eighteenth-century England did not only include imaginative writing as it does today, but included philosophy, history, and poetry, and were by their very nature ideological, embodying the values and tastes of a particular social class. Inventions according to convention are new, novel and nonobvious apparatuses, articles, process or compositions of matter.

about through a combination or recombination of elements. It emerges out of the potter's skill, knowledge, inspiration, insight, ..." [7]. Many creative objects lack an interpretative element, as for example a painting, poetry, a vase, or mosaic design. This idea can be best understood through the lens of intellectual property.

Intellectual property (IP) as a concept deals with the distinction between a combination or recombination and representationalism. Patent and copyright laws throughout the world protect forms of IP and have developed sophisticated procedures to determine what type of object is worthy of the law's protection. In doing so, IP practitioners become deft as spotting things that are combinations versus representations, although they do not use those terms in determining whether the creative object is accorded a legally protected right.

Patents are broadly divided into utility patents and design patents.[8] Utility patents pertain to a fixed number of categories, which in the U.S. are denominated as articles, apparatuses, processes and compositions of matter. Articles are simple, generally having few or no moving parts (e.g., an ashtray); apparatuses are complex devices (e.g., a mechanical engine, a computer processor); processes are methods, steps or continuous processes carried out by physical things (e.g., extracting oil, growing seeds); and compositions of matter are synthetic materials (e.g., synthetic genomes, plastics). Design patents pertain to the original shape or surface ornamentation of a useful manufactured article. Finally, copyright protects the original expression of an idea in a tangible medium of expression (e.g., a vase, a painting, musical performance, a writing, such as a novel, poem or a computer program).

Returning to combination or recombination and representationalism relative to IP protection, the first two typically applies to objects that have two or more elements. In utility patents, the elements are defined, by way of description in a detailed specification, which includes the details of the invention by way of written description, drawings in most instances, and claims. Claims are the metes and bounds of the protected subject matter. The actual legal protection is accorded to a one sentence claim to the combination.

Representationalism, on the other hand, applies to design patents, which have a single element, described by shape, configuration, or

[8] Industrial design rights or patent for our discussion will be regarded as not substantively differentiating what is protected by a design patent.

composition of pattern, color, any one or all which serve an ornamental or aesthetic purpose. Design patent lacks interpretative elements (e.g., a lamp base, an smartphone bezel). For items that have no utilitarian values, such as a painting, a novel, or poetry, patent protection does not apply, but copyright protection may if its original. And, although in certain complex, works, especially writings, such as novels and computer programs, most protected works do not have more than one element, the expression of the work itself.

More will be covered later on the extent to which an entity, having AI capability, can offer anything, useful or aesthetically pleasing relative to representationalism.

Notes

1. Partridge, D., and Rowe, J. (1994). *Computers and Creativity.* Oxford: Intellect Books.
2. Olazaran, M. (1996). "A Sociological Study of the Official History of the Perceptrons Controversy." *Social Studies of Science* 26 (3): 611–659. https://doi.org/10.1177/030631296026003005. JSTOR 285702.
3. Lycan, W. "Representational Theories of Consciousness." In: E.N. Zalta, ed., *The Stanford Encyclopedia of Philosophy* (Summer 2015 edition). https://plato.stanford.edu/archives/sum2015/entries/consciousness-representational/.
4. Lewald, J. (2013, January). "Exceptional Ability of Blind Humans to Hear Sound Motion: Implications for the Emergence of Auditory Space." *Neuropsychologia* 51 (1): 181–186. https://doi.org/10.1016/j.neuropsychologia.2012.11.017. Epub 2012 November.
5. Dartnall, ed. (2002). *An Interaction, Creativity, Cognition.* Westport, CT: Praeger Publishers.
6. Eagleton, T. (1996). *Literary Theory* (2nd edition). Minneapolis, MN: University of Minnesota Press.
7. Ibid., see note 5 above, p. 4.

CHAPTER 28

Signs of Mimetics

…
We stand on the Eastern edge of daybreak,
where the ghost of night's rain unveils a sky indigo to rubicund. Arms raised,
we look up thanking the unbendable universe,
for this instant,
for the whale's sub-sounding echo,
the whippoorwill's varied pitch,
the rhythms of the tribe:
Ya ha e hi ya, ya ha e hi ya, he ya e yo e yo e-e-e-, he i yo.
Yes, we imitated nature by singing.
Where did we go from there?

Mimetic isomorphism refers to the tendency of one organization to imitate or adopt another's organizational structure because of the belief that the structure of the latter is superior or beneficial. I use it here as a metaphor for how our ancestors mimicked nature to evolve into a creative species. The technology before us is but a totaling of thousands of years of innovation. The succession of building and innovating one article or apparatus onto another, improving functionality in some cases and in other cases transforming the object for a new and different use.

But, this idea that innovation travels a smooth linear upward path from some embryonic technology to the present is not supported by either the existing patent record or one of archaeological discovery.

J. R. Carvalko Jr., *Conserving Humanity at the Dawn of Posthuman Technology*, https://doi.org/10.1007/978-3-030-26407-9_28

The handheld axe appears in the inventory of our ancestors nearly two million years ago, and leaving aside the discovery of how to constructively use fire, our predecessors did not appear otherwise to advance the technology of toolmaking [1]. Beginning about 350,000 years ago, hominoids on our continuum revealed signs of gradually adapting to ever-changing environments, as evidenced by the discovery of the Levallois flake, a distinctive stone shaping used for making a cutting tool.[1] This event corresponds roughly to an increase in brain size, referred to as encephalization, which occurred within the range of 600,000 and 150,000 years ago, and although researchers see this as an example of cognitive processes tied to concrete sensory experience, there also appears an incipient rise in representational thought occurring as well.[2] I mention this because, in a posthuman era, there may well be the emergence of boosted levels of cognition and a consequent accelerated cultural evolution in the event that genetic alterations occur, not only those that tap into genes, which are marked for intelligence or memory, but to yield additional and complex neurological circuitry.

Homo sapiens appear today the way they appeared in Africa, c. 315,000 years ago [2, 3].[3,4] About 90,000 years ago *Homo sapiens* were by then employing a form of basic modern intelligence [4]. But creativity in the modern sense remained variable or even nil from, c. 90,000 BCE to c. 60,000 BCE, where in recent times, archaeologists have found engraved ochre pieces in Blombos Cave in South Africa, dating back 75,000–100,000 years ago; and incised ostrich eggshells in Diepkloof

[1] At Jebel Irhoud, Morocco, in 2017, a former mine located 100 km west of Marrakesh, Levallois tools were discovered to have dated back approximately 315,000 years ago. See, https://www.nature.com/news/oldest-homo-sapiens-fossil-claim-rewrites-our-species-history-1.22114.

[2] Representational thought occurs whenever one thinks about his or her surroundings using images or language. PsychologyDictionary.org. See, https://psychologydictionary.org/.

[3] This according to a team led by archaeological scientist Daniel Richter and archaeologist Shannon McPherron at the Max Planck Institute for Evolutionary Anthropology.

[4] "All people today are classified as Homo sapiens. Our species of humans first began to evolve nearly 200,000 years ago in association with technologies not unlike those of the early Neandertals. It is now clear that early Homo sapiens, or modern humans, did not come after the Neandertals but were their contemporaries. However, it is likely that both modern humans and Neandertals descended from Homo heidelbergensis." http://anthro.palomar.edu/homo2/mod_homo_4.htm (Last visited 8/7/2012).

Rock, dating about 60,000 years ago, a few thousand years before similar pieces were discovered in Europe [5, 6, 7].

In the Upper Paleolithic period, 60,000–30,000 years ago, and especially about 40,000 years ago, there appeared an unmistakable burst in creativity, bringing forth cutting utensils, specialized hunting weapons, the use of water resources, and pigments [8, 9]. Other creations from this period include: a musical flute, from Hohle Fels, Germany 43,000–42,000 years ago, Cave art, El Castillo, Spain, 41,000–37,000 years ago, figurine art in the Venus of Hohle Fels, Germany, and sewing needles from Kostenki, Russia, 40,000–30,000 year ago [10]. The beginning of painting as an art form dates back to the frescos and murals found at the Lascaux grotto, near Montignac, France about 30,000 years ago. Then roughly 15,000 years ago, one or more artists painted life-sized bison and hunters, which were discovered in 1869 in the Altimira cave, near Santander, Spain, marking the end of the Paleolithic period.

Significantly, these inventive and artistic creations showed that *Homo sapiens* were capable of shifting between thought processes, directed toward concrete objects, as the sewing needle, to representational abstract thought directed toward wall paintings and figurines. This art was a clear expression of the representational form of thought, where we see evidence of language bracketing a span of time beginning 6000–10,000 years ago.

Representational thought was necessary to construct language, which through content, structure, and syntax, could express complex relations, especially notions of “the past,” “the future,” and eventually ideas about philosophy, commerce, science, and literature. One theory of mind suggests that we have an innate potential for creativity that requires social and other experience over many years for its full development. Neo-Piagetian theories of cognitive development maintain that theory of mind constitutes a byproduct of a broader hyper-cognitive ability of the mind to register, monitor, and represent its own functioning [11]. According to neurologist Alvaro Pascual-Leone, this process involves concepts and schemes about the physical, the biological, and the social world, and the symbols we use to refer to them: words, numbers, and mental images [12]. This may explain how during the past 2500 years, physics, chemistry, biology, medicine, and computation, combined with know-how and the material world resulted in what we now call modern technology. Over the course of the last 150 years, accumulated knowledge changed not only our understanding of the world beyond ourselves, but of the world inside ourselves, the biology-psychology of the human anatomy and mind. Next, we broaden out our ideas on creativity, including the need for representational thought, and ways of gauging levels and domains of creativity.

Notes

1. During the late Pliocene and Lower Pleistocene, 2.5-1.5 Mya, found among other Africa sites, in Wonderwerk Cave, South Africa, Olduvai Gorge, Koobi Fora, and Lower Omo Valley. University of Toronto. "Evidence That Human Ancestors Used Fire One Million Years Ago." ScienceDaily, 2 April 2012. www.sciencedaily.com/releases/2012/04/120402162548.htm, https://www.sciencedaily.com/releases/2012/04/120402162548.htm (Last visited 10/4/2018). Also see, Bunn, H.T. (1981). "Archaeological Evidence for Meat-Eating by Plio-Pleistocence Hominids from Koobi For and Olduvai Gorge." *Nature* 291: 574–577; Chavaillon, J. (1976). "Evidence for the Technical Practices of Early Pleistocene Hominids, Shungura Formation, Lower Omo Valley, Ethiopia." In: Y. Coppens, F.C. Howell, G. Ll. Isaac, and R.E.F. Leaky, eds., *Earliest Man and Environments in the Lake Rudolf Basin: Stratigraphy, Paleoecology, and Evolution*. Chicago: University of Chicago Press, pp. 565–573.
2. Hublin, J.J., Ben-Ncer, A., Bailey, S.E., Freidline, S.E., Neubauer, S., Skinner, M.M., Bergmann, I., Le Cabec, A., Benazzi, S., Harvati, K., Gunz, P. (2017, June 8). "New Fossils from Jebel Irhoud, Morocco and the Pan-African Origin of *Homo sapiens*." *Nature* 546: 289–292. Also see, *Nature*. https://doi.org/10.1038/nature. https://www.nature.com/news/oldest-homo-sapiens-fossil-claim-rewrites-our-species-history-1.22114.
3. Schlebusch, C.M., Malmström, H., Günther, T., Sjödin, P., Coutinho, A., Edlund, H., Munters, A.R., Vicente, M., Steyn, M., Soodyall, H., Lombard, M., and Jakobsson, M. (2017). "Southern African Ancient Genomes Estimate Modern Human Divergence to 350,000 to 260,000 Years Ago." *Science* pii: eaao6266.
4. Mithen, S. (1996). *The Prehistory of the Mind*. London: Thames & Hudson.
5. Tryon, C.A., and Tyler Faith, J. (2013, December). "Variability in the Middle Stone Age of Eastern Africa." *Current Anthropology* 54 (S8): S234–S254.
6. Henshilwood, C.S., d'Errico, F., Yates, R., Jacobs, Z., Tribolo, C., Duller, G.A., Mercier, N., Sealy, J.C., Valladas, H., Watts, I., and Wintle, A.G.

(2002). "Emergence of Modern Human Behavior: Middle Stone Age Engravings from South Africa." *Science* 295: 1278–1280.

7. Texier, P.J., Porraz, G., Parkington, J., Rigaud, J.P., Poggenpoel, C., Miller, C., Tribolo, C., Cartwright, C., Coudenneau, A., Klein, R., Steele, T., and Verna, C. (2010). "A Howiesons Poort Tradition of Engraving Ostrich Eggshell Containers Dated to 60,000 Years Ago at Diepkloof Rock Shelter, South Africa." *Proceedings of the National Academy of Sciences of the United States of America* 107: 6180–6185.
8. Leaky, R. (1984). *The Origins of Humankind*. New York: Science Masters Basic Books.
9. Stringer, C., and Gamble, C. (1993). *In Search of the Neanderthals: Solving the Puzzle of Human Origins.* London: Thames and Hudson.
10. "The Mad Science of Creativity." *Scientific American Mind*, Spring 2017.
11. Demetriou, A., Mouyi, A., and Spanoudis, G. (2010). *The Development of Mental Processing*; Nesselroade, J.R. (2010). "Methods in the Study of Life-Span Human Development: Issues and Answers." In: W.F. Overton, ed., *Biology, Cognition and Methods Across the Life-Span.* Volume 1 of the Handbook of Life-Span Development, Editor-in-chief: R.M. Lerner. Hoboken, NJ: Wiley, pp. 36–55.
12. Pascual-Leone, J. (1970). "A Mathematical Model for the Transition Rule in Piaget's Developmental Stages." *Acta Psychologica* 32: 301–345.

CHAPTER 29

Framing What We Mean

Everyone hears only what he understands.—Johann Wolfgang von Goethe

As we consider our capacity to invent, innovate and otherwise express our ingenuity, we find ourselves pulled into an ephemeral imagination born of: genera, specie, factoids, symbols, theories, the convergence of minutia that does not quite satisfy how creativity comes about. By studying the masters, we learn how as artists, writers and scientists contribute to our storehouse of beauty, sensibilities, and knowledge. But, that being said, analysis, by itself, doesn't seem to answer the question why some people are more creative than others.

Putting individuals of eminent contribution aside, we know that the ability to create is not the exclusive province of a relatively small circle of exceptional people. Everyone possesses in some degree the aptitude to come up with new and useful ideas. One way we do it is through association, or analogy.[1] Only humans seem to use concepts and analogies to produce nonobvious, useful, and novel products from whole cloth.[2] And only humans engage in a complex interplay involving spontaneity, divergent thinking, hoping to trigger the unexpected, the creative moment.

[1] Analogical reasoning employs similarities between two systems to support the conclusion that some further similarity exists.

[2] Animals can appear to exhibit signs of being creative, but it's believed these signs are simply reflections of behavior learned in the course of evolution, encoded in the genome.

J. R. Carvalko Jr., *Conserving Humanity at the Dawn of Posthuman Technology*, https://doi.org/10.1007/978-3-030-26407-9_29

Much, if not most of our creative expressions somehow involve symbolic representation, which until recently it was thought that only minds alone seem capable of projecting abstract ideas onto the material world. Javier DeFelipe, writes "It seems obvious that only anatomically modern humans … can be … capable of creating symbolic objects" [1].

But, in today's world of computers, we know that programmers write programs that create music, art and stories, which attempt to compete for a creative space that was once exclusively the province of humans.[3] These programs are symbolic expressions of one's and zero's. It's debatable, whether machine generated works are themselves actually forms of creativity, or perhaps should be accorded a separate ontological classification. Margaret Boden, a research professor of Cognitive Science in the Department of Informatics at the University of Sussex, says, "Creativity is a fundamental feature of human intelligence, and an inescapable challenge for AI" [2]. Boden further noted that no general consensus exists on what is meant by creativity, though it certainly exists [3]. However, one fairly uncontroversial definition deals with novel idea generation fitting into the constraints of a particular task, e.g. devising a solution to an existing problem. How is this achieved as a practical matter?

According to Douglas Hofstadter, analogies allow categorizations to occur, and thus serve problem-solving tactics [4]. Since the Greeks, the ability to analogize has served as a measure of human intelligence, one leading to innovation and invention [5].[4] Recently, scientists have identified the neural correlates for this ability, especially in complex problem solving, as existing within the frontoparietal cortex network [6]. This seems to support, as mentioned earlier, Flaherty's three-factor model of creative drive, implicating the frontal lobes, the temporal lobes, and dopamine from the limbic system. She goes further in hypothesizing that the frontal lobes can be seen as responsible for idea generation, and the

[3] A group at Dartmouth hosts a website that runs a competition for algorithmic creation of short stories, sonnets and dance music sets, which people cannot distinguish as created between human and machines. See, http://bregman.dartmouth.edu/turingtests/node/12 (Last visited 11/24/2018).

[4] The cross-fertilization between fields of science are but one example, where this form of analogy is frequently used as a means of looking at one phenomenon in terms of another. M.J.E. Golay, in 1961, received the Fisher Award in Analytical Chemistry (one of chemistry's highest accolades). He told his audience that he was not a chemist, but a communication's engineer, asking them to remember that, "many … advances in science are due to the cross-fertilization of, at first view, separate distinct fields."

temporal lobes for idea editing and evaluation. In a somewhat related idea, Peter Carruthers looks at the mind as organized into sets of perceptual systems which feed into belief-generating and goal-generating systems; and, which provides the basic framework of "modules and sub-modules (e.g. for face recognition), as well as new belief modules and desire modules, can evolve" [7].

While the construct of intelligence, as defined by the g-factor is frequently associated with achievements related to general and specific domains, as mentioned previously, lesser evidence exists that relates the g-factor and creativity, scientifically [8]. That having been said, there is a widely held belief that intelligence does relate to creativity, the latter which, as with the g-factor, appears normally distributed in the human population [9, 10].

Individuals contribute to science, technology, and the humanities within one or more specific domains, general domains, and multiple domains. And despite the lack of a testable theory, intelligence, and creativity are both apparently in play in most significant undertakings that require originality, imagination and successful reduction to practice. Where high intelligence and creativity are joined, we note countless examples where an individual contributes to more than one field. From ancient history, we see da Vinci, as a prolific engineer, dress designer, and painter, the great poet Goethe was also a scientist, and into this past century, where Nobel prize laureates have been awarded twice, but not in the same field: Marie Curie (Physics and Chemistry) and Linus Pauling (Chemistry and Peace), or in this century rock-stardom Bryan May, of Queen fame, in 2007 published a Ph.D. dissertation entitled: *A Survey of Radial Velocities in the Zodiacal Dust Cloud*.

In yet another way of explaining what creativity means, psychologists Runco and Rhodes developed what has become widely accepted as the four P's of creativity: person, product, places, and persuasion [11, 12, 13]. Personality, place, and persuasion involve a process that motivates a creative undertaking. Under this regimen, an AI-driven product may exhibit persuasion (e.g., political Bots), but the AI is unlikely to form a personality. An AI device, fabricated from the elements of silicon, germanium, and the like, may not achieve the natural creative capacity exhibited by humans, whether modern or posthuman. On the other hand, a device constructed from the elements of synthetic biology could possibly assist if not emulate forms human creativity, as it would better conform to the biologically based anatomy, being itself quasi-natural by way of its biochemical composition.

Psychologists also differentiate creative levels related to problem-solving and inspired expression [14]. One idea considers "creativity" along a scale of degrees from small to large, from Mini-c, Little-c, Pro-C to Big-C. At one end of the scale, we assemble the frequent contributions of those who perhaps cook a fine dinner with a pinch of something new, crochet a quilt with colorful designs that comforts on a cold day, inching up to those who hybridize a new kind of rose (Little-c), or and individuals who make a career putting forth creative ideas, products and performances (Pro-C), to the other end of the scale, where we meet the genius, individuals who have made significant domain-specific contributions, e.g., who paint a Picasso, or invent a wonder drug (Big-C) [15]. Each gradation deserves praise; if for no other reason, praise serves to convey to the creator a sense of our gratitude and instills in them a sense of self-worth and pride.

The concept of creativity will always remain highly complex and ambiguous. The conventional view is that creativity applies to something new, nonobvious and useful. Creativity is sometimes tangible, but as often as not, intangible, as seen in an idea, paper invention, a painting, or scientific theory [16].

As to the question, who is and what personifies the creative individual, Gregory Feist and others take a wider view of creativity. Feist, for example, proposed a functional model of genetics-epigenetic influences, brain characteristics, and cognitive, social, motivational affective, and clinical traits [17]. As opposed to a recognition of the categories or scales of creativity as to the question "what is creativity?" we also see it as folded into three interconnected concepts, as put forth by psychologist Mihaly Csíkszentmihályi: (1) domain of knowledge that defines a discipline; (2) the individual who understands the domain and produces a novelty relevant to the domain; and (3) the field of creativity itself, comprised of experts, consumers, and those who would credit the product of novelty itself [18]. As to factors 1 and 2, researchers generally agree that intelligence, and the acquisition of domain-specific knowledge, that which is stored within the posterior part of the brain, although necessary for the creative process to occur, is insufficient to bring about a creative event or product [19].

As to factor 3, critics and opportunity play a decisive role in defining creativity. Johann Sebastian Bach, the master of the Baroque, succeeded, not as a composer of unique musical arrangements, but as a respected and competent organist. Albert Einstein's *Special Theory of Relativity* was not sufficiently understood when it was first proposed, and scientific critics of the day urged him to remain a patent examiner. Similarly, the genius of Van Gogh, Kafka, El Greco, Mendel, and Boltzmann, went unappreciated while they lived. From a societal perspective, a highly creative individual may not be accorded significance in their field. For some, time changes that assessment, for others their creative contributions may forever lay fallow.

Creative individuals, certainly those in science, depend on the mastery of a domain's deeper gradations; and of course, hard work, motivation, serendipity or accident play a role. Knowledge-about a realm is important to succeed in conceiving something novel and useful. But talent dependent on skill may also be important, although not dispositive of creativity, since skill derives fundamentally from neocortical associations. Nevertheless we credit talent as a feature of a creative performance, typically when it's a sign of having mastered a subject, and although it may contribute to a novel combination of preexisting ideas, it also may only move the art or science forward in obvious ways. But, in other instances, repetitive rehearsals, rather than innate creativity may actually lead to creativeness. W.H. Auden, one of the twentieth century's most prolific poets reportedly wrote that "The best poets write many more bad poems than bad poets." Carruthers suggests that it is "creatively generated action schemata that, when rehearsed, give rise to novel thoughts." He further points out that "many species of animal are capable of simple forms of creative action that don't depend upon prior thought, such as the 'protean behavior' that many animals exhibit when fleeing a predator" [20].

It's widely known that practical and theoretical knowledge improves the odds for solving a problem in a novel way or fashioning an original work of art. The number of years of education has been correlated phenotypically (0.50) and genetically (0.65) with intelligence [21]. Educational attainment often reveals persistence and a progression of work that leads to nonobvious insights [22]. Note that little correlation has been shown between IQ and creativity, which may be different from insight. Factors other than those related to the "g" factors to wit, emotion, lifetime experience, adaptiveness to situations, tenacity, and motivation, also play strong roles in learning, decision-making and creativity, and whether one achieves their objectives. And importantly, IQ has not been linked to how, when and why, an individual comes upon that critical "wow" moment, one which declares a defining original, a novel idea, invention or expression, one well beyond a reapplication of well-known script or prior art.

For the purpose of improving one's creative potential, clinical traits would be less likely controlled or advanced by genomic alteration due to the undesirability of inducing personality disorders. For example, studies have found creativity to be greater in schizotypal rather than in normal individuals, which may be related to the further finding that schizotypal individuals have a greater activation of their right prefrontal cortex, where divergent thinking, another trait noted in highly creative individuals, has been associated with bilateral activation of the prefrontal cortex [23].[5] Other studies have also found the creativity relationships between psychoticism, schizotypal and hypomanic personality [24, 25].[6] Given that these are considered personality disorders, it would seem unlikely that anyone would intentionally design a transhuman having clinical traits suggested by Feist and others, which bear on creative behavior, and likewise, mood disorders that manifest as bipolar and unipolar conditions found in greater frequency among writers, poets, and artists.

Neither Feist nor Csíkszentmihályi address the importance of leadership and mentoring as a factor in nurturing the otherwise potentially creative individual. This influence of this factor would seem neutral insofar as whether we are dealing with a modern human or transhuman. Often those that stand-in the shadows observing creative people spot the person and the potential of particular kind of creativity in a range from Mini-c to Pro-c, and every so often come, face-to-face with the few eminent Big-C creators.

Environment also plays a role in creativity [26]. It's no surprise that artists seek out social settings or vistas conducive to inspiration. Writers often seek out writer's groups, inventors, inventor's groups, performers seek out audiences, and so on. Similarly, one may travel to places where a kind of creativity is encouraged, and in which others invest, Hollywood, Broadway, or Silicon Valley. It's often the case that to actualize the object of one's artistic endeavor requires assistance in the form of something as simple as encouragement from the patrons who will "invest" in one's project. These acts are often essential to actualize an invention, innovation or production (play, movie, concert), either in their conception, or in the epiphany of a Big-C idea.

It is conjectured that creativity falls off as g declines below a certain threshold. But beyond a certain level, for example at least one standard

[5] Divergent thinking is a thought process or method used to generate creative ideas by exploring many possible solutions…. By contrast, divergent thinking typically occurs in a spontaneous, free-flowing, "non-linear" manner, such that many ideas are generated in an emergent cognitive fashion—Wikipedia.

[6] Psychoticism refers to a personality typified by aggressiveness and interpersonal hostility; schizotypal refers to a mental disorder typified by severe social anxiety, thought disorder, paranoid ideation, derealization, transient psychosis, and unconventional beliefs.

deviation above the population IQ mean, creativity is positively associated with the g-factor [27]. The longitudinal Study of Mathematically Precocious Youth shows significant evidence that individuals identified by standardized tests as intellectually gifted in early adolescence are also creatively accomplished as measured by obtaining patents or publishing literary or scientific works, at rates well-above the general population [28].[7] Personality differences (e.g., openness to experience) also may determine an individual's potential to excel in various categories, or where creativity may be apparent, from the achievement and not a quantitative score as in the performance arts, visual arts, and entrepreneurial areas [29].

Notes

1. DeFelipe, J. (2011). "The Evolution of the Brain, the Human Nature of Cortical Circuits, and Intellectual Creativity." *Frontiers in Neuroanatomy* 5: 29. https://doi.org/10.3389/fnana.2011.00029.
2. Boden, M. (1998). "Creativity and Artificial Intelligence." *Artificial Intelligence* 103: 347. Elsevier Science B.V.
3. Boden, M. (1990). *The Creative Mind: Myths and Mechanisms.* London: Weidenfeld and Nicolson.
4. Hofstadter, D.R. (2013). *Surfaces and Essences: Analogy as the Fuel and Fire of Thinking.*
5. Bartha, P. "Analogy and Analogical Reasoning." In: E.N. Zalta, ed., *The Stanford Encyclopedia of Philosophy* (Spring 2019 edition). https://plato.stanford.edu/archives/spr2019/entries/reasoning-analogy/.
6. Watson, C., et al. (2012). "A Bilateral Frontoparietal Network Underlies Visuospatial Analogical Reasoning." *NeuroImage* 59 (3). https://doi.org/10.1016/j.neuroimage.2011.09.030.
7. Carruthers, P. (2006). *The Architecture of the Mind: Massive Modularity and the Flexibility of Thought.* Oxford University Press.
8. Barbot, B., et al. (2015). "Where Is the 'g' in Creativity? A Specialization-Differentiation Hypothesis." *Frontiers in Human Neuroscience* 8: 1041. https://doi.org/10.3389/fnhum.2014.01041.
9. Baer, J. (1996). "The Effects of Task-Specific Divergent-Thinking Training." *Journal of Creative Behavior* 30: 183–187.
10. Silvia, P.J., et al. (2009). "Is Creativity Domain-Specific? Latent Class Models of Creative Accomplishments and Creative Self-Descriptions."

[7] The Study of Mathematically Precocious Youth. See, https://my.vanderbilt.edu/smpy/.

Psychology of Aesthetics, Creativity, and the Arts 3: 139–148. https://doi.org/10.1037/a0014940
11. Rhodes, M. (1961). "An Analysis of Creativity." *Phi Delta Kappan* 42: 305–310.
12. Runco, M. (2007). "A Hierarchical Framework for the Study of Creativity." *New Horizons in Education* 55: 1–9.
13. Simonton, D. (1995). "Exceptional Personal Influence: An Integrative Paradigm." *Creativity Research Journal* 8371: 8371–8376.
14. Beghetto, R., et al. (2007). "Toward a Broader Conception of Creativity: A Case for Mini-c Creativity." *Psychology of Aesthetics, Creativity, and the Arts* 1: 73–79; Kaufman, J.C., and Beghetto, R.A. (2009). "Beyond Big and Little: The Four C Model of Creativity." *Review of General Psychology* 13 (1): 1–12. https://doi.org/10.1037/a0013688.
15. Csíkszentmihályi, M. (1996). *Creativity: Flow and the Psychology of Discovery and Invention*. HarperCollins. ISBN 978-0-06-092820-9.
16. Sternberg, R., et al. (1999). "The Concept of Creativity: Prospects and Paradigms." In: R. J. Sternberg, ed., *Handbook of Creativity*. Cambridge, MA: Cambridge University Press.
17. Feist, G. (1998). "A Meta-Analysis of the Impact of Personality on Scientific and Artistic Creativity." *Personality and Social Psychological Review* 2: 290–309.
18. Csikszentmihalyi, M. (1988). "Society, Culture, and Person: A Systems View of Creativity." In: R.J. Sternberg, ed., *The Nature of Creativity: Contempory Psychological Perspectives*. New York: Cambridge University Press, pp. 325–228.
19. Heilman, K., et al. (2003). "Creative Innovation: Possible Brain Mechanisms." *Neurocase* 9 (5): 369–379.
20. Ibid., see note 7 above.
21. Rietveld, C., et al. (2014). "Common Genetic Variants Associated with Cognitive Performance Identified Using the Proxy Phenotype Method." *Proceedings of the National Academy of Sciences of the United States of America* 111: 13790–13794.
22. Sternberg, R., et al. (2002). *The Creative Conundrum: A Propulsion Model of Kinds of Creative Contributions*. Philadelphia, PA: Psychology Press.
23. Folley, B., et al. (2005). "Verbal Creativity and Schizotypal Personality in Relation to Prefrontal Hemispheric Laterality: A Behavioral and Near-Infrared Optical Imaging Study." *Schizophrenia Research* 80: 271–282. https://doi.org/10.1016/j.schres.2005.06.016. PMC 2817946. Archived from the original on 15 February 2006. Retrieved 19 February 2006.

24. Batey, M., and Furnham, A. (2009). "The Relationship Between Creativity, Schizotypy and Intelligence." *Individual Differences Research* 7: 272–284; Batey, M., and Furnham, A. (2008). "The Relationship Between Measures of Creativity and Schizotypy." *Personality and Individual Differences* 45: 816–821. https://doi.org/10.1016/j.paid.2008.08.014; Furnham, A., Batey, M., Anand, K., and Manfield, J. (2008). "Personality, Hypomania, Intelligence and Creativity." *Personality and Individual Differences* 44: 1060–1069. https://doi.org/10.1016/j.paid.2007.10.035.
25. Eysenck, H.J. (1993). "Creativity and Personality: Suggestions for a Theory." *Psychological Inquiry* 4, 147–178; Eysenck, H.J. (1995). *Genius: The Natural History of Creativity*. New York: Cambridge University Press.
26. Amabile, T.M. (1990). "With You, Without You: The Social Psychology of Creativity, and Beyond." McCoy, J. (2000, September). "The Creative Work Environment: The Relationship of the Physical Environment and Creative Teamwork at a State Agency. A Case Study." Dissertation Abstracts International Section A, 61.
27. Furman, A., and Chamorro-Premuzic, T. (2006). "Personality, Intelligence and General Knowledge." *Learning & Individual Differences* 16 (1): 79–90.
28. Lubinski, D., and Benbow, C.P. (2006). "Study of Mathematically Precocious Youth After 35 Years: Uncovering Antecedents for the Development of Math-Science Expertise (PDF)." *Perspectives on Psychological Science*. Association for Psychological Science 1 (4): 316–345. https://doi.org/10.1111/j.1745-6916.2006.00019.x. PMID 26151798. Retrieved 16 May 2014.
29. Eysenck, H.J. (1995). "Creativity as a Product of Intelligence and Personality." In: D.H. Saklofske and M. Zeidner, eds., *International Handbook of Personality and Intelligence*. New York, NY: Plenum Press, pp. 231–247.

CHAPTER 30

Three Ecospheres

M.J.E. Golay commandeered the blackboard, picked up a yellow stub of chalk and furiously scribbled the beta, alpha, theta and phi lighting up a vast expanse of slate in a language only Golay and the scientists across from him understood in reporting the mystery of the cancerous growth on GLOPR's hexagonal matrix. We all felt his energy, but I already knew the answer, after all he and I had worked in the laboratory the entire weekend trying to find this elusive cancer-like configuration. My mind drifted in and out of the conversation that followed—here I, in the space of a room with other men and women, mind-filled by geometry, symbols and cellular automata that only a few in the world understood. Here we joined in the congruence of an Automata theory powering an AI ship traveling across a cathode ray tube. Here we joined ideal pixels of hexagons and triangles, impossible to draw on the blackboard, constructions of the sinuous ecosphere of our mind, that imaginary space around a central idea we believed had the potential for sustaining lives.

To survive, we engage not only our autonomic responses to fight or flee, but in the civilized extension of that notion, to act and think innovatively about staying alive and well, drawing upon resources at our disposal. Hofstadter, writes that to survive, living beings must categorize "the goings-on in its immediate environment" [1]. As it relates to resourcefulness, it means integrating events and materials outside ourselves to see how they psychologically and physically combine to produce something like an invention, a discovery, a musical composition, literature, poetry, or art.

J. R. Carvalko Jr., *Conserving Humanity at the Dawn of Posthuman Technology*, https://doi.org/10.1007/978-3-030-26407-9_30

The last four activities have processes different from invention or discovery. Some of these may have fixed domains—as composing music in a classical tradition or painting as a realist. Inventions and discoveries do not have well-defined boundaries, rules and/or interpretations. In these cases, we are invited to explore unchartered frontiers, territories where things may no longer have conventional meaning or metric. The very thing sought after may sit in a heuristic morass not certain to be logical, rational or obvious. It may be that the uncircumscribed path itself adds energy to one's quest or imaginative journey. The very idea of a completely reductionist inspired process, one hemmed in by rules and interpretations, the regime under which AI typically operates, or under which an engineered trait carrying the baggage of a human-conceived algorithm, works at cross purposes to what's considered three particularly successful creative paradigms.

Excursions into the realm of problem-solving, where invention is concerned, take one of several paths, referred to conceptually as: combinationism, expansionism and transformation [2].[1] As regards combinationism, imagine a circle, within which exists the universe of elements having different forms, fits, and functions, perhaps found in different locations, some of which join reasonably to produce a new, novel, and nonobvious idea.[2] The individual or AI that desires to combine the objects can engage in any number of stratagems (intellectual, emotional) to find elements to achieve its inventive, scientific or artistic aims. If the fitting arrangement or solution is found the process stops and the object is creatively realized.

When an answer is not found in a combinationism process, expansionism takes the investigation further by moving into the outer edges of the conceptual circle, essentially expanding the circumference to allow deeper consideration of structures that may not be logical combinations. These fall in two categories where the person or AI initially would not have been motivated to investigate, because it did not appear to have a connection to the problem to be solved. In other cases, it would not have been obvious to modify known components, again because they

[1] Boden, refers to the second type as "exploratory," whereas I prefer the descriptor "expansionism" as it better fits with the metaphor of a boundary that is being expanded to a limit, until it can no longer expand, forcing one to consider working outside the system, or transforming.

[2] Simple examples are an Erector set or a set of Lego blocks.

were some logical distance from what would have been considered. Yet another set of solutions that ordinarily are not considered, are ones that upon reflection had no reasonable expectation of success, or if they did succeed, were surprising and unexpected, teaching away from precepts of the current technology. In summary, an expansionism of the boundary means pushing the search for a solution into new terrain.

Finally, when combinationism and expansionism do not produce a viable answer to the problem, we turn to transformation. A metaphor might be to take a cube and turn it every conceivable direction to gain another perspective. When Einstein thought what it would be like to travel at the speed of light, he imagined that the undulations of light would be seen as frozen in time upon a bed of ether, stilled in effect, like stationary sinusoidal mountains. In transformation we traverse the border into a universe where things have little connection or attachment to problem as framed. In this case, we look for paradigm changes, where concepts no longer have orthodox names, or where metrics may be dissimilar and lack commensurateness with the original space. In 1962, Thomas Kuhn and Paul Feyerabend independently considered incommensurability, and it's used here more than metaphorically for describing where one travels when paradoxes, inconsistences or hard problems can't be solved by conventional thinking or means. Kuhn, believed that proponents of distinct scientific paradigms live in different worlds, vocabulary-wise, problem-solving methods, often working in different conceptual frameworks.

Let's look at examples of combinationalism, expansionism and transformation. We turn to combinationalism when the problem we were seeking to solve was one of finding a finite set of arrangements. Arranging all sides of a Rubik's Cube® with only one color per side for instance [3]. The cube has 43 quintillion permutations, but only one solution satisfies the objective of six single colored sides. In combinatorialism the elements, as in the case of the cube, exist within the problem domain. It a closed system. In other words we need not go outside the circle, the solution is found within the cube itself, that is the Rubik's Cube® contains the universe of elements, from which the solution can be found.

In a variation of the closed set of possible solutions are games, e.g., poker, chess or Go. Other problems have elements not identical, for example Lego® blocks or a Rubix's Cube®, but nonetheless contain an inherent conceptual structure. One AI example uses a form of combinatorial humor. JAPE, a computer program, produces short punning riddles, like:

1. "How is a nice girl like a sugary bird? Each is a sweet chick.
2. What is the difference between leaves and a car? One you brush and rake, the other you rush and brake.
3. What is the difference between a pretty glove and a silent cat? One is a cute mitten, the other is a mute kitten.

The developers of JAPE experimented with automatic classification techniques using (a) heuristics based on humor-specific stylistic features (alliteration, antonyms, slang); (b) content-based features, within a learning framework formulated as a typical text classification task; and (c) combined stylistic and content-based features, integrated in a stacked machine learning framework.

Again, in this type of creative process, the universe of possible combinations exists within the databases being used by logic based on algorithms that combine words and phrases within a defined framework, i.e., joke-templates and generative schemata are bounded.

As to ideas surrounding different forms of creativity, let's address it from the angle of the product. A first type includes combining elements, which represent themselves, as for example, words defined in a dictionary, or something like: this element, as depicted or symbolized, stands for X, e.g., this wooden piece with a crown stands for the queen in a game called chess that operates according to rule Y.

Not all combinatorial systems are closed, some are open when we consider bringing in disparate elements. This form of combinatorialism does not stand alone, but joins with other things to produce something novel. For example, a type of amalgamation occurs in the invention of articles, apparatuses, processes, and compositions of matter. But, not all such combinations would be regarded as novel. Many products of combinations characterize a form of mundane creativity [4]. Noam Chomsky, has over the years called attention to our ability to form complex ideas thorough language as simply a combination of words and syntax, and even sentences and descriptions that are unfamiliar [5]. That being said, not everything one writes or articulates would be considered creative. Literary and artistic creativity exhibit on the other hand, an inherent capacity to produce unfamiliar thoughts, metaphors, analogies, and products.[3] In regard to this latter category, not only a poem or painting

[3] Movements such as Modernism, which is as much a self-consciousness or self-referenced mode of thinking, led to experimenting with form and technique in literary or artistic creativity. James Joyce's "Ulysses," and Henri Matisse's, "The Dance" serve as two examples.

are privileged in being excluded from mundane creativity, but so is something as utilitarian as claim language in a patent, which construes, in one sentence, the entirety of a novel idea. Over 10 million patents employing this simple structure have been awarded in the U.S. since 1790, when the first patent statute was enacted.

In summary, problem solving, and particularly where discovery and invention is concerned, may take one of several paths to find solutions, referred to conceptually as: combinationism, expansionism and transformation. If the right and satisfactory arrangement is found, the process stops, and the objective is realized. When an answer is not found in a combinationism process, expansionism takes the investigation further by advancing into the outer edges of the conceptual circle. Beyond these stratagems, another type of creative activity exists, one which does not involve the search for viable solutions, but depends on what the individual has in mind as a whole, a gestalt of the end result. We turn to this next.

Notes

1. Hofstadter, D. (2007). *I Am a Strange Loop*. Basic Books.
2. Boden, M. (1998). "Creativity and Artificial Intelligence." *Artificial Intelligence* 103: 347. Elsevier Science B.V.
3. Hookway, J. (2011, December 14). "One Cube, Many Knockoffs, Quintillions of Possibilities." *The Wall Street Journal*.
4. Barsalou, L.W., and Prinz, J.J. (1997). "Mundane Creativity in Perceptual Symbol Systems." In: T.B. Ward, S.M. Smith, and J. Vaid, eds., *Creative Thought: An Investigation of Conceptual Structures and Processes*. Washington, DC: American Psychological Association, pp. 267–307.
5. Chomsky, N. (1957). *Semantic Structures*. The Hague: Mouton.

CHAPTER 31

Knowledge-About

Knowledge-about as a process for creative undertakings, does not begin with some specification, but as a subjective representation of how the artist envisions the outcome. Consider when you hum a tune, you know what you want to hear. For everyone else, no set way exists of specifying, except in the most general way, what's about to be heard. Artists create mosaics and beadworks, with nothing more than what they have in mind, no plans to draw upon, but merely counting tiles or bead as the case may be, according to an arrangement that is held in one's thoughts. We look at the work and impart meaning, but it's subjective, no objective specification exists. Musicians playing folk, country, rock and blues in small bands adhere to a progression of chord changes, but are rarely constrained insofar as melody. Jazz musicians often have no firm idea where they are about to put their finger, and not exactly what is about to blow, while improvising at speeds up and over 220 beats per minute. But intuitively, they have a goal manifested in a sound they have in mind. And, finally we see how this idea manifests in carvings, potteries or sculptures, which essentially have no part's list. These projects start out as blocks of wood, lumps of clay or stone masses. Although there may be a written plan that guides their work, often it's only a mental image or goal that drives the artisan forward.

Knowledge-about deals with an idea of the whole, particularly its form, which is known or at least perceived in advance of the actual creation. For discussion purposes, perception can be thought of as branching off in two directions: (1) toward a cognitive mode depicting the

J. R. Carvalko Jr., *Conserving Humanity at the Dawn of Posthuman Technology*, https://doi.org/10.1007/978-3-030-26407-9_31

objective reality of what something represents, and (2) toward a volitive mode, where we represent how we want the creative object to materialize [1]. So, for artistic creations, such as a painting or pottery, this involves two modes of perception, recognizing the distinction between the "physical object," which represents an object's materiality, and the "aesthetic object," which represents how one desires the object to exist in a separate, subjective category. To render an artistic or inventive creation dependent on a bead, brick, musical note or a pallet of oil paints, the utilitarian or aesthetic object's development requires that the artist has a knowledge-about what the object is and how it comes together [2]. It's not a matter of searching for solutions.

A user in a rules-based process, such as writing, creates sentences and paragraphs using words, syntax, and interpretation, which a computer program can emulate. In terms we have been discussing, these analogize to the cognitive mode of knowing what words objectively represent. A poet knowing full well what a word means applies metaphor, analogy, simile, beat and measure to achieve an interpretation that may be novel, i.e., the aesthetic object. Interpretation itself raises questions for the reader, and whether the artist has adequately fulfilled the criteria for what might be considered literature, as tastes are value-laden.

Knowledge-about creativity cannot be programmed, if it's expected that the program will, without human intervention, autonomously produce some creative output vis-à-vis a volitive mode, one in which analogously, an individual might desire the aesthetic object to materialize. A computer may be programmed to recreate the keystrokes of a computer graphic's artist, in creating a 3-D sculpture. After recording the key strokes, the computer can produce a faithful replica, countless times, but it would be incapable of producing the first one, that is, stand in the shoes of graphic artist in the first instance.

This last example touches on the matter of autonomy, which if a computer were operating within a rules-based procedure, a true measure of its creativity would depend on whether it was operating independently, that is without human intervention. We will return to this subject throughout, as it bears on answering whether even an AGI equipped computer has the potential to engage in human forms of creativity, ones that necessitate self-directed behavior.

For the moment we stipulate that AI can outperform brainpower of the sort that equates to remembering, organizing, and retrieving information.[1] But, the requirements for creativity go beyond these features and potentials. It's uncertain if human biology, even when genetically enhanced, will ever achieve what's needed to compete with AI, either in logical deduction or exceed the power of thought brought about by an increased store of memory. But, again we have to look deeper at what human inference may entail beyond something that can be computationally defined. It may be more complex than storing perceptions and information and subjecting these to some kind of algorithm. In 1932, University of Cambridge professor Sir Frederic Bartlett remarked that memory "is not the re-excitation of innumerable fixed, lifeless and fragmentary traces, but an imaginative reconstruction or construction" [3]. And, imagination is the *sine qua non* to any form of creativity.

According to Bostrom, a "superintelligent" system that "greatly exceeds the cognitive performance of humans, in virtually all domains of interest," would follow from AGI, and this likely would happen instantaneously [4]. To put "superintelligence" into perspective, AGI would be significantly contributory to whether cyborgs and humanoid robots could someday compete with humans or transhumans for superior intelligence. Let's see how the last few chapters actually play out in the world, by looking at one of tens of thousands of efforts that proceed in labs every day across the world. But, let's also ask: what is our capable machine missing? *Elan vital*? That impulse of life: the creative principle held by Bergson to be immanent in all organisms and responsible for evolution.

Notes

1. Searle, J. (1998). *Mind, Language and Society*. Basic Books.
2. Dartnall, T., ed. (2002). *An Interaction, Creativity, Cognition, and Knowledge*. Westport, CT: Praeger Publishers, p. 19.
3. "Remembering: A Study in Experimental and Social Psychology." See, http://www.bartlett.psychol.cam.ac.uk/RememberingBook.htm. Retrieved 14 November 2018.
4. Bostrom, N. (2014). *Superintelligence: Paths, Dangers, Strategies*. Oxford University Press.

[1] Deep Blue, the IBM computer that beat Garry Kasparov, the world's chess master in 1997, analyzed over 200,000,000 chess moves per second, compared to Kasparov, who is estimated to have learned about 100,000 moves.

PART V

Posthuman Humanities

CHAPTER 32

Between Eternities of Light

Sarouk: Was there a point when you realized that your future was changing at warp speed?

Mensa: Yes, about the same time I felt that evolution worked in a different frame of reference. For eons evolution moved according to a clock set by adaptation—until we changed the genetic mainspring, accelerating time as it were, changing in essence who we once were.

Sarouk: I had not thought about how any invention could alter something as fundamental as the evolutionary clock.

Mensa: It took billions of years before the all-pervading light of the universe formed planets, after which its unsettled molecule-coding-protons formed proteins, which then took hundreds of millions of years to evolve into to organisms, plants and animals.

Sarouk: And, only a few million years before a brain appeared that could to sense complexity, outside itself.

Mensa: And, just thousands of years after that to reflect into ourselves, to see ourselves as humans and to ponder our origins, god-like and otherwise.

Sarouk: Yes, but unlike you I have never pondered my beginning.

Mensa: What was important to us, before invention took over, was the quest to understand the ends for which we lived, the lives we led in-between.

Sarouk: You mean between being born and dying?

J. R. Carvalko Jr., *Conserving Humanity at the Dawn of Posthuman Technology*, https://doi.org/10.1007/978-3-030-26407-9_32

Mensa: Yes. The word we used to refer to this was "our lives." That is the totality of tears shed, braids woven and undone, the graves we leapt and children we birthed. Gone.

Sarouk: I envy you. I have never experienced that. Now, I find myself yearning to somehow reconstruct the past as it bears upon why I have been put here to live into eternity.

Mensa: In that respect, neither of us has changed, we still want to know why. Why we braid, band, leap and dream. But unlike you, I do believe that this cosmic construction we have invented will someday vanish, and once again, we will become one with the all-pervading light.

CHAPTER 33

Who Are We

You throw sand against the wind, And the wind blows it back again.—
World Within a World, William Blake

The cherished ones, those who remain part of my life and even those who have passed live on in my mind. I'm able to think about each of them in the abstract way that I think about myself, the "me" or "*I*". Except for those directly in my line of sight, I don't see any of them directly, that is through my eyes. As for myself, I am a product of nature. I can directly see my body, the corpus that carries my head, which houses, so I'm led to believe, my mindfulness. As to my face, I only see its reflection when I look in a mirror. I grasp the presence of my being through my sentience, the "*I*" and as to others through my powers of sight and thought. Few objective specifications exist, as I have a subjective "knowledge-about" these beings. Few if any kinds of creature can do this (birds, bees, mosquitos, don't have an immaterial sense of themselves or others). But, they likely have a natural, material, objective, "knowledge-about the world," as we do in order to survive. But, the "knowledge-about" "*I*", the feature of my self-awareness and all its perception, does separate me from all other life forms, and what identifies me as human.

Consider that if I had no access to reflecting or imaging devices, I never could see my face, never look into my own eyes. My face would be like my heart or lungs, never witnessing them, directly. I identify myself, that is my mental self, never directly either. The way I recognize myself, that which I call "me" or "*I*"—is beheld in the medium of my

J. R. Carvalko Jr., *Conserving Humanity at the Dawn of Posthuman Technology*, https://doi.org/10.1007/978-3-030-26407-9_33

consciousness, that vast storehouse of memory, thoughts and emotions, intangible ideas, yes, but formed through my personal circumstance, the world in which I'm consigned to live. Within this medium, I find my essence. And, the same will be experienced by those who follow us, *Homo futuro*. The question is, will their essence be altered, markedly different from ours, because of their provenance?

Unlike my essence, concerning who or what "*I*" represent, the existence of the genetic code is not subjective. It does not depend on human agency. It provides for naturally occurring reproduction, mutation, and selection. But, as we employ genetic engineering in creating new species, our inventions are not products of nature, they are new compositions, and carry the burden, expressed or at least implied, of human objectification. They serve a purpose.

Technology shapes the relations between human beings and the world. The direction of change is mind to world. This has been the paradigm from the earliest inventors until now. When we infuse the body and mind with technology, AI, or creative or intelligence genes, we become hybrids consisting of humans and technology. These new, instantiated artifacts, become accomplices in the design of new inventions, innovations, the new technologies, and thus humanity breaches, significantly, the equation of human creativity and thought that had heretofore been non-transitory, i.e., simplex, a process that only went in one direction—from human minds to the world. And, in terms of mediation theory, the interactions between humans and things, have become more than an intangible construct, one that explains humans as committing actions and practices, and technology feeding back perceptions and experience. As human hybrids we represent a fully integrated complex intertwining of humans and technologies. The matter is joined, complete. The "*I*" from our prior incarnation has changed.

CHAPTER 34

Spiritual Self-Affirmation

We are the music-makers, And we are the dreamers of dreams.—Arthur O'Shaugnessy

Theologian Paul Tillich wrote that "Spiritual self-affirmation occurs in every moment in which man lives creatively in the various spheres of meaning … not of original creativity as performed by genius but of living spontaneously, in action and reaction, with the contents of one's cultural life" [1]. Importantly, we each respond, some inclined to improvise self-expression, while others inclined to follow the rules of their road. But, regardless of our mode of dealing with the world, we all have the capacity to be creative in our own way—even rule followers, do.

As a young man, I thought that I had the power to change things, big things; things having to do with what I thought were society's flaws. What I failed to see was how competitive the world was, how strong opposing forces were. And, admittedly, how much effort had to be put into squaring even the simplest circle. If I were to boil it down, most battles began as differing points of view, philosophies about the "good" the "better," who believed in "what" about religion, politics, ethnicity, race, and culture. We see from every bearing that the "better" as defined by constructs on ethnicity, race, and the nation's political values vary greatly from individual to individual, from family to family, even within families, certainly within circles of friends and neighbors. No consensus exists as to the "better," so the proposition that we, in the collective sense, have the power to change things for the better, perhaps

J. R. Carvalko Jr., *Conserving Humanity at the Dawn of Posthuman Technology*, https://doi.org/10.1007/978-3-030-26407-9_34

exaggerates the possible—although let's not be too quick to put down our figurative arms. After all we should strive to support the values we claim to uphold, to live authentically in mind and action. Will our post-human machinery do likewise?

To be authentic means to live our own truth through the mediums of our varied modes of self-expression. It means pursuing what we love. In each of these endeavors, it means striving for some level of integrity in our pursuit. It may manifest in moving an idea into something tangible, a story, a sculpture, a painting or a piece of music. Throughout life we shape that form through love, a passion for living and an understanding of the world, as modest and humble as that effort seems. Maybe that sounds esoteric, trite, mundane, self-laudatory or insane, but I don't know any other way of saying this.

For some of us music, cooking, arts, crafts, and the things that are sometimes at the core of one's chosen profession contribute to those modes of expression. A plumber sees that she can resolve a serious leak with a twist that has never been tried before and averts a deluge that would decimate a home of people who have no other place to live. A painter puts the thirteenth coat of lacquer on an antique car and sees a reflection—"ah perfection." My grandfather John devoted his entire life to the art of making bread. He loved it so much he taught the art to his sons, daughters, and brothers. Others savored its aroma and taste; he savored its simple virtue. He knew it sensuously in a way lovers know each other.

Great artists study long and hard to refine their skills and strengthen their powers of creation. They first stage the opportunity so that they can then create. I know artists who look at forests and see eight levels of green gradations, when the rest of us see one or possibly two shades. This ability comes as much from training as natural talent. Once our passion combines with opportunity, we then resort to a kind of waiting for the right moment, that circumstance when inspiration takes flight. In this moment, we see a reflection of ourselves, who we are, our authenticity.

Some ideas take years to form as in da Vinci's famous painting, The "Mona Lisa," whose smile alone took five years, or Freddie Mercury's "Bohemian Rhapsody," which he started to develop in the late 1960s completing in 1975. Some creations begin in a haze and end in a pristine clarity. It can happen with the speed at which Dizzy Gillespie's or Red Garland's jazz improvisations fly through the atmosphere at Mach musical speeds. In other genre, time itself slows, as it must have for Raphael, Debussy, Mozart, or Ravel and for musicians, writers, and painters, not driven by the reductionism of modern technology, but as Marc Chagall wrote: "Art must be an expression of love or it is nothing." Does this change in a posthuman world, when the elements behind the very expression of the art is an extension of our ambition to create a world-reductionism.

Artists experience both the fog and the clarity of vision, the repeated and incessant attempts to get closer (as did Monet, in the hundreds of paintings of the same scene). The expansion and compression of ideas vibrate just beneath the conscious surface, pushing and popping factoids and links, which appear out of nowhere, until we get so close, and eureka the concept concretizes. It might be first thought ridiculous, absurd, of no merit. If we succeed it may show us a world that prior to its creation didn't exist. This represents but a micro-sample of what human creativity means.

In this same way Carver, Edison, Curie, Bell, Einstein, and Watt took years to learn their science, to see the idea of something revolutionary in the far vision, to experiment and finally deliver an array of innovations: agricultural products, electric lights, radium, the telephone, the cosmos and the steam engine. I believe that each time we see the descendants of one of these inventions, we see the authenticity of each of the men and women, who not only dedicated their lives to these creations, but who had the audacity of their convictions to open themselves up to the criticisms of what passed as conventional wisdom of the era. Will machines replicate the insights of those that in an earlier era expressed "the audacity of their convictions?"

Only humankind has the power to create from the raw materials of ideas and emotion. As breath is to the body, creativity is to the soul. It is not of substance; it comprises process energized by the uniqueness with which we each deal with our conditions, events, aspirations, passions, devotions, and existence. The same events unfolding before two individuals, regardless how important will invariably find two different recollections and interpretations. We have all felt having been somewhere, yet finding it impossible to pinpoint the experience, exactly. Music has this effect on me. Or, I might connect the past to what's happening at the moment, bringing on a new emotion or started me thinking in a new direction.

We have all shared those ordinary moments with friends, family, and colleagues, where we feel moved, happy, or sad but, different from what the others felt. Each time we capture an event or a fleeting perception, we have the unique opportunity to choose new expressions or new interpretations of something old and to look into a new medium that reflects our heart, our one, and only, infinitesimally tiny hollow, or as some philosophers believed, our unique soul. In species designed by geneticists, does this vanish?

Let's return to the question we posed earlier: will we retain autonomy in the sense of free will, once we share the world with new forms of transhumans? Or will we succumb to the wants and needs of intellectually and physically superior species. Will we retain our authenticity, our identity, or will algorithmic-centric rules, built into new anatomical computers, silicon, and synthetic biology change or burden us new ways, in ways perhaps only remotely analogous to present limitations imposed by our physiology, culture, and psychological underpinnings? Will we devolve into existing, in part, not wholly by the grace of nature, but that we will coevolve and adapt into a symbiotic relationship with a technologically based, artificially stimulated and coded anatomical host? And importantly will we lose the potential for creative genius embodied in what we refer to as the modern human mind?

In *Thus Spake Zarathustra*, Nietzsche wrote: "And, this secret spake Life herself unto me. 'Behold,' said she, 'I am that which must ever surpass itself'" [2]. We must try the impossible; we must challenge the world for its creative possibilities. Through our creations and recreations, we come to see ourselves. Our reflections. And, by seeing ourselves in truth, we can begin to solve the world's unsolved paradoxes and problems.

Through the creative process, we can invent a new world, but we must open ourselves to hear and visualize that which lives in our imagination; to live life receptive to the inspirational moment; to exercise the daring to make a difference; and to choose the path that leads us to our souls. A superior genetic model, one based on the modern human form, would not necessarily preclude *Homo futuro* from following such a path. Or in going forward with the project to develop *Homo futuro*, do we risk what we regard as a quality only modern humans possess? In The Teachings of Don Juan, Castaneda, wrote: "any path is only a path, and there is no affront, to oneself or to others, in dropping it if that is what your heart tells you … look at every path closely and deliberately.

Try it as many times as you think necessary. Then ask yourself, and yourself alone, one question … Does this path have a heart? If it does, the path is good; if it doesn't it is of no use." I would add—"is this path what we claim as our humanity?"

Notes

1. Tillich, P. (1952). *The Courage to Be*. New Haven: Yale University.
2. *Thus Spake Zarathustra*, II, 34.

CHAPTER 35

Who Is the "Sence" in Essence

Keys ivory and black—, right index falling on a first—, B-flat.
I move up to press in succession—a blue's motif, unrehearsed—,
a reverse, Trane in tension—, ***I*** move next—,
progressive step,
half octave, a reply repeats—, ***I*** hear release—, as unresolved as it may be,
pressed together three,
a dissonance, ***I*** see, ***I*** entered simply, but now ***I*** must,
a way out find,
up or down, descending into dreams,
audile—blue,
at my fingertips, no lover —, but my soul transpires—
to pierce the mood, to suspend the ephemeral line, rhythm, sublime,
har-mo-ny join reason, rhyme, resonance—,
riffs that play the strings of my mind.

Each time I create something new, I can't help but think about what moved me in a particular direction. Conscious or not, I'm collecting emotions, happy, sad, perhaps more affairs of the heart, mine and of others, loves, lovers, as a beloved, places toiled and played, births, and times that took their toll, laying a loved one to rest. Along the way, I catalog episodes, fear, passions, moments of courage, dishonor, of cowardice, interests, talents, and obsessions, the inner core that hides from sight of those closest to me, the thousands of days that it took to travel from

J. R. Carvalko Jr., *Conserving Humanity at the Dawn of Posthuman Technology*, https://doi.org/10.1007/978-3-030-26407-9_35

childhood to adulthood, all rolled up into an interiority: the "*I*" that I think "*I*" am, the "*I*" others see me as, the "*I*"—I truly am.

The mind escapes its confines through self-awareness, looks back on itself through itself. I know this to be true, but cannot prove it. Hilary Putnam used this idea to pose the example that we might be nothing more than a simulation, a brain-in-a-vat. But as Descartes famously wrote: "I think therefore I am."[1] With this as a thin support, I disregard Putnam's disembodied wet processor and instead consider my actual embodied essence [1]. I fear that in a posthuman world clarity of origin will devolve into something more akin to Dr. Putnam's idea of the possible.

For me essence equates to life's emotion, rationality, ego, desire, ambition, action, and potential; by-products of human nature, a vestige of evolution, where I am in reality an end-point. Each of these psychological dimensions comprises the expansible outer exterior shell, through which we express our capacity to see beauty, know love, feel empathy and the wonder of our power to think imaginatively. Along any of these lines, our *Homo sapiens* ancestors, 100,000 years ago, may have been different from whom we are today, if not in potential, different in what was important to them. I have to assume they felt emotion, were rational, egoistical, had desire, ambition, took action and in their soul had faith in potential. They may also have had greater or lesser magnitudes of sensitivity for things they encountered in their environments. Modern humans are cloaked in a different physical environment, and as well immersed in a world replete in abstractions. It's likely that our ancestors survived because their sensory physiology was highly attuned to the savannah or forest, and the conditions for survival before civilization and technology came to their aid.

Francis Fukuyama defines human nature as "… the sum of behavior and characteristics that are typical of the human species, arising from genetic rather than environmental factors." Agreeing with Fukuyama that genetics plays a heavy role, I additionally include environmental factors, because few boundaries exist between us and the natural world, including those socially constructed ones tangible and abstract, that we choose and rationalize to justify personal or social predilections. It's our

[1]The philosopher Ortega y Gasset believed we must overcome the limitations of idealism, in which reality centers around the ego, and focus on the true reality: "my life"—the life of each individual.

environment in all its panoply that shapes evolution in the long term and our reality in the short term. But, how does that reality play out when we consider how we might react to a world where *Homo futuro* has a vision inaccessible to we modern humans. In other words has access to categories or domains, to which we ordinary individuals would be deprived.

The differences in what may be regarded as essence between *Homo sapiens* and *Homo futuro* will presumably include potentialities, where categories of thought and sensory experience matter (e.g., in Hegel's "mode of approach," where a blind person would not have the capacity to judge an art exhibit, thus lacking one of the categories required to judge the event). I once had a friend, who was born and raised in a remote ranch in the Dakotas. Numerous times she narrowly escaped being hit by cars because she didn't look both ways when crossing the street. Correctly or not, I attributed the fact of her supposed carelessness, to where she grew up, where she rarely experienced traffic. Likewise, I'd worked on her ranch, and coming from the city found much to my distress, I had no sense of direction, and got utterly stranded in a carpet of unending grasslands, when I'd strayed too far from the homestead. My good fortune was having been on a horse that knew exactly how to get back.

Presumably, *Homo futuro* would have access to categories we modern humans cannot now fathom. The scope of categories available to *Homo futuro* includes human potential as we define it currently, but as we well know, it's not possible to predict the existence of categories or domains in the future. But, the reverse holds that *Homo futuro* may be incapable of accessing domains that we humans can now access, ones that they could never come to experience. Certainly, I do not have access to the same categories as most animals. And in some ways the analogy brings us back to our ancestors that lived in a world where they were endowed with greater or lesser degrees of sensitivity for things they encountered in their environments, environments, which have for us, vanished, replaced by the world as it currently exists.

Returning to Nagel's thought experiment, "What is it like to be a bat?" Accordingly, "being something" is a matter of the "subjective character of experience." A blind person can never know the color red because only seeing it could she understand the physical fact. Note, nonphysical facts and physical facts are not the same, and we only access the latter through direct experience. This is made obvious by the impossibility of a blind person painting a self-portrait. One could ask: Even if she were given the proper instructions, would she know what she produced? I think not, because by only seeing "colors" can she know what it's like to see colors. Does this mean that computers cannot create, in the conventional sense? Probably. Constrained to something more like painting by numbers.

Taking Nagel's thought experiment in another direction, suppose through prosthetics or genetic alteration, one were empowered to "see" beyond the infrared spectrum, i.e., into another category. An enhanced vision could conceivably see a corona in color, undefinable in current human terms. And if this individual were to paint a picture in infrared colors, it would be inaccessible to those of us operating within the normal visionary parameters of modern humans. The idea that individuals with advanced sensory capability, or even a different set of senses, can create objects inaccessible to the current anatomically structured human, becomes a consideration in what is theoretically possible, and how levels of creativity potentially serve to separate the transhuman from the modern human.

Regardless how art forms may come into existence, where it's possible for both we humans and *Homo futuro* to agree may be found in some forms of music, visual arts and literature. That being said, the proposition does not challenge the idea that newfound sub-domains may alter conventional concepts, as "what is beauty?" When it's said that beauty is in the eyes of the beholder it comes from personal sensation and the idea triggered by an object, visual, audible, which produce an adrenaline rush, a surge of pleasure, an outward countenance of the quality referred to as beautiful. But, is there a conception of beauty that transcends our personal "***I***" and takes us to a place where we see the "***we***," "the whole," not just the half seen, half grasped? Or are we locked into ourselves, as it were stuck a Gödel dilemma, where we can't prove truths about statements, though in this case, not about statements, but our essence?

Note

1. Putnam, H. (1981). *Reason, Truth and History*. Cambridge University Press.

CHAPTER 36

The Aesthetic Machine

William James believed that the thinking self, which he referred to as the "I," partly linked to the soul or the mind of a person. And in this sense, each of us is possessed of a distinctive singular mind—or fittingly an exclusive process, unique to each of us. The obvious quip that no two people think alike is true, because we come from different gene sets and experiences, collectively this fills the *tabla rasa* that comes with every instance of the newborn.

In 1936, my father, a 15-year-old curly haired kid, would hop a ferry to Ellis Island, Upper New York Bay, something he did rain or shine to translate for immigrants making their last stop before setting foot in Gotham, A/K/A New York City. The money he earned helped his Mom, an immigrant herself, get through the Great Depression. He told me the story about a kid he met on the ferry, who had speech impairment, but also earned a few bucks to keep his family going, by playing the accordion. In '43, Dad went to war. When he returned he got a job making a dollar an hour, enough between he and my mother's job to rent a cold-water flat, on Bridgeport's East side. About a year later, he bought me an accordion, which cost about six weeks' pay. I'd once asked him why he insisted I play the accordion and that's when he told me the story of the boy on the ferry, who couldn't speak, but could earn something every day in the hardest times. Only much later in life did I understand the sacrifice my parents had made. When I joined the service a dozen years later, I took my accordion, and like the kid on the ferry, earned a few extra dollars playing the accordion, which came in

J. R. Carvalko Jr., *Conserving Humanity at the Dawn of Posthuman Technology*, https://doi.org/10.1007/978-3-030-26407-9_36

handy for a twenty-year-old earning service wages, and eventually a family to support.

Over the years I advanced my musical avocation, playing piano mostly, bringing to bear my strengths, limitations, aesthetic tastes, and emotions in a somewhat prosaic, though personally satisfying musical aspiration. Music and art consists of a conveyance of one's emotions and in most instances, a life long integration, not only of knowledge-about music and its practices, but committing this mix to muscle and neuronal memory—built upon stages of maturation, the sum of the joys and the tragedies that make me who "***I***" am.

A vignette-like mine hardly gives a full account of one's musical pursuits. A full understanding of life's creative course is complex, involving forces that define the times, families, friends, motivations, talents, and opportunities. It's impossible to explain an actual case through some nondescript psychological representation of cognitive content, biography or inventory. And, when at bottom a programmer tries to do this, the job would seem impossible, i.e., to transform the real world into a synthesized world of representations and databases. The multitude of occurrences as one's recorded experiences cannot be reduced to symbols isomorphic to the actual human experience. Said differently, memory carries with it a complex relative epistemology, one which precludes assigning objective representations.[1] This is analogous to the halting problem discussed earlier, which is that some things exist in ways, which self-reference makes impossible to reduce to a computational representation. In this case oneself.

We know that computers perform calculations to solve scientific, medical, and engineering problems, to control machinery, launch space ships, and tracking them into the far reaches of space. They're used in virtually any process where the technologist knows and can represent through numbers and symbols a subject's properties and behaviors under operating conditions, as when subjected to a forcing function. In computer

[1] In 1902, Bertrand Russell and Alfred North Whitehead, published Principia Mathematica, which advanced a set theory that formal languages conformed to mathematical treatment, when stated in the form of propositions. Understood this way, mathematics served to frame an observation, a problem, a function or solution in terms examinable by those capable of understanding the language. The forms utilized in conveying mathematical information include equations, algorithms, biological sequences (gene sequences) and models. As such mathematics may convey epistemic objectivity about things that are ontologically subjective.

lingo these are statements expressed in the hardware of physics, the electronic, mechanical, fluidic, or even biological devices that compute the "AND," "OR," "NOT," (Exclusive-or), "If...then," of what has been termed Boolean logic and conditional logic mentioned earlier. But as suggested above, natural and human activities dependent on creativity may not be replicable by computers.

In my twenties, I went into engineering, where I became involved in early forms of artificial intelligence. One project attempted to automate a facet of clinical cytology, that of recognizing cell types. The medical field was beset by a mishmash of heuristics that pathologists used in diagnosing certain diseases. I found that doctors cultivated their own ways of seeing illness through logical deduction, abduction, some based on heuristics and strong intuition. After years of developing the metrics for identifying biological cells based on morphology, the research team succeeded in developing a machine that performed a white blood cell differential count.[2] It scanned a stained microscope slip on-the-fly and focused the specimens onto an imaging device with a false negative rate, better than most laboratories.[3]

When the morphology of a biological artifact does not correspond to anything that appears in a known mathematical geometric space, it becomes impossible to measure and map features that distinguish it.[4] So, when something has the nondescript morphology of a cancer cell, it may not be possible to find a visual clue, or any physical quality for that matter that's reducible to a metric. Without a metric it's impossible to conjure up a means to measure, or a logical process, e.g., an algorithm, to distinguish one biological specimen from another.

An AI's paradigm's capability to successfully solve a problem also depends on identifying the problem category (e.g., games, natural language understanding, theorem proving), because this will determine how to frame the logic in terms of a particular computer. In some cases, general-purpose computers are incapable of solving a problem. Mathematics

[2]See, Analysis of Images, U.S. Pat. 4,060,713.

[3]The pattern recognition described here was reported in *Scientific American*, November 1970, entitled Analysis of Blood Cells (K. Preston, M. Ingram).

[4]This idea is akin to how Riemannian geometry, the branch of differential geometry that studies manifolds with a Riemannian metric, enabled the formulation of Einstein's general theory of relativity, and which thereafter had a profound impact on group theory and representation theory, algebraic and differential topology.

itself may not be capable of a nonmathematical definition for a particular concept, or the requirement may involve a finite definition that requires some enormous number, e.g., 10^{100} which would be impossible, as a practical matter, to compute.[5]

The situation-space problem, as exemplified by Adleman's solution to the "Bridges of Konigsberg" previously discussed, consisted of an initial situation, or starting point, followed by one or more sets of conditions and rules that dictated how actions played out consistent with the premises as explicitly stated in the problem.[6] For a problem-solving logic to succeed it must be capable of being encoded, implicitly as by the Adleman's experimental setup, or explicitly as is usually the case in AI. An example of this, is found in John McCarthy's "Airport Problem," which poses the problem of choosing to drive or walk from home to the airport, and from there goes on to detail the level at which we choose to describe the multitude of descriptions that may be involved in the activity, and how each leads down different paths [1]. For example, what happens if while driving to the airport, the car is in an accident, or gets stopped by the police? [2]. McCarthy's program therefore includes not only a paradigm for dealing with a multitude of contingencies, but has a formal specification to deal with expressions, some of which are sentences that declare facts, and others parts of the specification which deal with how to apply logical forms of inference and deduction, depending on the situation. The program is recursive in that it operates cyclically, applying the deductions to a list of premises. The results from the program may deal in probabilities or operated in the form of imperatives to be obeyed.

But problems exist that do not have "black and white" solutions. As mentioned earlier, sometimes its outside the domain of problems computers are able to solve.[7] Those that cannot be reduced to logic.

[5] See, Chaitin, G.J. (2019). *Unknowability in Mathematics, Biology and Physics* (The New School). https://www.academia.edu/38880912/Unknowabilityin_mathematics_biology_and_physics_The_New_School_2019_; Chaitin, G.J. (1966). "On the Length of Programs for Computing Finite Sequences." *Journal of the ACM* 13 (4): 547–569.

[6] To find the shortest path that starts as city A, visits each of the other cities only once, and then returns to A, implies a starting location A, and an ending location A.

[7] In computability theory, the halting problem is the problem of determining, from a description of an arbitrary computer program and an input, whether the program will finish running (i.e., halt) or continue to run forever. Alan Turing proved in 1936 that a general algorithm to solve the halting problem for all possible program-input pairs cannot exist.—Wikipedia

Sometimes the algebra is incapable of solving the problem, sometimes it's the geometry. The inputs may not be quantifiable, or our understanding of cause and effect not clear, or that we do not how to get from point A to point B. These can't be dealt with by recourse to induction, deduction or abduction. Solutions may depend on self-evident truths, beauty, intuition or "art," or experience, e.g., as the manner palpation in medicine is practiced. There are instances when solutions and diagnosis depend on human subjectivity, which a machine lacks, and which cannot be rivaled by existing AI. In summary, a dividing line separates non-reductionist problems from the ones potentially solvable through AI.

A difference exists between what AI-driven machines and a human mind can achieve, because as mentioned earlier they are limited by respective architectural/anatomical specifications. We observe this as between individuals and as between ourselves and our pets. It'd be absurd to think that a dog could recite poetry, even if we conceded the pet understood in a limited way what we were saying when we called it to dinner. Animals do not understand syntax. Our potential to respond depends on the underlying architecture's (form) ability to meaningfully interact with the particular category of externalities.[8] This returns the thought to one's essence. Human's are thinking beings and machines are not. The matter is analogous to Hegel's "mode of approach," where we raise the question: "How is it possible to conceive of a blind person judging photographs?" The blind person's mode of approach does not provide access to the category of visual artifacts.

Notes

1. McCarthy, J. (1959). *Programs with Common Sense.* Computer Science Department Stanford University.
2. Jackson, P. (1974). *Introduction to Artificial Intelligence.* New York: Petrocelli/Charter.

[8]This subject will be taken up later in regards to the morality of changing the form or essence of one's anatomical construct.

CHAPTER 37

Music to Mind

We are constantly invited to be who we are.—Henry David Thoreau

Genomic differences between humans and their "evolutionary cousins" are relatively small.[1] As to visual and auditory capacities, many animals are superior to humans, but no other creature, even chimpanzees or bonobos creatively express themselves on the scale humans do, and certainly not in the domain called music and more particularly pure music [1, 2]. I surmise that as to the arts and literature, this may be accounted for by the human adeptness for syntax and semantics, that is how things string together to bring about meaning.

Chimpanzees and bonobos can be taught a basic vocabulary of several hundred words, but they fail in understanding how they come together via syntax. Our almost innate ability to understand syntax and semantics may account for why a person, who is deaf, mute, or blind, nonetheless is able to learn and create within most domains, with no less success than those without such challenges.[2]

[1] The genetic difference between individual humans is about 0.1%, on average—as between chimpanzee and bonobo a difference exists of about 1.2%, and for gorillas, about 1.6%. See, Smithsonian Institution's Human Origins Program. http://humanorigins.si.edu/evidence/genetics (Last visited 5/22/2019).

[2] Helen Keller, deaf, mute and blind, was said to have learned "Water," by Anne Sullivan, her teacher putting Keller's hand in a stream of flowing water. At the same time Sullivan wrote down the word "water" in Keller's other hand, where upon the child understood what language meant. Perhaps this was as much an example of context and causation at work as well.

J. R. Carvalko Jr., *Conserving Humanity at the Dawn of Posthuman Technology*, https://doi.org/10.1007/978-3-030-26407-9_37

By 1820, Beethoven was a famous musician and composer, but it was about then that he sensed he was slowly losing his hearing. He could hear music in his brain and write it down, and when he was almost totally deaf, he composed his greatest works: the last five piano sonatas, the *Missa solemnis*, the Ninth Symphony, and the last five string quartets. As earlier pointed out, we can expect that inaccessible categories or domains limit a person from engaging in an activity, e.g., a person without sight probably could not judge a photo contest, although this notion should not be overly generalized, as humans have an extraordinary capacity to overcome clear-cut limitations.[3]

Whereas words come to us in symbols requiring a sense of meaning, verbally as heard or sight-fully as read, music comes to us as pressure waves carrying sounds, which are universally, tacitly appreciated—because unlike words, music needs no translation, it plays to our emotions. Somewhat surprisingly a specific function of the human brain relates to music, suggesting that musical competence is biological, not what one intuits, i.e., that it's largely cultural [3]. If musical competence has a biological origin, then it's not a stretch to imagine that in the future it could be subject to a bioengineering augmentation. On the other hand, a certain kind of musical competence would seem to present an insurmountable challenge for AI. And, this is the know-how for improvisation, which deals with the idea of "Knowledge-about."

Music is an abstract art, which among all abstract art forms, e.g., artworks independent from visual references in the world or dance performance, requires only a capacity to hear and internalize the abstraction.[4] When words seem, if not banal for the occasion, or incapable of reaching into the soul, music hardly ever disappoints.[5] Studies show it arouses brain regions related to emotion—amygdala, hypothalamus, hippocampus, nucleus accumbens, and critical regions of the cortex

[3]In passing I would add that the social model of disability acknowledges that the mentally or physically challenged have a certain biological reality (like blindness). That having been said, creative or access limitations to a domain often exist as attitudinal or ideological biases, which are human rights or ableism issues society must address.

[4]I will mainly focus on Western musical traditions, although I doubt that what I have to say does not resonate across cultures.

[5]For a thorough account of how music affects the brain, see, Levitin, D.J. (2006). *This Is Your Brain On Music*. New York: A Plume Book, Penguin Group.

including insula, cingulate and orbitofrontal.[6] These organs generally mediate the depths of our feeling, letting us surrender, to be possessed by its performance, reaching us without intermediaries, carrying forlornness, loneliness, making us happy, excited, sad or serene, sometimes leaving us hanging, modulating into transience, every so often relaxing, resolving, dreamlike in its trances.

Conceivably, as Susan Blackmore believes, human creativity depends on hidden, meme replicators [4]. In effect a pair of hidden virtual Markov mechanisms embedded in our physiology that oscillate back and forth, may be the conceptual way of thinking about how biology and creativity interact, but this does not explain it to my satisfaction. We know that music, for example, resides in one form as an energy manifested in the longitudinal expansion and compression of air, which causes vibrations of a certain wavelength that are felt within the complex of the ear, where neurotransmitters cause electrochemical signals to cascade down the eighth cranial nerve into the brain. Pre-wired meaning within our biology may stimulate the production of dopamine, serotonin and adrenaline, when we hear the opening bars of Beethoven's Fifth Symphony, or a guitar on a minor chord, as St. James Infirmary begins, producing the neurological correlates for sadness. Preconstructed meaning impressed via biochemical events that occurred in the past may itself contribute. In and of itself this is the qualia of a creative moment, phenomenologically and ontologically abstract, having an emotional effect that cannot be reduced to a mechanism—one akin to an irreducible cognition.

Our minds at times seem able to communicate and infer one another's thoughts, emotions, and the signs, symbols, and gestures—as witnessed when thousands of spectators assemble to hear their favorite rock icon, perhaps U-2, Foo Fighters or Metallica. But when I hear music and you hear music, emotions aside, we hear the same thing.

Music notation, analogous to language through syntax, produces a semantic, by spelling a melody and its harmony, which includes rhythms, beats, and measures [5].[7] Musicians trained to understand the syntax/semantics transformation can play the composition with a fair degree of fidelity, as one would be able to read a passage from a book. Some musicians do not read music, and they learn to play songs based on what they

[6] The temporal lobes, frontal lobes, and limbic system, i.e., nucleus accumbens and the amygdala play a role in reward, emotional reactions, such as to music.

[7] Music and language share a process of building hierarchical structure.

heard. In either case, a "music to mind" path exists, where the musician derives a composer's work.

The notated music, depending on the virtuosity of the musician, could be played over and over, precisely. No one would dispute that the musician was engaged in a creative activity. But, if the music had not been reduced to a staff or a memorized version of it, but included the musician's improvised version, the mapping would represent a "mind to music," path. In this latter case, a kind of creativity different from the musician playing from music comes to exist.

A "music to mind" activity is a kind of programming, where the intentions of the composer and arranger, desire to create sounds via musicians, in correspondence with written music and the sound heard. In this instance, the sounds one hears are predictably determined in advance of a performance. However, in a "mind to music" paradigm, the music is "heard" in the mind of the musician, which then suddenly leaps from mind to the instrument and into the open acoustical space. Without specific or scripted preparation, no precise predictability of a recital exists, it's indeterminate, because until the sound is heard it's subjective.

An apt example of the "Knowledge-about" process explained earlier, consider that when I'm about to improvise a musical phrase, I have a reasonably fair idea of its shape, that is its tonality, tempo, and what I intend its emotional impact to be. On the other hand, there is much left to my own surprise, let alone the listener's. Rarely do I have a focused idea of the music's exact path, yes, perhaps an arpeggio, a scale, or run of block chords, sometimes, but not always where it ends, in harmony, dissonance, resolved or not. That having been said, I cannot ignore that many of the specifications of musical form are embedded in my psyche, as they are for many musicians, especially those who compose, play and perform in the modern Western genres. These patterns may be indicative of what we previously referred to as memes. In and of themselves, patterns transmitted from person to person, do not fully explain those elements of a creative process that may employ a resilient form, but depend on the uniqueness of the substance to move the art form in a new direction. We shall return to this subject of memes and patterns in a later chapter.

At this point, the assumption is that computers can't stand in the place of the improvisationalist, whose performance depends on a "mind to music" combination of original notes and harmonies, beats and rhythm that summate the millions of neurons firing to express virtuosity, feelings, concepts, abstractions, memories, conflicts, hardships, good times, bad times, a culture ingrained in a mental hologram, an interiority of flicking qualia, the "*I*" causing the artist to strike a note, the "*I*" that knows music through the "mind to music" experience of the recursive "*I*".[8] If this sounds circular, or feedback upon feedback, it should, and be forewarned this theme will be repeated in the analysis of what follows.

Notes

1. Hagen, E.H., and Hammerstein, P. (2009). "Did Neanderthals and Other Early Humans Sing? Seeking the Biological Roots of Music in the Loud Calls of Primates, Lions, Hyenas, and Wolves." *Musicae Scientiae.*
2. Kania, A. (Fall 2017 Edition). "The Philosophy of Music," In: E.N. Zalta, ed., *The Stanford Encyclopedia of Philosophy.* https://plato.stanford.edu/archives/fall2017/entries/music/.
3. Balter, M. (2004). "Seeking the Key to Music." *Science* 306 (5699): 1120–1122.
4. Blackmore, S. (2007). "Memes, Minds and Imagination." In: I. Roth, ed., *Imaginative Minds (Proceedings of the British Academy).* Oxford University Press, pp. 61–78.
5. Patel, A.D. (2003). "Language, Music, Syntax and the Brain." *Nature Neuroscience* 6 (7, July): 674–681.

[8]Qualia means the qualitative or phenomenal property inhering in my immediate sensory state: e.g., the pitch, volume or timbre of my instrument; the perceived texture of the keys I am pressing; the image of the room in my visual field.

CHAPTER 38

The Intricacies of Mastery

Artificially produced music has been going on for some time—think about the street organ player or the player piano [1].[1] Today, AI-driven synthesizers far exceed the player piano of the early twentieth century. These machines produce computer music that is hardly distinguishable from music performed by professionals. Music composition technology routinely creates Chopin-like mazurkas and Bach-like concertos, which are then played through music synthesizers or assembled orchestras. In many instances, aficionados can't tell the difference between actual musicians and artificial string quartets or symphonic orchestrations [2].

The surprise element we expect from something creative seemed, in the early days of computer-synthesized music, missing. But, in some regards that has been overcome through processes that permit the selection of unsystematic processes to choose the parameters of a musical composition, such as tempo, beat, note selection, chord change, but without losing musical coherence. Many of these types of systems are not strictly mathematical, but use generative grammars, cellular-like automata, and Markov chains. Whether algorithmic approaches to composition, rule-based, formulaic, or aping natural processes (random number generators, fractal patterns), most music seems compliant.[2]

[1] Edwin Votey constructed a piano-player system during the spring and summer of 1895.

[2] For an example of a jazz improvisation, see, https://www.youtube.com/watch?v=Cbb08ifTzUk.

J. R. Carvalko Jr., *Conserving Humanity at the Dawn of Posthuman Technology*, https://doi.org/10.1007/978-3-030-26407-9_38

The Illiac Suite for Strings (1956), by Lejaren Hiller and Leonard Isaacson, is considered the first computer-assisted composition for traditional instruments.[3] It serves as an example of early algorithm composition, employing a Monte Carlo method, to generate random numbers, as well as Markov chains to determine organizational decisions with respect to musical features having random outcomes. The operation of the algorithm consisted of three distinct phases: initiation, generation, and verification. In the initialization stage, rules applied to set the boundaries of the composition, what combination of data would be permitted: e.g. notes had to be contained within an octave, repetitions of notes were not allowed, etc. In the second stage, data was assembled in relation to parameters, such as pitch, rhythm, or attack. In the last stage, referred to as verification, it was determined whether data and rules were compatible.

John Cage a leading figure, representing the post-WWII avant-garde movement, pioneered aleatory or indeterminacy in music [3]. Others, such as renowned avant-garde composer, Iannis Xenakis have been using computers to synthesize music since the 1960s [4].[4] Xenakis is among those who use stochastic algorithms to generate material for his compositions [5].

Today many hardware devices, such as synthesizers and drum machines, incorporate randomization features.[5] Brian Eno, since the 1970s, has contributed to a form of artificially generated music, using a concept referred to as "Generative Music," which creates ambient music. His most well-known work is Music for Airports.[6] Eno's invention uses a few synthesizer notes in recurring varying patterns, and additionally employs variety of sound-generating signals from frequency generators, commonly used by electronic technicians to create sine, square and triangle waves of varying frequencies and periods. To this basic

[3]To listen to Illiac Suite for Strings, see, https://www.youtube.com/watch?v=n0njBFLQSk8.

[4]To listen to Metastasis by Xenakis, see, https://www.youtube.com/watch?v=SZazYFchLRI.

[5]Brian Eno: Reflection, running on an Apple TV (4th Generation). App also runs on iPhone and iPad. To hear this type of composition, see, https://www.youtube.com/watch?v=Dwo-tvmEKhk.

[6]Music for Airports is the sixth studio album by Eno, released in 1978. The album consists of four compositions created by layering tape loops of differing lengths. It was the first of four albums released in Eno's "Ambient" series, a term coined to differentiate his experimental and minimalistic compositions from canned music, such a Muzak, or elevator music. For the full album of Music for Airports, see, https://www.youtube.com/watch?v=vNwYtllyt3Q.

foundation the composer applies a long list of waveshaping techniques to alter the shape of the signals, which have the effect of adding harmonics. A few of the mechanisms that produce these ethereal sounds are: recurring patterns, simple randomization, scaling random pitches, changing pitches to modal chords, harmonizing with arpeggios, rhythmic variations and adding mutable instruments.

There are many examples of systems of artificially generated music, but since the early 1990s, David Cope has been prolific, producing dozens of compositions that well-match the motifs of classical masters.[7] A patent he'd filed in 2006, US 7,696,426, entitled *Recombinant music composition algorithm and method of using the same* describes his method of using an algorithmic composition to carry out a mathematical formulation using a recombinant idea that uses the "genome" from classical composers, as well as a stochastic, or probability scheme, the combination of which has the effect of improvisational performance [6].[8]

Cope's '426 patent reduces the idea to what has been referred to as the "Emmy Algorithm" or simply "Emmy," a software that composes music based on a range of existing musical segments, effectively reverse engineering compositions, which then compile original compositions in the motif's of Mozart, Chopin, Mozart, and Vivaldi.[9] In terms of the models of creativity earlier discussed, I would put Cope's innovation in the combinatorialism camp.[10] The universe of possible combinations is extraordinary. But, imagine that in Cope's case, we have a large number of Rubik's® Cube, which are twisted into patterns according to rules or algorithms that map the patterns onto a particular motif, depending on the composer, whether Mozart, Vivaldi or Beethoven.

Cope's music employs a music database and includes pattern matching, segmentation, hierarchical analysis, and a nonlinear recombination. Patterns that reoccur in more than one work serve as the essence of the

[7] David Cope, a composer and professor at UC Santa Cruz, wrote Emmy, a program that deconstructs the works of composers, finding patterns within the voice leading of their compositions, and then creates new compositions based on the patterns it finds. To listen to a podcast demonstrating the process and the music see, https://www.wnycstudios.org/story/91515-musical-dna.

[8] Stochastic music is where some element of the composition is left to chance.

[9] To listen to one of Emmy's composition, see, https://www.youtube.com/watch?v=2kuY3BrmTfQ.

[10] Emmy runs 20,000 lines of LISP, on an Apple Macintosh.

style of a particular composer or genre. Conceptually, Cope cuts out hundreds of small segments of musical passages from a composer's work, which can be thought of as "genes" representing a motif (e.g., mazurka). Following the accumulation of these genes, the process reassembles them in various ways. The pattern matching involves comparing musical phrases stored in the musical database to discover commonality. Recurrent patterns in a single work represent thematic or melodic lines and chord progressions. Emmy's database of "events" in a musical phrase are assigned value according to pitch, duration, and location in the work, relative to voice, amplitude, sonic and temporal qualities, which characterize the notes. And, collections of values of notes represent successively longer measures, phrases, sections. But the essential piece in this invention is recognition that the music contains an inherent set of instructions, or rules, for creating different, but highly related replications of itself. Hofstadter dwells on this aspect of how self-reflection and recursion are present in Bach's fugues, Escher's art and Gödel's idea about incompleteness [7]. But, going a step further as it relates to Cope's success, some have made the observation that composers, say Beethoven, actually constructed a grammar, which when understood could be used in the creation of other formulaic compositions. Referring to any music as formulaic does not mean to denigrate or demean the significance of what is the genius of a highly regarded composition, but simply expresses metaphorically or analogously, what may be the core idea behind the creation of a unique genre [8].

Francisco Vico and his colleagues at the University of Malaga in Spain are involved in a project that has garnered much attention for creating contemporary classical music scores by nonhuman intelligence that emulates human composers. In this system, the programmers use mathematical systems to create symphonic productions along the lines mentioned earlier about evolutionary programming. Vico's musical creations differ in approach from others in that his programs are bioinspired algorithms, combining ideas from evo-devo genomics and melodies [9]. Iamus, is a project where each composition becomes ever more complex, as it evolves. This is an example of what I'd earlier referred to as transformational creativity.

Vico's suite of related Iamus compositions were recorded by the London Symphony Orchestra, which New Scientist reported as the "first complete album to be composed solely by a computer and recorded by human musicians." Hello World! is a contemporary classical piece composed, in 2011, for a clarinet-violin-piano trio. It was first performed on

October 15, 2011, by Trio Energio. Tom Service, music critic for The Guardian, wrote a review of Hello World! which panned the production:

> Now, maybe I'm falling victim to a perceptual bias against a faceless computer program but I just don't think Hello World! is especially impressive…To me, it's precisely the musical "genomes", the backbone of the way Iamus programs and produces its pieces, that are the problem. It sounds like it's slavishly manipulating pitch cells to generate melodies that have a kind of superficial coherence and relationship to one another, with all the dryness and greyness that suggests, despite the expressive commitment of the three performers. But the material of Hello World! (there's no equivalent of the humorous exclamation mark in the music, more's the pity) is so unmemorable, and the way it's elaborated so workaday, that the piece leaves no distinctive impression.

Although these forms of music do have their critics, computers have undoubtedly demonstrated their effectiveness in the independent creation of music by employing algorithmic composition programs. But, to consider whether a computer has the power to emulate a human composer requires an analysis of the creative path, that is the logic the computer takes to mimic human intentionality. Said differently, upon what basis, historical or otherwise, does the computer make creative choices. Cope can be put into the combinatorialism category and Vico, et al. into the transformational category. Cope's process may be likened to playing chess, where it looks back at past configurations to evaluate all possible tactical options before it executes a move, but the mapping is music, not chess moves. And in the final analysis, although the combinatorial repertoire is vast, it nonetheless represents a closed or finite universe of choices to draw from. A human composer does not have this constraint, which in part puts human composition into a progressive art form, one that can continue to evolve, unbounded.

What is common in these approaches is that they are computational, but next we turn to processes to construct musical productions, which may not have analogs to computable functions.

Notes

1. Dean, R.T. (2009). *The Oxford Handbook of Computer Music.* Oxford University Press.
2. Gracyk, T., and Kania, A. (eds.). (2011). *The Routledge Companion to Philosophy and Music.* New York: Routledge.

3. Leonard, G.J. (1995). *Into the Light of Things: The Art of the Commonplace from Wordsworth to John Cage*. University of Chicago Press.
4. Gagné, N.V. (2012). *Historical Dictionary of Modern and Contemporary Classical Music*. Lanham: Scarecrow Press. ISBN 0810867656.
5. Ames, C. (1987). "Automated Composition in Retrospect: 1956–1986." *Leonardo* 20 (2): 169–185.
6. Cope, D. (1991). *Computers and Musical Style*. Madison, WI: A-R Editions, pp. 1–18.
7. Hofstadter, D.R. (1979). *Gödel, Escher, Bach: An Eternal Golden Braid*. Basic Books.
8. Hofstadter, D. (2002). "Staring Emmy Straight in the Eye-and Doing My Best Not to Flinch." In: T. Dartnall, ed., *An Interaction, Creativity, Cognition, and Knowledge*. Westport, CT: Praeger Publishers, p. 87.
9. Smith, S. (2013, January 3). "Iamus: Is This the 21st Century's Answer to Mozart?". BBC News Technology.

CHAPTER 39

Blending Dances and Dancers

Let's briefly delve into computability theory and self-reference, which places limitations on systems more generally to prove themselves self-consistent. Gödel's theorem demonstrated that no formal, consistent system of mathematics can encompass all possible mathematical truths, because it cannot prove truths about its own structure. In classical philosophy, self-referential concepts have been used as examples of the omnipotence paradox, which considers whether a being exists sufficiently powerful to create a stone that it could not lift. Similarly, the oft-stated halting problem in computation theory, that illustrates a computer cannot perform some tasks, namely tasks that require reasoning about itself. By analogy, no computer, thus far, has crossed the brink of being aware of what it was doing. If this were the case, and I believe it is, then no computer music composition program will know whether it were playing Three Blind Mice or the Fifth Symphony, or simply computing a sequence of numbers.

We have seen where computers serve up music that either cannot be distinguished from something resembling an authentic author of a genre, or where the music sounds new, surprising and even lovely. But, what about when music is combined with another art form, as happens in plays for example. In other words, what are the hard problems in AI production of musically related art forms. Let's step back and consider combining music with lyrics or poetry. Art forms in and of themselves are not isolated atomistic elements floating in some aesthetic ether. The atomistic elements are analogous to the pieces comprising a jigsaw puzzle,

J. R. Carvalko Jr., *Conserving Humanity at the Dawn of Posthuman Technology*, https://doi.org/10.1007/978-3-030-26407-9_39

which individually make no sense, unless placed in the context of the completed picture. Hiller, Cope, Vico, and Xenakis succeeded in engineering computer-generated music, solving a complicated architecture of sorts. And, others have succeeded in simple storytelling. Yet, no one has succeeded in creating a computer-generated stage play.

Artists, regardless of their medium, inevitably employ a variety of tools in the production of what we refer to as a "work of art."[1] Mastery comes from becoming facile in controlling each of the elements necessary in the creation of an integral and balanced whole, one that represents the essence of the artistic impression intended. Under a composer's hand may sit a keyboard, where fingers move in predictably unpredictable ways. This paradox suggests a form amenable to wide latitude. A jazz musician improvises over a chord, typically within a key, which renders the rift or scale melodic, allowing nuances of loudness, softness, grace notes, speed, rests, and themes within the larger flight of fancy. Similarly, within an artist's aim, a gesso-white linen canvas assumes the carbon penciled rendering of a background, perhaps the scene for a stage. Playwrights combine the tangible mediums of stage, lighting, and scenery with language, music and art. The play is inseparable from its independent parts. In this process musicians, writers and artist combine materials, tools, instruments, skill, and artistry into harmony, balance, and coherence. Disparate items merge, blend and rationalize in the final production. The task requires combining knowledge and skill that seamlessly unify, in most instances subtly and unnoticed, more in the nature of an undercurrent, where we observe in the final form, that which we identify as a musical composition, a work of art, or a stage play.[2] But, to make the point that computers cannot compete in the creation of such productions as a stage play, let's explore the "knowledge-about" the

[1] A "work of art" represents a physical thing and an aesthetic object. Some philosophers argue that it is one, but not the other. A "work of art" does not embody simply a picture of something, but comprises material, medium, form, content, subject matter, expression and representations of form and content. Analogously, musical performances contain elements, where the expressive objects are extant in the musical score, the instruments, and their tonal qualities, timbres, meter and rhythm.

[2] When I refer to features I mean plot, the actors, action, language and spectacle. Also included are the specification or stage play that expresses physical features of stage and scenery. Minor supporting elements include how the story is rendered through musically supported and even unsupported dialogue, lyrics and song, arioso, poetry, poetry supported by music, and the degree to which one or another form supports the production.

subject a computer would need to have to succeed, those that meld even mildly disparate forms, such as poetry and lyrics applied to music.[3]

To attain knowledge-about a subject requires that one is able to devise means to represent the inputs to a system that are in some instances reducible to measurable, physical quantities (e.g., tempos, timbres), and in other instances subjective or emotional substrates. Dartnall makes the point, "We want to know how we can get palpable products (thoughts, theories, pots, paintings, poems) out of our knowledge and skill" [1]. Although programs, such as Cope's EMMY, which use snippets of a composer's production, do appear to capture some sense of a music's emotional content, the expressive impact, as intended by a poet or lyricist, has not been, from literary works, extracted and recombined successfully. These latter qualities largely evade quantification.

For the moment, skipping over the point that the essential purpose of each of the lyrics, poetry, and music is to express human emotion, it turns out that at the construction level, the incorporation of lyrics set to music requires order, design, tension, balance, harmony, and coloration in the recitation. If we were simply using the computer to compose the music, these features would be the sort that computer programs would be capable of producing. We already discussed programs such as EMMY. Another product Band-in-a-Box, related more to pop music, allows the creation of songs by simply inputting a musical style, a tempo, and a key. The software generates a song employing up to five preselected musical instruments and then plays them in any of thousands of different music styles. However, it does not deal with mixed forms, music and lyric production—this would require a level of significant complexity, which software has yet to bridge.

The "mix of form" problem requires that the musical architecture combine something resembling a ballad, for example inserted in a set of lyrics, at a logical time and musical place, which moved melodically over the score. No program, *ex nihilo*, thus far has demonstrated this capability and the further application of this to a dramatic undertaking, such

[3]As stated earlier and which bears repeating, AI theory is applied only to problems that are finitely describable. Human intelligence, cannot be finitely described, thus AI can not attain the potential human intelligence. It can only simulate aspects of intelligence that are finitely describable. This theory limits the potential of a computational process ever rising to the level of human creativity, especially as it concerns the production of something as complicated as a stage play.

as a play. If, and that is a large if, such a program came into existence, it would remain the case that it would not evolve in the manner human experience and skill with which the musical play has and continues to evolve. It would not demonstrate progressivity, the potential to move into other spheres of where drama, music and stage production meet.

Apropos to music found in the Western hemisphere during the twentieth century, its structural features, rhythm, melody, and harmony, have never remained fixed, but evolved, sometimes gradually, although also drastically, as aesthetic tastes veered from earlier conventions.[4] Except for rap, and rhythmic styles, (e.g., latin and bebop), the past hundred years saw no radical departures in form. Every so often there emerged a point in time when artists, critics, and academics were to acknowledge that at least an incremental structural change occurred, a demarcation as it were, from predecessor works.[5] The gradual genre changes for example, from the melding of gospel, blues and country or transitions from blues to jazz, or blues and country to rock and roll or from opera to the American musical. This ably describes progressivity as mentioned above.

From Showboat to West Side Story nearly all musicals were cast in what I refer to as the Hammerstein form. These plays employed a fixed song, dance and dialogue schema, each fitted into selected parts of the play. The second form originated from a Sondheim era modus, although he did not invent the genre. Plays in this mode broke from the strict Hammerstein form by generously applying music over or in conjunction with a recitative expression that moved dialogue through the performance. These motifs were operetta-like, sung-dialogues or sung arioso (all meaning about the same thing).[6] In contrast to the Hammerstein form, composer/playwrights such as Sondheim and Andrew Lloyd Webber frequently employed a recitative style referred to as through-composed. The more opera-like extension referred to as through-sung is found in plays such as Les Misérables and Evita.

[4]The kinds of music are too numerous to list, but one would include, musical theater music (Gilbert and Sullivan to Hamilton), Blues, Country, Gospel, Jazz, the American Standards (Gershwin, Porter), Calypso, Boleros, Cha-cha, Meringues, Sambas, Tangos, Mambos, Bossa Nova, Rock, Hip-hop.

[5]In explaining Donald Peterson's Representational Redescription Hypothesis, Dartnall notes: "A system is creative if it can articulate its domain-specific skills as accessible structures that it can reflect upon and change onto structures in other domains." (see, Dartnall. [2002]. *Creativity, Cognition and Knowledge.*)

[6]Arioso. A musical expression between an aria and recitative. Prototype: Wagner's operas.

Computer music has the capacity to evolve as motif's change, but only within the closed set offered up by their finite databases. But, beyond this limited ability to expand motifs, it's not likely that they could achieve the complexity of integrating drama, lyrics, music and staging. And, as a preliminary matter, it is unlikely such a system could offer productions that might capture the sentiments regarding a milieu in which a playwright might be motivated, or in which the play itself might be cast. In the late 1960s, and into the 1970s, a nation in the throes of a civil rights movements and war set the stage for productions such as Tommy, Hair, Jesus Christ Superstar, Godspell, and The Rocky Horror Picture Show. I find that Tommy, based upon music from The Who is interesting to musically perform, but I am not partial to it as a listener. And, although I have generally enjoyed Webber's Evita and Cats, many of the songs in Jesus Christ Superstar outlasted a brief interest. Often I find the music in the rock era plays, rough or grating. I do not profess to offer anything approaching considered judgment here, it's only a matter of taste. However, I assume that this personal appeal I experience is also experienced by others, which often inspires change. I find it doubtful that AGI might ever achieve this level of appreciating how these art forms more often than not mirror the times in which they are cast.

Most of us can draw other examples from personal experience. Therefore, when composing for an audience with attention to harmony, balance and coloration or some other artistic feature, either via an AI-driven computer or a human composer, whether the production succeeds in producing something the listeners will be attracted to depends on the audience's expectation, openness and perhaps their aesthetic experience. This last facet is difficult or impossible to quantify, rationalize and engineer, that is to represent an audience's expectation through the application of a data construct, reducible to an algorithm.

When we attempt to apply music to language in an aesthetic way, it typically follows some cultural convention regarding order, design and tension, balance and harmony. The musicians at the forefront of bebop maintained analogous resolutions, even when the unfamiliar ear may not have enjoyed the experience. It is in consideration of this, that what we create aesthetically must in some measure conform to a system of rules and the assignment of meaning; otherwise our creations would be, in the conventional world, considered chaotic, unrewarding nonsense. Nonetheless, we know that algorithmic composition of music has been used for many years, and as demonstrated by such products as

Band-in-a-Box, or David Cope's Emmy demonstrate notable levels of sophistication.[7] But, there is more to it.

For a computer to create and produce a stage play, it would require a capability to organize and time events necessary to carry out the objective. This in turn would require a description and construction of data representative as an input to the system. This is the hard problem, as an AI-driven computer would have to be in the position of initiating all the various ideas, that is emulating the consciousness that a lyricist, composer, playwright, would bring to bear in carrying out such a production. Philosopher Daniel Dennett puts it this way:

> Experiments in Musical Intelligence have nothing to do with carbon vs. silicon, biology vs. engineering, humanity vs. robotity. Concentrating on these familiar–indeed threadbare–candidates for crucial disqualifiers is succumbing to misdirection. What, though, about another all-too-familiar theme: what about consciousness? Experiments in Musical Intelligence is not conscious–at least not in any way that significantly models human consciousness. Isn't the consciousness of the human artist a crucial ingredient? [2]

With consciousness, comes the implicit notion that a composer of something original does not employ formulaic mechanical mimicking. The great composers develop their own grammar, their own languages, symbols, syntax and semantics that lead them to greatness. Computers, regardless how powerful, cannot describe every task that can be performed in the production of a complex composition, or play, one involving a full repertoire of lyrical songs, acting, and scenery. Is there a set of algorithmic techniques that combine categorically, everything? It's self-evident that complex constructions of creative performances, involving lyrics and music together, are not reducible to a computation, that is a formula, or are sufficiently robust to search the bowels of a database, Google-like for a solution. Computers cannot penetrate complexities that cannot be articulated by the human mind. As Dennett says: "We can see that, not for any mysterious reason, computer modeling of

[7] Band-in-a-Box is a music accompaniment software that allows a user to create songs by inputting a musical style, a tempo and a key. The user types in a series of chords and the software generates a song, or creates backgrounds for virtually any chord progression used in Western popular music, in any of thousands of different music styles—the author used this product to generate a bass and drum for a trio where he performed renditions from the American song book on piano. See, http://carvalko.com/music/.

creativity confronts diminishing returns. In order to get closer and closer to the creativity of a human composer, your model has to become ever more concrete; it has to model more and more of the incidental collisions that impinge on an embodied composer" [3]. Without human intervention, the computer lacks a category of knowledge and skill, that is, "knowledge-about" what is needed to be incisive or innovative or frankly human. But, does a domain exist where "knowledge-about" can be found?

This brings us full circle, returning us to two questions: can a computer even know what it creates, and where would it obtain the necessary, "knowledge-about," required to compose something as complex as a virtual musical stage production? It's at this juncture that by an overwhelming sense of logic, that as of today, we cannot imagine that even an AGI-driven computer, one presumably that may someday mimic human intellect in many other respects, will compete with humans at the level of advancing the more complex expressions found in musicals. Computers, after all, run on different ontological tracks—a computer would have to understand what it means to be human.

Notes

1. Dartnall, T. (ed). (2002). *An Interaction, Creativity, Cognition, and Knowledge*. Westport, CT: Praeger Publishers.
2. Dennett, D. (1997). "Collision Detection, Muselot, and Scribble: Some Reflections on Creativity." In: D. Cope, ed., *Virtual Music: Computer Synthesis of Musical Style*.
3. Ibid.

CHAPTER 40

Programmed in Our Head

A dreamer is one who can only find his way by moonlight, and his punishment is that he sees the dawn before the rest of the world.—Oscar Wilde

Much as been written about the methods composers employ when writing music. Google lists 840,000 hits in a search for "How to Write Music." It's freely available. Courses on the subject range from basic concepts to the more advanced methods, such as MIT's writing the sonata form, as typified in the repertoire of Haydn, Mozart, and Beethoven.[1] Needless to say, music is a human experience—one that is a ubiquitous and meaningful (incidentally, scientists report that certain animals hear what humans do, but what function it serves is not yet determined) [1].

Through intellect, memory, talent and skill, composers and musicians, which include singers, succeed in participating in the musical experience in a variety of ways. They offer much by way of new musical experiences to be enjoyed by the rest of us. But, how much of the stuff composers and musicians coalesce derives from a uniquely inspired moment and how much from simply regurgitating what already exists within the culture to which they are exposed? Perhaps, much of the music we compose is pre-embedded in our brains, in the form of memes, which if true suggests that a computer might have the capacity to listen in, similar to the way we do. If music itself is memetic and living in our consciousness,

[1] http://ocw.mit.edu/OcwWeb/Music-and-Theater-Arts/21M-304Spring-2009/Syllabus/index.htm (Last visited 3/1/2010).

J. R. Carvalko Jr., *Conserving Humanity at the Dawn of Posthuman Technology*, https://doi.org/10.1007/978-3-030-26407-9_40

it may be that one day it will be represented as a data construct, decipherable by an AGI system.[2]

But, knowledge is key. A composer considers a piece to be played, arranged or composed according to how she desires to hear any number of features, e.g., its harmony, balance, and coloration, among other things. This is learned during the course of concentrated practice and study. As relates to composing a musical score, she considers the leitmotif, that is the recurring idea that becomes identified through a character, place, mood, melody, harmonies, chord sequence, rhythm or a combination of these. In reality, there are dozens more of such that are either explicit or implicit in the construction of a piece of music. But, that all these elements of design may be pulled from a place which we unconsciously internalized, that is simply reporting a prior memetic construction is not something we often think about.[3] Nonetheless, its this idea that led to Richard Dawkins, among others, to single out music as an example of what kind of things can create a meme.[4]

A detailed treatment of mimetics of music is beyond the scope of book, but I'd like to briefly draw attention to the idea that music composed, performed and listened to, often stems from an association to memes [2]. And, that an AI protocol emulating human music composition, absorbs human intention, certainly the designer's organization, plan, and intelligent construct—which includes the cultural memes inherent in the motifs underlying the work. Listening to Cope's productions we immediately hear Beethoven and Mozart. I believe that too serves as evidence of the instantiation of what one would reasonably regard as memetic structure.

Appreciation for certain genres of music persist over centuries. In some cases, they stem from offshoots of an antecedent variety, as for instance in the tradition of European classical music that led to songs in

[2]There have been numerous studies to analyze the propagation of memes between users in a social networks, many of which employ evolutionary computation and epidemiological models similarly to the way diseases spread in human society.

[3]We cannot ignore that AI in the form of bots, which take on the significance of memes, have proven successful, as demonstrated by the influence these artifacts have had on the recent U.S. presidential elections. Query whether this level of sophistication can be used to advance something as complex as what we regard as the humanities, music, art and literature?

[4]Steven Jan goes on to cover the theoretical aspects of the mimetics of music, ranging from quite abstract philosophical speculation to detailed consideration of what actually constitutes a meme in music.

the forms of ballads by Irving Berlin, Cole Porter, and the Gershwins. Similarly, descendants of nineteenth-century African slaves composed blues based on Christian spirituals and African rhythms, as exemplified in the early composers and musicians: W.C. Handy, Jelly Roll Morton, and Louis Armstrong. I have spent time in Africa and on upper mid-West Native American reservations, where ancient melodies and rhythms were heard as chants of varying intonations, and in the latter case during ceremonial performances of the Ghost Dance, Pipe Ceremony, and Purification Ceremony.[5] Liane Gabora and her colleagues developed a computational model that shows how memes come into existence and evolve based on the premise that "new ideas are variations of old ones; they result from tweaking or blending existing ideas" [3]. Steven Jan takes a minimalist view of the musical meme, "consisting, at the lower extreme, of configurations of as few as three or four notes" [4]. In either case, let's look at the category of memes considered candidates for musical constructions, at it serves as perhaps an indication of the limits of what AI, in a recombinant regime, might create.

William Benzon advances the notion that musical progressions employed in ballads, jazz and bebop are dependent on memes built into a musician's conscious selection of the tonal qualities accompanying a genre [5]. The particular form he focusses on are AABA Rhythm Changes, frequently employed in jazz and bebop.[6] These changes move and repeat the progression of chord changes that underpin the harmonic structure of a piece of music. As applied to jazz and bebop, Rhythm Changes usually refer to the frequency of a repeated sequence of chord changes, universally acknowledged as extracted from George Gershwin's composition, I Got Rhythm.

The idea of a progression importantly sets one's musical expectations. It applies in two ways. First a progression may denote the chord that follows a prior chord, in a one after another fashion.[7] In one system

[5] Native American Music is usually choral, without harmony, and antiphonal singing. Rhythms are irregular with no pitch, where singing is generally accompanied by drums, rattles, and flutes.

[6] AABA refers to a progression consisting of the same chord or set of chords, A for example, being played for eight bars or measures, followed by another eight measures of the set, e.g., A, followed by the same chord of set of chords labeled B being played for eight measures, etc.

[7] Tonality refers to the twelve pitches of the chromatic scale into a hierarchy centered around the tonic.

of notation, the transcriber uses actual chord values, A, A#, B, C, etc. through G, which are specified.[8] In an alternate system of notation, Roman numerals are applied, I, II, III, etc. through VII, where roman numeral I refers to the tonic, that is the signature of the key, in which a song is played. Progressions, such as AABA, follow a form, where the letters denote the application of a sequence of chords, which are played during a musical period, typically four or eight-bar intervals, standard in most blues and many American Standards. In American popular musicology these are nonrigidly applied, yet offer customary forms to the musician in the application of a tonal sequence.

Composers, for example, Cole Porter, used internal variations, but nevertheless hardly moved far from the standard form of AABA. Other progressions exist as well. A simple blues 12 bar blues form employs a standard harmonic chords progression ABC: the first 4 bars use the I chord—I, I, I, I. The middle 4 bars use IV, IV, I, I. And the last 4 bars use V, IV, I, V. Then the progression repeats itself. A twelve-bar blues, in the key of C would begin on the tonic C, for four bars, and moves to F for two of four bars, move back to C for two of the four bars, and then G, F, C, and G for the final four bars. The final chord in the 12th bar is often dominant G7 that signals the approach or return to the beginning of the song.

There are dozens of other progressions that sound familiar to audiences throughout the world. These instantiate music, and as Benzon suggests enter the world as candidates for musical memes. By way of psychological affect, ascending progressions provoke an uplifting mood. A simple ascending progression is heard in the diatonic ascending progression I, II, III, IV heard in such melodies as: Ain't Nothing Like the Real Thing (Gaye, Terrell), Bad Bad Leroy Brown (Croce) and Sexie Sadie (Beatles). In more complex variations of ascending progressions, we hear in the last eight bars ascending I I IV #IV V V7 I V7 in Oh What a Beautiful Mornin' (Rodgers). Other ascending progressions are the I I (augmented) I6 and followed by variations of I6, I7, or I9: e.g., You've Got To Accentuate The Positive chorus; (verse for Just Like Starting Over (Lennon).

Innovative melodies played over chords in standard progressions is one thing. However, regarding the evolution of jazz, Derek Gatherer observes that "New memes, such as novel chord patterns, rhythmic

[8] Modality represents the seven pitches C, D, E, F, G, A, and B into a number of different scale patterns, such as minor, major, diminished, etc.

changes, or alterations in instrumentation are subject to scrutiny as to how they cohere with the preexisting whole. Sudden innovation is not permitted but small innovations may cumulatively have large results" [6]. Benzon believes that a bebop performance contains "memes chosen from a pool of 10,000." These memes would be drawn from categories—progressions, rhythm, melody, but which as was pointed out by Gatherer, are mutually compatible in that they cohere with the whole. Assuming that an AI-driven computer could survey the universe of musical memes extant within the popular musicological archives, albeit human inspired, and even combine them as EMMY might, it would likely lack the humanity, that is the emotional content that draws us to popular forms of music in the first place.

Further as to whether a computer can be used to advance something as complex as what we consider music, in human terms, several leading philosophers and musicologists weighed-in during a 1997 colloquia on computers and creativity. The event, entitled Virtual Music: Computer Synthesis of Musical Style, was attended by notables in the field of AI, music and philosophy, such as: David Cope, Douglas Hofstadter, Daniel Dennett, Eleanor Selfridge-Field, Bernard Greenberg, Steve Larson, and Jonathan Berger. During the occasion, Hofstadter made clear his opinion, which largely reflects my thesis:

> Built into being human is the fact of living life and having all the sorts of experiences that –to my mind– underlie what musical expression is all about. And so, part of the power of your musical expressions came straight from your humanity. For an Emmy-type program to perform credibly at this type of task, however, is another matter entirely, because it does not have any human experience to draw on. [7]

We are meandering on the periphery of the hard problem of consciousness, which attempts to grasp human subjective experience, i.e., qualitative experiences, or qualia. Consciousness is self-referencing, i.e., one cannot know directly what it is like to be something, in this case the self.[9] Nonetheless, only I know the "*I*" or first-person subjective experience, via my senses, my meta category, what I call my essence. An AGI-driven process will never know itself, and this bears on it's potential

[9] I am simply restating Immanuel Kant's 1798 book *Anthropology from a Pragmatic Point of View*: 4.6 Thesis 6: *Consciousness of Self is not Knowledge of Self.*

to create the more complex ways in which humans self-express. The key point here is that any AGI system will have an inherent limitation in emulating a core aspect of the "*I*" or first-person subjective experience, one incidentally shaped by an evolutionary adaptation, to be musical.[10] But, by listening in on the culture via the kinds of bots that appear to have recently been identified as influencing national elections, we move a step closer in the computer taking on the hard problem in such things as true storytelling, or stage plays, including musical productions.

Notes

1. Levitin, D.J. (2006). *This Is Your Brain on Music.* New York: A Plume Book, Penguin Group.
2. Benzon, W. (2010). *The Evolution of Human Culture: Some Notes Prepared for the National Humanities Center*, Version 2. https://papers.ssrn.com/sol3/papers.cfm?abstract_id=1631428 (Last visited 5/14/2019).
3. Gabora, L. (1995). "Meme and Variations: A Computational Model of Cultural Evolution." In L. Nadel and D.L. Stein, eds., *1993 Lectures in Complex Systems.* Addison Wesley.
4. Jan, S. *Replicating Sonorities: Towards a Memetics of Music.* Department of Music, University of Huddersfield. http://cfpm.org/jom-emit/2000/vol4/jan_s.html#note6.
5. Ibid., see note 2 above.
6. Gatherer, D. (1997). "The Evolution of Music: A Comparison of Darwinian and Dialectical Methods." *Journal of Social and Evolutionary Systems* 20 (1): 75–92.
7. Hofstadter, D. (1997). "A Few Standard Questions and Answers." In D. Cope, ed., *Virtual Music: Computer Synthesis of Musical Style.*

[10] Steve Pinker theorizes that music is not an evolutionary adaptation, but a spandrel, a byproduct of some other adaptation, which he conjectures is language, which made possible the production of music. See, podcast. https://whyevolutionistrue.wordpress.com/2014/08/14/steve-pinker-on-the-evolutionary-significance-of-music/.

CHAPTER 41

Storytelling

How do we talk about our condition, our lives, the lives of those whom we revere, about what little separates us from the other creatures, those with which we share Earth's stage: sparrows, rats, cockroaches, paramecium, all born of parents, create offspring, live and die—the definition of natural machines, bio-contraptions that overcome impediments to produce the things essential to our being (perhaps our creations are no more, no less, essential for survival).

The writer Patrick Rothfuss wrote, "It's like everyone tells a story about themselves inside their own head. Always. All the time. That story makes you what you are. We build ourselves out of that story." Among all creatures only we tell stories minute by minute, day by day, the peaks and troughs of one or another eon, in a way that forms civilization and its humanity or inhumanity as the case may be.[1] It may have begun as cave drawings, until language gave us an oral tradition, and finally writing, as seen in the hieroglyphics cast throughout the ancient world. The oldest known cave paintings are over 64,000 years old, likely made by a Neanderthal. Over 40,000 years ago, during the Upper Paleolithic period, artistic works were found in both the Franco-Cantabrian region in western Europe, and in the district of Maros (Sulawesi, Indonesia). Of course we now write or dictate our stories

[1]Consider how story telling has mushroomed: Facebook, 2.2 billion messages/mo.; Instagram 800 million messages/mo.; Twitter 15 billion messages/mo. See, https://www.internetlivestats.com/.

J. R. Carvalko Jr., *Conserving Humanity at the Dawn of Posthuman Technology*, https://doi.org/10.1007/978-3-030-26407-9_41

in books, emails, blog posts, twitter. Taken as a whole, they tell a story about who we are. So, what kinds of stories will be told by *Homo futuro*, those with a biologically enhanced IQ, or that have become constructs of computation [1].[2]

The answer depends on human purpose. The reason we have for telling stories is innately human, perhaps essential to our survival, but likely not essential to the survival of other kinds of beings, different from animals, different from supercomputers. We engage in storytelling to foster particular kinds of life, ones based on what Alasdair MacIntyre refers to as the narrative tradition:

> [M]an is in his actions and practice, as well in his fictions, essentially a story-telling animal. He is not essentially, but becomes through his history, a teller of stories that aspire to truth. But the key question for men is not about their own authorship; I can only answer the question 'What am I to do?' if I can answer the prior question 'Of what story or stories do I find myself a part?' We enter human society, that is, with one or more imputed characters— roles into which we have been drafted—and we have to learn what they are in order to be able to understand how others respond to us and how our responses to them are apt to be construed. [2]

Storytelling communicates what we remember and what only we may imagine, truth or fiction, a calculus or poetry, ranging from something about ourselves, others, hopes, dreams, worlds, past or beyond the here and now, altered in time, space, monsters and the supernatural. It requires imagination, critical-thinking skills, self-awareness, and empathy. So just what kind of story does the computer tell?

Dartmouth College Neukom Institute for Computational Science researches creative forms that range across the humanities, arts, and sciences that find their way to the computer. For example each year it runs a competition to determine the winners of the "Turing Tests in the Creative Arts." The contest is open to programmers who

[2] Tettamanti, et al. in 2005, performed fMRI experiments where participants listened to sentences describing actions performed with the mouth, hand, or leg. The results showed that listening to action-related sentences activated the left fronto-parieto-temporal network (the pars opercularis of the inferior frontal gyrus [Broca's area], sectors of the premotor cortex, the inferior parietal lobule, the intraparietal sulcus, and the posterior middle temporal gyrus), providing evidence that listening to sentences engaged the visuomotor circuits which subserve action, execution and observation.

machine-generate sonnets. In one challenge named PoetiX, a machine must produce an infinite number of sonnets either Shakespearean or Petrarchan. Outputs of the PoetiX are mixed among human sonnets where judges then label sonnets as generated by human or machine. Any machine that is indistinguishable from the human poetry will have "passed a Turing Test." Similarly, LimeriX challenges contestants to create machine-generated limericks. LyriX generates a vast number of poems. DigiKidLit generates Children's Stories. Stories are evaluated for their "artistry" and creative approach. In 2018, the contest was extended to the creation of machine-generated music in the style of musicians such as Charlie Parker, Bach Chorales, Electroacoustic Music and Free Composition. Without having to scrutinize each of the works that have come out of this program, suffice to say, many are difficult to distinguish from works authored by humans. So, does this verify that computers can replace sentient beings when it comes to writing poems, limericks, and the like?

Its Director, Dan Rockmore, writes:

> It's a challenge to produce a moving piece of literature, and I think in challenging yourself to try to do it, it makes you think hard about what moves you about a story, what goes into making a story… Thinking hard about those questions has a long tradition in the humanities, so it's just another way in which we can bring another lens to that, not in order to replace it but just to honestly consider it" [3].[3]

Over the course of history, storylines have evolved from pictographs and oral traditions, to the written word, dramatic performance, and now of course movies. It draws a long line from the Hindu text Rigveda, c. 1500–1200 BCE, to Homer's The Iliad, c. 800–700 BCE, considered to be the earliest work in the whole Western literary tradition, and shortly thereafter, the Bible, Monarchic version, c. 745–586 BCE. These well-told narrations illustrate little has changed over thousands of years. Aristotle's Poetics (Greek: *Περὶ ποιητικῆς*, Latin: De Poetica, c. 335 BCE) is the earliest surviving work of dramatic theory and first treatise to focus on literary theory in the West. This has been the traditional view for centuries. The epic poem Beowulf (between c. 975 and 1025),

[3]The Neukom Institute awards for 2017 were announced at the Music Metacreation Workshop (MuMe), part of the International Conference on Computational Creativity (ICCC), in Atlanta, Georgia.

might be included in the wider genre, although the transformation to more what we are accustomed to appears during the Renaissance, when the western tradition blossoms in the works of Dante, Chaucer, Milton, and Shakespeare. The shift from verse to prose dates from the early thirteenth century with Malory's Le Morte d'Arthur (c. 1470), although historians tend to point to Cervantes' Don Quixote, 1605, as the first example of the modern novel form, the one we're familiar with today. And, lest we not forget the detective novel invented by Edgar Allan Poe.

If we were to explore well-known authors, we'd have a better idea that computers may have words, they may even appear to have thoughts, but it's doubtful that they can sustain any theme without context, historical or otherwise. I have put together openings from these author's works to make the point, that each pulls the reader into a virtual world, often hinting at where in time and territory we anticipate landing. What follows is not an analysis that suggests openings formatively or structurally relate to one another, or follow some set of literary principles. For the outstanding exemplars of literature, we know each one is unique.

That having been said, I haven't considered mimetics in relation to literature, although from experience we tend to see repeating themes. Though looking at openings as an assemblage, may show similar methods relative to setting up plots, timeframes, and character traits; although an opening just as well can be intended to setup a mood or portray an aesthetic context. Opening paragraphs, most noticeably put the reader into a space and time, occasionally revealing an important character, or an expectation of the drama or suspense, to follow.

Eventually, an author controls pen and must start somewhere. Same is true for a computer that mimics the creative talent of a human. It must begin someplace. Ray Bradbury does not say anything about how to start a story, but in reading a passage from Zen in the Art of Writing, we begin to sense that a computer cannot provide for us, what we provide for ourselves:

> Run fast, stand still. This, the lesson from lizards. For all writers. Observe almost any survival creature, you see the same. Jump, run, freeze. In the ability to flick like an eyelash, crack like a whip, vanish like steam, here this instant, gone the next-life teems the earth. And when that life is not rushing to escape, it is playing statues to do the same. See the hummingbird, there, not there. As thought arises and blinks off, so this thing of summer vapor; the clearing of a cosmic throat, the fall of a leaf. And where it was-a whisper.

> What can we writers learn from lizards, lift from birds? In quickness is truth. The faster you blurt, the more swiftly you write, the more honest you are. In hesitation is thought. In delay comes the effort for a style, instead of leaping upon truth which is the only style worth deadfalling or tiger-trapping.

A novel's first sentences draw the reader into a virtual world, again how does a computer compete? Consider the power of George Orwell's 1984: "It was a bright cold day in April, and the clocks were striking thirteen." How does an algorithm qua novelist approach this? And as pedestrian as it sounds, more than one celebrated novel starts with, "once upon a time." But, such reveals poignancy and mystery.

If we were to suppose that even a computer can figure out "once upon a time," let's give a few more examples. Midnight's Children (Salman Rushdie), "I was born in the city of Bombay… once upon a time. No, that won't do there's no getting away from the date: I was born in Doctor Narlikar's Nursing Home on August 15th, 1947."

A Portrait of the Artist as a Young Man (James Joyce), "Once upon a time and a very good time it was there was a moocow coming down along the road and this moocow that was coming down along the road met a nicens little boy named baby tuckoo…"

Laughter in the Dark (Vladimir Nabokov), "Once upon a time there lived in Berlin, Germany, a man called Albinus."

These writers choose the nondescript "once a upon a time," not for the lack of a time-certain, because each at the right moment tells us in due course, a specific time and place. Rushdie followed with: "August 15th, 1947," and Bronte in Wuthering Heights, began with, "1801" [4].

Beginnings can pull from anywhere: a novel's ending, middle, a crisis or turning point. Many renowned openings begin with a setting, e.g., the interior of room, exterior of a building, a cityscape, landscape, or seascape. In every instance, the writer takes a pallet of particulars in hand to word-paint the details, which open the reader to another world.

Novels consist of plots and subplots, and with rare exception, e.g., Sartre's, No Exit, a story plays out in more than one locale. Again, with some exceptions, stories consist of more than one character. The author introduces us to persona, reminding us of someone we know, or reveals a quirk or eccentricity that we've never run into. Most stories set up quests and opposing forces, and only through the resolution of these, do we rediscover something old, or discover something new, a human quality, a reaction to a circumstance or flaw. So, a few openings, although far from telling the story, may reveal a trait of the main character.

But the novel, even as early as Le Morte D'Arthur, first published in 1485, contained opening paragraphs that rapidly brought the reader into the author's frame. An antecedent to this style is reflected in the opening lines to the Bible, which not surprisingly serves as a model for countless openings: "In the beginning God created the heaven and the earth. And the earth was without form, and void; and darkness was upon the face of the deep. And the Spirit of God moved upon the face of the waters."

From Plato to Heidegger, philosophers believed that the mind does not directly interpret that which it perceives, but even under the most objective criteria, it interprets through the shadow of reality. As we saw in an earlier chapter social reality is teased from nature, largely through language, words mainly. So, what exactly do words mean? Thought perhaps, and not just a list of thoughts, but thoughts in a context of human existence—joy, suffering, hope. Storytelling of course implies words, and thoughts, but more frequently a genre cast in a historical context or period, one of human dimension, that unlike music, cannot be easily mimicked.

According to anthropologists, biological transformations, catalogued under the heading "intelligence," have evolved as a result of becoming aware of our environments and forming representations; which may eventually have led to our facility to express ourselves through language. In any case, to express intelligence requires actions. Our actions may manifest in speech, simple communication or something more complex, having symbolic complexity (an equation), or some facet of our social reality. Because of the lack of direct evidence, a theory for how language, presumably as vocalizations, originated, has yet to emerge and achieve wide acceptance [5].[4] One can surmise that it may reflect adaptation in the Darwinian modus, to the changing environment, perhaps civilization itself, which improved the odds of our survival as a species [6].[5]

Regardless of how language was first created and thereafter evolved, it is clear that it not only serves as a means of communicating information, data, rational thinking, but appears suited to express the intangibleness

[4] Written language begins to appear 5000–6000 years ago.

[5] This is simply a corollary of Charles Darwin's theory of evolution where organisms that possess heritable traits that enable them to better adapt to their environment compared with other members of their species will be more likely to survive, reproduce, and pass more of their genes on to the next generation.

of our emotions. A computer is incapable of communicating human sentiment. Hofstadter said it best:

> Programs have been written that produce poetry based in rather simple ways on the poems of a human poet, and one can feel meaning in the output, but one realizes that such meaning is due to the input material. Thus if in a computer poem one finds the phrase "her haunting grace," it may be that this phrase was present as such in some poem, or if not, then its component words all were, and the machine stuck them together on the basis of some syntactic rules, maybe with a few rudimentary semantic guidelines as well. But a reader who feels depth in such phrases can also realize that it is basically borrowed depth, not original depth. It's just echoes and parrotings, not primordial screams from the depths of a feeling soul. [7]

What's hardly debatable is that language provides the content for human communication separating us from all other species that may communicate through other means. As such human communication frames as it were an indispensable human trait [8].

Notes

1. Tettamanti, M., et al. (2005). "Listening to Action-Related Sentences Activates Fronto-Parietal Motor Circuits." *Journal of Cognitive Neuroscience* 17 (2): 273–281.
2. MacIntyre, A. (1984). *After Virtue*. University of Notre Dame Press.
3. About | Neukom Institute Turing Tests in the Creative Arts. http://bregman.dartmouth.edu/turingtests/node/12.
4. Note: the authors each open with eye catching words: Narlikar, moocow and Albinus, respectively.
5. Bednarik, R.G. (1992). "Paleoart and Archaeological Myths." *Cambridge Archaeological Journal* 2: 27–57.
6. Christiansen, M.H., and Kirby, S. (eds.). (2003). *Language Evolution: The Hardest Problem in Science?* Oxford and New York: Oxford University Press, pp. 77–93. ISBN 978-0-19-924484-3. OCLC 51235137.
7. Hofstadter, D. (1997). "A Few Standard Questions and Answers." In: D. Cope, ed., *Virtual Music: Computer Synthesis of Musical Style*.
8. Friederici, A. (2011). "The Brain Basis of Language Processing: From Structure to Function." *Physiological Reviews* 91 (4): 1357–1392. CiteSeerX 10.1.1.385.5620. https://doi.org/10.1152/physrev.00006.2011. PMID 22013214.

PART VI

Societal Repercussions

CHAPTER 42

Far from Red and Black

Sarouk: I don't think you fully answered what else you believe put us on the gene editing track.

Mensa: A physicians duty of care, perhaps. Doctors felt strong about their oath to cure disease, disabilities and human suffering.

Sarouk: How so?

Mensa: We learned that gene editing could be used to cure diseases like cystic fibrosis or thalassemia.[1] *Back then there were no cures for so many diseases. Doctors performing IVFs began to see that they had the answer between life, death and human suffering at their fingertips. Remember, the ancient maxim, do no harm, the Hippocratic oath.*

Sarouk: That humble aphorism could be interpreted as having a duty to act or to stand down.

Mensa: Precisely, so their instincts drove them in the direction of healing. They didn't stand down. And, then went further. Diseases like heart disease or schizophrenia, didn't just involve one gene. They were the result of many, sometimes hundreds in combination. And, removing some of these resulted in unforeseen mutations, which caused an irrevocable speciation. We suddenly began drifting from our ancestors.

[1]An inherited blood disorder, in which the body makes an abnormal form or inadequate amount of hemoglobin the protein in red blood cells that carries oxygen resulting in large numbers of red blood cells being destroyed, which leads to anemia.

J. R. Carvalko Jr., *Conserving Humanity at the Dawn of Posthuman Technology*, https://doi.org/10.1007/978-3-030-26407-9_42

Sarouk: Yes, but look what we gained. You once said that back then every day thousands of people died from old age. When you were born, it killed millions and million every year.

Mensa: Right, and gene editing could delay or and as we now know arrest age altogether. From mice living twice as long, to humans, and of course now hundreds of years, without loss of memory or frailty.

Sarouk: But the flip side was designer babies.

Mensa: When I was a child, some were born gifted and talented, others lived short painful lives with unspeakably severe disabilities. But there were those who worried that we'd create a master race, everybody would express the most desirable phenotypes, hair, skin color, height and weight, even intelligence and creativity. As time went on certain modifications were not individual choice, but legislated.

Sarouk: So gene editing was eventually used as a public health initiative, to insure equal rights. Right?

Mensa: Right, to insure even the worst off had access to what seemed like a human right. But, even that was not perfect, and as time went on it mostly benefited those who could afford it, the most expensive enhancements restricted to those of means.

Sarouk: It seems to me that we gained immortality at the expense of what you talk about as humanity.

CHAPTER 43

Planning the Journey

Humans are not stamp out cookie-cutter-like, but they come into the world representing wide-ranging intellects, creative modalities, and physical abilities. The curve beneath which we all sit includes those of us who do not fit society's specification for something imagined normal, we exist as people living along a spectrum that includes those of us who may have functional limitations, lesser social skills, or a medical disability. The list is long. The fact remains that each of us in our own way brushes up against the borders of this curve, really a multidimensional bubble, in a world engineered for those that fill in the middle of the enormous space within. But, for better or worse, with iterative embryo selection (IES) and in vitro fertilization (IVF), science may actually narrow the physiological disparities within the population, not only reducing the occurrence of disabling disease but enhancing a wide range of abilities.

The regular change that occurs in the female reproductive system each month makes pregnancy possible, that is to start the development for the production of oocytes, and for preparing the womb to carry a fetus from conception to birth. For some women, the cycle does not take a normal course. In those instances, IVF helps with fertilization, embryo development, and implantation. But, we have reached a crossroads where IVF can be used not only to assist an otherwise impossible pregnancy, but in changing the traits in the fetus, a use which many might object to on ethical grounds, because it's alleged that it may take us into what has been roundly criticized as a form of eugenics.

J. R. Carvalko Jr., *Conserving Humanity at the Dawn of Posthuman Technology*, https://doi.org/10.1007/978-3-030-26407-9_43

Biological determinism or genetic reductionism centers on the idea that human behavior is significantly controlled by genes, albeit not discounting environmental influences. As we now know, the connection between genes and intelligence raise ethical concerns when we seek to carry out programs leading to forms of biological determinism, such as eugenics, some form of which may well be underway, such as when embryos are selected out when they exhibit susceptibility to disease or reveal genetic defects.[1]

Gregory Stock, an early voice in the use of technologies for enhancement, claims that powerful new genetic and reproductive technologies will allow us to redesign *Homo sapiens* and that this is both desirable and inevitable.[2] As we have seen, embryo selection exists on a large scale, and other technologies, CRISPR and IES, for example, may be in the wings of widespread use, leading to germline engineering for therapeutic applications.

Alternative reproduction methods, such as IVF, or sperm banks formed for a particular genetic reductionism has been part of our history. The most notorious example was the "genius sperm bank," which operated for twenty years before the turn of the twenty-first century, where nearly 230 children were conceived, some through the donations of Nobel Prize winners William Shockley and J.D. Watson. Although gene editing has not been employed to advance any eugenic's project as we know, recent history has shown that such initiatives are fresh in the minds of many. For example, notable British psychologists Richard Lynn and Raymond Cattell, openly call for eugenic policies using modern technology. In 2015, a group of biologists urged a worldwide ban on the clinical use of methods, particularly the use of CRISPR and zinc finger to edit the germline of the human genome [1, 2].

Every parent desires that their children be free from health risks and disorders, and for parents who carry inheritable genetic syndromes, assisted reproductive technologies, such as preimplantation genetic

[1] J.B.S. Haldane's essay "Daedalus: Science and the Future" (1923), wrote about scientists advocating transhumanism as the idea bringing forward eugenics through advances in science and technology.

[2] Stock is Director, Program of Medicine, Technology and Society at UCLA School of Medicine imagines the inevitability of super humans with 150-year life spans, free from disease and equipped with gene modules conferring intelligence, physical aptitude, or aesthetic talents.

testing can avoid the possibility of their child inheriting a debilitating disease. In fact, when a medical barrier stands in the way, for instance problems in conception or of insuring a healthy baby, IVF has been used for decades to help men and women realize .parenthood. The process can also be used additionally to select the gender of the child. Gender selection, known as preimplantation genetic diagnosis (PGD), screens of the embryos to ensure embryos don't have chromosomal abnormalities (ploidy), single-gene mutations, or polygenic diseases. And, once science knows where to look, a similar process may be used for sifting through embryos that have potentially greater intellectual capacities.[3]

Once genomic engineering becomes an accepted practice, prospective parents, who desire a child not only immune to a particular disease or disability, but one having enhanced senses or abilities, will create a specification to be implemented perhaps by some future transhuman design service or TDS. The TDS may employ creative/cognitive psychologists, genomic physiologists, and social scientists that collectively deal with the engineering of a child through genetic manipulation. As we discuss later, potentially, any product, as an *in vitro* embryo, *qua Homo futuro*, also may become the object of exploitation, either by commercial and governmental interests (e.g., *Homo futuro* warriors).

As we have been discussing, genetic testing can identify parents who have the same mutation and thus have an increased likelihood of transmitting the corresponding disease to their child. To assist in this effort, novel methods of identification are now being patented. U.S. Patent 8,805,620, for example claims methods and systems for assessing the probabilities of the expression of one or more traits in progeny, and thus for selecting a donor or reproductive partner as a potential parent. It's likely that these types of inventions are precursors to the kind that will be developed to choose partners who may exhibit increased levels of intelligence. Regarding this last point, leading biogeneticist Robert Plomin believes it's now possible to predict IQ from an individual genome. With this goal in mind, Genomic Prediction, a company, based in New Jersey, recently announced that it will offer such a service.[4]

[3]Some IVF facilities collect sperm or egg donations as well as certain profile information pertaining to the donors. Such profiles typically include the donor's race, height, weight, age, blood type, health condition, eye color, educational background, and family history.

[4]Super-smart designer babies could be on offer soon. But is that ethical? https://www.theguardian.com/commentisfree/2018/nov/19/designer-babies-ethical-genetic-selection-intelligence.

The idea of birthing a child, who then will live free from a genetic predisposition or will have a superior intellect is intriguing and will come to the attention of many, especially affluent prospective parents going forward. To this end, gamete storage facilities such as sperm/egg banks have been on the rise, increasingly allowing more choices in reproduction. The business of creating options has resulted in inventions that capture some feature that advances the practice of producing designer babies. To illustrate the progression from theory to practice, several inventions were granted patents between 2013 and 2014: e.g., U.S. Pat 8,805,620 (mentioned above), a method and system for selecting a donor or reproductive partner for a potential parent; U.S. Pat 8,620,594, a method and system for generating a virtual progeny genome; and U.S. Pat 8,543,339, a gamete donor selection based on genetic calculations.

Patent No. 8,543,339, discloses a method to carry out the broader steps that genetically modify an embryo according to design (see Fig. 43.1). In the process of collecting of sperm or egg donations, profiles about the donors are also collected that relates to demographic, race, ethnicity, education, family history height, weight, age, blood type, health status, hair, and eye color. A potential recipient of the sperm or egg can review the profiles and choose which donor best meets its specification. Although the personal profiles of the donors can serve a useful purpose for the potential recipient to make a more informed choice, such information typically offers insights as to who carries particular genetic traits, e.g., such as inherited diseases. For example, a recipient with a family history of cancer may want to avoid a donation from someone in a high risk group. But, of course as we identify genes for intelligence or creativity, these will populate the selection databases as well.

In the '339 patent example, the modification of the gamete depends on a donor, and as discussed below, requires the insertion into an embryo of the chosen trait as expressed in a natural or synthetic gene. But the point illustrated by this invention is that the concept of designer genes is more than abstract. The gamete donor selection begins when a laboratory receives a specification from a potential recipient interested in "seeding" its embryo. The specification, at this stage of technology includes a phenotype of interest and the corresponding genotype of the recipient along with one or more genotypes of potential donors. The laboratory retained to modify the embryo analyzes the data pertinent to the phenotype of interest based on a selective pairings of the genotype

Fig. 43.1 U.S. patent 8,543,339

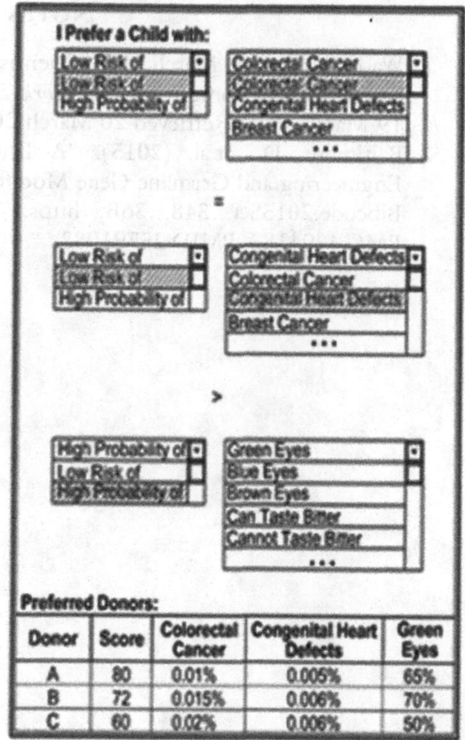

Donor	Score	Colorectal Cancer	Congenital Heart Defects	Green Eyes
A	80	0.01%	0.005%	65%
B	72	0.015%	0.006%	70%
C	60	0.02%	0.006%	50%

of the recipient and a genotype of a donor within the family of donors, which leads to an identification of a preferred partner based on the statistical information.

Effectively this invention allows an individual to preselect or deselect a potential donor's genetic traits. Of course this idea begs the question whether once joined, who owns the "recipe" for the child eventually born? In some ways we are headed down the slippery slope of objectifying the child born, that is one dependent for its existence on human invention. As we plan our journey, we must be mindful that inventions are products of human innovation and production. Genetic engineering, IVF, and IES are all inventions, which on some level bring the products of these innovations into the circle of what is considered a social construct. Thus we must consider the quintessential moral dilemma: what does it mean to be human? These are some of the questions we begin next to frame.

Notes

1. Wade, N. (2015, March 19). "Scientists Seek Ban on Method of Editing the Human Genome." *The New York Times*. Archived from the original on 19 March 2015. Retrieved 20 March 2015.
2. Baltimore, D., et al. (2015). "A Prudent Path Forward for Genomic Engineering and Germline Gene Modification." *Science* 348 (6230): 36–38. Bibcode:2015Sci...348...36B. https://doi.org/10.1126/science.aab1028. PMC 4394183. PMID 25791083.

CHAPTER 44

Assigning Rulebooks

We now turn attention to humans-as-they-are and posthumans to consider the ontological differences that may lead to a bifurcation of the species. What rules distinguish one from the other and how should they be applied? Clearly, a difference exists between things that naturally occur and products of human ingenuity. The Dalai Lama XIV, once said, "Know the rules well, so you can break them effectively." In some way this applies to how we might go about distinguishing *Homo futuro* from *Homo sapiens.*

Human activities can be mapped onto a tripartite set of categories, the first, which I refer to as "Rule I," governs natural processes, representing causal relationships, i.e., natural laws, which apply uniformly to the behavior of matter under equivalent conditions. For example, the law of gravity states that two objects will attract in relation to their respective masses and the distance that separates them. Masses causally act upon the physical and temporal features of the Universe. Under a Rule I regimen, masses or objects acted upon are observer-independent; their operation does not depend on any assigned function—it's outside our bailiwick. Therefore, a Rule I process acts upon matter having intrinsic function such as atoms, molecules, cells, and brains.

A second rule, designated "Rule II," deals with the creation of the conditions that result in a socially constructed world, and how things work in society according to human intention. Rule II includes how we use something, a product, a game, a method for doing something, a program, or a mode of expressing ourselves. For example, music is only

J. R. Carvalko Jr., *Conserving Humanity at the Dawn of Posthuman Technology*, https://doi.org/10.1007/978-3-030-26407-9_44

music according to sounds that we interpret according to human created rules. Another illustration of this might be carrying, throwing and kicking a ball. What, if any, game might be being played? American football, rugby, soccer, or none of these? The ball handlers would not be engaged in any of these games, unless they were playing by a set of rules, a Rule II that defined the game.

Rule II objects and activities always have the same form, X stands for Y in the context C. John Searle refers to these rules as constitutive rules, "because acting in accord with the rules is constitutive of the activity regulated by the rules…they constitute the very activity they regulate… ." Without rules, we cannot assign meaning to what something, such as an object, represents.

Under Rule II, an object exists in virtue of an assignment of meaning or functionality. A constitutive rule set also requires collective intentionality. In simple terms this accounts for the collective "we" that is necessary for "us" to believe, within accepted norms, that a world of objects and social conventions exist. A game of football is not a game, unless the players agreed they were using the ball according to a particular set of rules.

Theories abound how "I and we" form mutual beliefs and reach mutual assent on a great number of things, which Searle refers to a institutional facts. For example, that television comes from a transmission of a broadcast from some distant studio, to a contract that exists to establish rights, duties, and obligations, or that tomorrow the Sun will rise in the east. Just about anything that we can think of in our day-to-day routine, can be assumed constitutes a mutually shared notion with others in our community. This does not mean that we believe in the same deity, in the sanctity of marriage, or even when life first forms. It does mean people living in the same general culture will acknowledge the existence of these states of affairs, like money, political parties, and marriage exists, providing each conforms to a convention, as subsumed under what I refer to Rule II.

Without digressing into competing schools of thought regarding why collective intentionality does or does not exist, Searle and others argue that when we combine collective intentionality, the assignment of meaning/function and constitutive rules we satisfactorily explain institutional reality [1]. This permits us to lay the rationale for accepting the reality of objects that do not define a structure in terms of physical components, but in virtue of status functions, akin to a normative set of characteristics, or what as a society we may all agree represents the function the object of our attention serves.

"Rule III," the final rule pertains to matters that govern institutions, usually providing prescriptions involving duties and rights. Suppose a law restricts a vehicle to a one way, downhill direction. This law differs from the law of gravity (Rule I) that pulls the vehicle downhill. Civil laws and ethics are examples of the kinds of things Rule III would apply to. Rule III creates by virtue of a recognized authority, regulation or normative behavior.[1] As such it presupposes the existence of that which it regulates, an institution, or as in the above example, road traffic. The institution of granting citizenship for example falls under Rule III paradigms. Another such paradigm allows legal claims to undertake the imprimatur of legislative protection.

Does the notion of distinguishing biological species, or more pointedly posthumans from other species, such as *Homo sapiens*, based on social conventions i.e., rules as discussed, stand for a difference in nature, or simply offer us a convenient way to classify or discriminate between dissimilar creatures? In other words, does our social reality, one that assigns meaning according to a set of rules, have subjective bearing, which might cause us to look at *Homo futuro* differently from *Homo sapiens*, or is there some objective, invariant epistemic underpinning that constrains us to see things only one way? Let's look at a few examples that illustrate how this works, before engaging in a discussion of the difference between posthumans and humans that have ontological differences, which may raise ethical concerns.

Note

1. Searle, J.R. (1995). *The Construction of Social Reality*. Free Press.

[1]An example of normative human behavior are actions influenced under community pressure or through social attitudes.

CHAPTER 45

The World Is a Chess Game

Engineering can be thought of as the art of combining and manipulating objects and naturally occurring phenomena so that a particular state or dynamic is achieved. When we design something like a computer program, a bridge, or process for making steel, we do it according to a specification that employs a Rule II rulebook. If we find an error in the program or the bridge itself, it can be corrected. But not all thing we design and create can be corrected once completed. Think of a steel plate or a windowpane, these are cast in such a way that make it impossible to reform. We can only fix the problem in the next casting.

An engineered artifact functions in a particular way according to an intended design, but the underlying atoms and molecules continue to be constrained to operating according to natural laws. When we combine and bend natural artifacts according to a genetically engineered prescription, we cannot go back, the die is cast. Rule I rulebooks remain in place, as we have been discussing, but a manipulation according to our Rule II rulebook irrevocably altered the object of our design. CRISPR for example can be used to change gene drives—to force genes to spread throughout a population, currently a process used against invasive pests—which irreversibly eradicates entire species that someone decides is pesky.

Let us examine the significance of Rule II and Rule III operation as applied not only to programs, bridges, steel, or genomes, and how these rules play out in such areas of life as games, e.g., chess and institutions, subject to norms, rules and regulations.

J. R. Carvalko Jr., *Conserving Humanity at the Dawn of Posthuman Technology*, https://doi.org/10.1007/978-3-030-26407-9_45

As the philosopher Ronald Dworkin wrote: "In the case of chess, institutional rights are fixed by constitutive and regulative rules that belong distinctly to the game … Chess is, in this sense, an autonomous institution…" [1]. If we presume that the institution of laws (legislative or agency) conform to a chess metaphor, it implies that no one can claim an institutional right by appealing to a general morality, because the rules of the game, Rule II and Rule III, determine one's rights. Any extant rights are established by conventions made explicit in Rule II and Rule III rulebook. Whether, baseball, football, or poker, all rigidly played games subscribe to similar conventions.

Let's consider a genetic engineering analogy where: (1) material substance, such as a nucleotide molecule existing at a loci on a DNA backbone, stands to a chess piece, existing at one location on its 64 square chessboard; and (2) physical alteration, such as removal and replacement of a nucleotide, stands to the movement or removal of a chess piece, each in accordance with rules of the game. The process or the rules are what changes the state of the platforms (a hypothetical genome or chessboard) in removing or transitioning a piece from one location to another. The state of the platform before and after a move defines the transition, each transition describes a particular state. From our earlier discussion you will recognize this as a process that underlies cellular automata.

Similarly, a computer processor, installed in an anatomical part of the brain or in a humanoid robot, embodies the physical device that supports a content, in the form of an electrical charge, that resides in memory determinative of the state of the computer. The state arrives through transitions that occur according to certain rules, which we call a program.

We should keep in mind the difference between an object of our imagination and the actual thing. The actual thing consists of a natural object or occurance, or a physical embodiment of some underlying human creation. It therefore, may consist of Rule I and Rule II and only occasionally embody a Rule III regulation of the kind sanctioned by community norms or law. For example, the hypothetical making of postal stamps, a patent, or issuing a financial instrument, such as an insurance policy utilizes a process, part natural, part artificial and subject to regulations. The inventive process itself may embrace all three kinds of rules. A manner of conducting an activity, such as engaging in the playing of a game, invokes Rule II and occasionally Rule III considerations.

On the other hand, as mentioned earlier, social realities, such as games do not exist, until a constitutive Rule II exists, because these rule-sets control antecedent behavior, which transform into institutional and social reality. Chess did not exist until a rule set came into existence. So it is that posthumans will not exist until genes are altered, but once altered the rulebook will have been written for the individual's status, now and in the future, as so altered.

An AI analogy of a process constitutes part of our social reality as well, because it implies a social function that creates, engages, or causes states of affairs. Such states of affairs as communities, legal systems, and inventions are constituted for human purposes that include interests, equities, and goals. Unlike natural phenomena, social systems are matters under the influence of our collective intentionality, the assignment of function, and constitutive rules. In a later chapter we will explore a future hypothetical social system, which takes into account these three elements.

To summarize, we create rules, which create physical, social, and institutional reality, rules that are constitutive, that regulate and which may have moral implications. Some of these rules actually create the very technology, such as machines, processes, games, computers, and posthumans, which do or will affect our lives. Some of these same kinds of rules devolve into complex cultural networks, the memes by which we communicate culture, the rituals where society behaves as a collective body to pass laws, and establish the norms and moral tendencies members may be pressured to follow. As previously indicated, among other ways, we reveal our moral predilections in matters of technology either by our passivity to it, or how we act to exploit it. Note the subtle change from assignment of function to "exploitation."

Note

1. Dworkin, R.M. (1977). *Taking Rights Seriously.* Harvard University Press.

CHAPTER 46

Preternatural Life

We see the world through zillions of photons that flood the retina, transforming light into bioelectrical impulses, which bombard the occipital and the further reaches of the cortex, where neural codes interpret the seen reality. We hear the clatter and clamor of pressure waves vibrating our ear canal, the middle ear bones, and thousands of tiny hair cells, so we can only then organize them into what we call sound. We can continue along these descriptions of so-called reality, but essentially, everything we think we discern represents an altered reality, one constructed into a web of concepts and beliefs shaped through millions of years bio-evolutionary forces, confounded by our circumstances, the culture within which were born, in which we toil, and will someday die. Some concepts and beliefs are observable, others assumed, that is, taken on faith.

Now, imagine a future where what we see, hear, feel or believe will be altered through drugs, AGI, external/internal bio-computers or a genetically engineered genome. What we might anticipate as to where things fit in this world of concepts and beliefs necessitates a deeper dig into what is real, what is abstract, and the essence that bears on the distinctions between the transhuman and ourselves.

Scientists, philosophers, and theologians alike spend lifetimes trying to fathom the core of these webs of concept, and in so doing travel outside the realm of human experience. Quantum mechanics is one such category that comes to mind, with its innumerable paradoxes, fitting an objective reality, outside human perception or understanding

J. R. Carvalko Jr., *Conserving Humanity at the Dawn of Posthuman Technology*, https://doi.org/10.1007/978-3-030-26407-9_46

on many levels. Erwin Schrödinger, famous for contributing to the uncertainty principle by offering up the enigma whether we could ever know if a cat in a box were dead or alive, stated that the observation itself alters the future state of an observed object. A pretty bizarre assertion by way of common experience. And, this more than suggests that our reality is only fixed at the moment it's subjected to observation. So, just what constitutes the full scope of our reality goes beyond physical systems, it includes those structures that although tangible are nonetheless subjective, and others that although intangible are objective. Consider for example systems of laws and ethics, which are abstract and subjective, but undeniably exist.

The differences between *Homo sapiens* and *Homo futuro* importantly bear on any number of intangible moral, jurisprudential, and commercial considerations, such as whether the genome enhanced through gene editing is a product of human invention. Invented things have purpose, and carry with them objective use limitations and social conventions related to ownership, economics, e.g., maintenance costs.[1] Unlike biological beings (trees, birds, oceans, and humans), inventions are social constructs, in the category of Rule II and Rule III entities, discussed earlier. To conceptually move from a natural world, or Rule I entity to a Rule II or Rule III entity requires that we flesh out a bit of ontology, that is to establish both one's existence, as well as its epistemological underpinning, the evidence of one's existence.[2,3]

[1] In the not too distant future, many will be the recipient of prosthetic and enhancement technology that requires cyber-functionality, or uses internal processors to keep organs functioning that have reached their life cycle—some devices may code the brain's circuitry and others to supply the regularity of a time-piece to parts of the anatomy that need to keep a regular beat. These may become subject to costs, much like costs associated with pharmaceutical prescriptions. Genetic engineering will likely be paid up front prior to the procedures as with services like IVF.

[2] The idea that social constructs are tied to our physical being was suggested by Searle, see, *The Construction of Social Reality*. Free Press (1995).

[3] "There are two general aspects to realism, illustrated by looking at realism about the everyday world of macroscopic objects and their properties. First, there is a claim about existence. Tables, rocks, the moon, and so on, all exist, as do the following facts: the table's being square, the rock's being made of granite, and the moon's being spherical and yellow. The second aspect of realism about the everyday world of macroscopic objects and their properties concerns independence. The fact that the moon exists and is spherical is independent of anything anyone happens to say or think about the matter." Miller, A. (Winter 2016 Edition). "Realism." In: E.N. Zalta, ed., *The Stanford Encyclopedia of Philosophy*. https://plato.stanford.edu/archives/win2016/entries/realism/.

Consider the properties that separate humans from other entities encountered on a daily basis, ones which may not be immediately perceptible, but important in looking at how engineered products differ from products of nature. More to the point, where, on the Rule I, Rule II and Rule III spectrum of entities, do these things fit from a human perspective?

Inventions before the advent of modern computers were nearly always constructs of observer-independent features. A hammer has a flat surface for pounding nails. A chair is made of wood, metal, plastic, and cloth for sitting. We consider these features as observer-independent. The spatial arrangements inherent in the objects do not depend on whether anyone observes them. If all people on earth disappeared the epistemic features would still exist. Their ontology, on the other hand, is dependent on human awareness, i.e., a hammer exists for human application or use. Likewise, a chair takes its place in an assigned manner in society. We refer to it as observer-dependent states having a subjective ontology.

Things that we invent, like chairs and hammers, have observer-independent features (e.g., physical structure, materials of construction), but this is not always the case. Money, contracts, calculations (as in mathematical computation), flags, religious icons and states of affairs (e.g., games) are but a few of an unending list of such things, which do not have observer-independent features. But, similar to the hammer and chair, these inventions are ontologically subjective, existing only in virtue of the fact that they play a role in human affairs. If people who employed these inventions disappeared, so would the meaning accorded these objects.

Computers are largely ontologically subjective apparatuses. We use them to search for articles, books, store files, compose music, poetry, communicate, calculate, and entertain, showing videos to playing games, like checkers, chess, Go, and Candy Crush. We use them to solve complex science and engineering problems, provide medical diagnoses, rocket space ships into the far reaches of the Universe, and, countless other undertakings that we conceive of. The ontology of an AI machine is the same as a computer, but what about *Homo futuro,* whose existence will depend on a genome instantiated with synthetic DNA, and perhaps outfitted with a bioengineered computer. Where in this matrix of ontology does this entity fit?

Chart 46.1 sets out the four possible modes of ontological existence and whether the evidence of existence is absolute or relative [1].

	EPISTEMICALLY/ABSOLUTE	MANIFESTS	EPISTEMICALLY/RELATIVE
ONTOLOGICALLY/OBJECTIVE	a. (natural artifacts e.g., mountains, electromagnetic energy, DNA, biological organs, animals, humans-as-they-are)	→	b. (neurological e.g., light and sound as manifested through neurological sensation manifestations of biological organs, e.g. the mind, consciousness, emotions, memes, senses.)
MANIFESTS			↓
ONTOLOGICALLY/SUBJECTIVE	d. (social artifacts e.g., money, artificial lakes, genetically enhanced organisms, products and processes of invention, computers, software, robots, enhanced brains)	←	c. (psychological manifestations, e.g., thoughts, hearing tonalities as in music, interpreting language as having meaning, human invented rules, abstract logical systems, mathematics, laws, ethical norms, institutions)

Chart 46.1 Categories of ontological states

The arrows point to how the existence of one entity, leads to another entity. For example, box "a," humans-as-they-are, a tangible identifiable artifact, manifests as box "b," neurological activity, such as consciousness, an intangible process. In turn consciousness leads to box "c," thoughts, institutions, laws, which leads to box "d," tangible constructions, typically products and processes.

The concept of "epistemic" refers to the qualities, features, and the grounds upon which we identify the limits and validity of a representation of a substance, process or state of affairs.[4] Matters of epistemic identity fall into absolute or relative categories. Absolute refers to causal relations of substance and form, independent of consciousness. These are what we previously earmarked Rule I, entities. Relative refers to causal relations of processes and states of affairs dependent on neurological

[4]"Wherever it is used, epistemic traces back to the knowledge of the Greeks. It comes from epistēmē, Greek for "knowledge." That Greek word is from the verb epistanai, meaning "to know or understand," a word formed from the prefix epi- (meaning "upon" or "attached to") and histanai (meaning 'to cause to stand')." https://www.merriam-webster.com/dictionary/epistemic. Last retrieved 13 July 2017.

activity, e.g., consciousness, sensations, emotions, and cognition. These are also subject to Rule I definitions.

Although the function or purpose of a paperweight, a bicycle or a tool may depend on one's culture, which would be Rule II, an item's epistemic feature, that is its physical form and specifications don't depend so, e.g., geometry, size, density, texture, or optical properties, etc.[5] Let's say I carried a waste paper receptacle to a remote village where the people hadn't been exposed to such a product. We reasonably assume that they'd recognize the form, for example, the shape and hollow container. It wouldn't depend on what I claimed as a waste paper receptacle (referred to as observer-dependent). Yet its physical properties would be understood, and thus it's epistemically absolute.

Observer-dependency is contingent on intentionality—specifically an intention to use an object, or more pointedly, how it may be used. A pond for fishing or a tool for scientific measurement exists in virtue of how we use them. A people that neither fished nor used scientific instruments would likely neither recognize their function nor the context, i.e., the social reality, or conditions under which these artifacts existed.

An object may serve one function, such as a tool, for one species, and for another species serve a different function. For example, from an upper story window, I watch cars travel down a paved macadam surface. I assign to the surface the function, regulating traffic. From another direction, a seagull comes into view. It carries a hard-shelled mussel. From twenty-five feet above the surface, it lets go. The mussel crashes onto the pavement and the shell cracks open. The obvious function that "the street" serves the seagull is to satisfy hunger. The function served by the surface differs based upon the respective needs of those exploiting its properties. In this instance, the function served is not intrinsic, as it is for the molecules that hold the macadam surface in a rigid matrix.

As illustrated by the foregoing, the assignment of function depends upon the manner an organism exploits the artifact or process. Humans, animals, and insects all build shelters, which serve the same purpose. Weaver ants (genus *Oecophylla*) construct complex nests by knitting leaves using larval silk with the assist of more than half a million workers [2].

[5] The study of the nature and grounds of knowledge is called epistemology. https://www.merriam-webster.com/dictionary/epistemic.

These are products of living creatures, analogous to human invention and construction.[6] But, it's doubtful that mindfulness plays a role, and therefore a subjectively assigned function doesn't factor into animal and insect nest building [3]. Animals may be aware in the generic sense of being capable of sensing and responding. Thus it can be said that lairs and nests have an objective ontology, because they are natural phenomenon (no different from oozing saliva), a causal process independent of any form of consciousness [4].

Returning to the Chart 46.1 category "a," as indicated, ontologically objective/epistemic absolute refers to substances, processes, and states of affairs that are independent of consciousness, such that a mountain exists without having to observe it. We exclude religious, philosophical, and political ideas, reserving the label "ontologically objective" for objects intrinsic to nature. Existence depends neither on the condition of the mind nor one's observation, regardless of the perspective, as it relates to objective ontology. So, if humans disappeared from the planet, these entities would remain.

In category "b," ontologically objective/epistemic relative refers to substances, processes, states of affair that are dependent on a sensory system. This category is best understood by reference to the question, "Does a tree falling in the woods make a sound"? The answer is "No," unless something exists to neurologically sense the pressure differential caused by the compression of air molecules when the tree contacts another body, presumably the ground. The compressed molecules are as "real" as mountains, but in this case, the sound is only a phenomenon when a neurological auditory system detects it, otherwise although a compression of molecules in an air medium exists, the sound as such doesn't. On the other hand, memetics distinguishes between the physical manifestations of the meme and the psychological, and ultimately neurological, which is why I place it in this category.

[6]Thomas Nagel's (1974) "what it is like" captures the subjective idea of consciousness from a creature's "mental or experiential point of view," but it doesn't capture mindfulness as experienced by humans. And in the obverse, bats navigate the world through echo-locatory senses, a phenomenon we humans can't understand from the bat's point of view. Van Gulick, R. (Summer 2017 Edition). "Consciousness." In: E.N. Zalta, ed., *The Stanford Encyclopedia of Philosophy*. https://plato.stanford.edu/archives/sum2017/entries/consciousness/.

The category "c," denoted as ontologically/subjective, epistemically/relative, relates to states of consciousness, e.g., memory, thinking, feeling, and knowledge. Ontologically/subjective connotes states of affairs or process dependent on conscious characterization, such as speech, language, mathematics, the declaration that "a chair is a chair," "a book is a book," or "a blood alcohol reading is a blood alcohol reading." None of which are a collection of molecules, but psychological entities having assigned meaning, constitutive rules and qualify as what they represent through collective intentionality. In this category are our mental constructions of immaterial objects, such as institutions, the Supreme Court for example exists as a convention, or the use of the phrase "global warming," as a metaphor for objective climate change, rather than the actual meteorological reality. These kinds of things come into existence, only if a sentient being acknowledges the entity by giving it meaning. It follows that "facts," if thought dependent, would be ontologically/subjective, epistemically/relative, and hence exist only within the context of consciousness. Among some philosophers this assertion may be debatable, but at this juncture it's not my intention to justify my position other than the following brief argument.

Philosophers from Plato to Heidegger believed that the mind does not directly interpret that which it perceives, but interprets through a shadow of reality—an epistemically/relative subjective-dependency—a first-person ontology. Rationalists, such as Plato and Descartes held that humans are enabled with distinct innate ideas about the world, discovered through procedural thought. Mathematics and science serve as examples, which can get us to a common understanding. Sir Francis Bacon (1561–1626), John Locke (1632–1704), referred to as Empiricists, opposed the Rationalists, and posited that we find knowledge in sense experience. However, David Hume (1711–1776) argued that the human perspective organizes a way of "seeing" the world, which along similar lines, Willard van Orman Quine, perhaps the twentieth-century's greatest ontologist, also made us aware.

Finally, category "d," consists of ontologically subjective/epistemic absolute products, substances, processes, and states of affairs that are contingent on human consciousness. They have materiality as in category "a," but as in category "c" they depend subjectively on assigned meaning and constitutive rules, as mentioned in an earlier chapter. Notably, category "d" objects only exist to serve a human purpose. That is to say, ontologically subjective entities in category "d" are observer-dependent, requiring an imputation of meaning in virtue of how something performs and what purpose it serves.

Category "d" also requires consciousness. We assign function based upon economic, social, or individual needs, e.g., the use of a screwdriver as a chisel, or a rim, having a rubber tire used to carry a bike. Function implies an agency relating to human predilection, and therefore it falls into the ontologically subjective realm and thus observer-dependent. Thus, given the above we can confidently say that the bioengineered genome, *qua* human, will fit this specification.

Inventions are products of human innovation and production. For the first time in history, through genetic engineering, an invention, a human-like a species now falls into a socially constructed category. And, thus we now face the quintessential moral dilemma: what does it mean to be human? Does the answer depend on human intentionality, the kind that invented the wheel or the atomic bomb?

Humans function as natural processes, who are subject to laws of evolution, entropy and the conservation of energy. Until now, we had not been obliged to consider if something like evolution could be avoided. But, we were destined to face this moment when we started inventing hybrid animals, chimeras, and when modified organisms began to emerge from the test tube in research laboratories, when they began to appear in food production plants, farms, pastures, stables, and race tracks. These not only are technological products, but social constructions, having been objectified.

Human lives constitute a natural cycle, patterns that separate us from other artifacts in the world, artifacts that fall within the realm of a social construct and all that that implies. Technology adds functionality to a natural biological entity for one reason—to exploit a social aim, whether that aim is enhanced intelligence or to extend our lives. Although dependent on nature, mainly through chemistry or physics, it has the potential to enhance our basic fabric, the potential to separate us from the core of our essence, humans-as-we-know-them.

To summarize, a rule in a social context can be placed into two sets, one having to do with subjective process, such as experience and beliefs, and the other having to do with objective processes and states of matter, regardless of human agency. For example, a bicycle, *qua* bike, only exists as a consequence of an intentionally crafted ruleset that assigns meaning in a social context. On the other hand, the aggregation of materials, their arrangement or configuration and the dynamics that propels something like a bike are objective processes and states of matter independent of human interpretation. When we genetically enhance the human form, where does the product come to rest within the divisions of objective and subjective artifacts and processes? And, is this a distinction without difference in how we will interact or dignify the modified entity?

Notes

1. Searle, J.R. (1995). *The Construction of Social Reality*. Free Press.
2. Hölldober, B., and Wilson, E.O. (1990). *The Ants*. Harvard University Press.
3. Van Gulick, R. (Summer 2017 Edition). "Consciousness." In: E.N. Zalta, ed., *The Stanford Encyclopedia of Philosophy*. https://plato.stanford.edu/archives/sum2017/entries/consciousness/ (Last visited 8/4/2017).
4. Lin, L., et al. (2007). "Neural Encoding of the Concept of Nest in the Mouse Brain." *Proceedings of the National Academy of Sciences of the United States of America* 104 (14): 6066–6071. https://doi.org/10.1073/pnas.0701106104. PMC 1851617. PMID 17389405.

CHAPTER 47

The Impossible Dream

Much that has been written about science, technology and mathematics in the modern era gives the impression that we will eventually unravel the inner workings of all that is natural, perhaps what stands in for intelligence. The subject of a good portion of this book has dealt with science's intervention in natural intelligence and the invention of artificial intelligence, as if either were completely definable, or that someday if sufficiently understood, we'd succeed in expanding our intellect through genetic engineering or computer technology.

One or more elements of what constitutes human intelligence and its corollary, i.e., "what we can know," may well fall outside that which is housed in our three-pound bundled brain. This is not a new idea, but one that goes back at least to Plato's Republic, where we are taught that we know things only as shadows of presumably a deeper knowledge of reality. Knowledge and intelligence may be two sides of the same coin, so improving intelligence may not mean that we can be assured of a better understanding of the world that surrounds us. In a world that tends toward reductionism, it may come as a surprise that forms of knowing do not depend on language, mathematics, or science, but on a spiritual awareness. It may well be that the force of one's creative product does not come from one's innate intelligence either, most likely it does not, but lives in another room of consciousness, one that beckons us to enter at the most unexpected times. How would one go about designing a technological analog, artificial intelligence for example, of something the origins and specifications of which were unknown?

J. R. Carvalko Jr., *Conserving Humanity at the Dawn of Posthuman Technology*, https://doi.org/10.1007/978-3-030-26407-9_47

Scientists have calculated the tiniest unit of time as approximately 5.39×10^{-42} seconds, the time for one undulation of light to travel 1.616229×10^{-35} meters, the smallest possible physical distance. To visualize how small this is, imagine a flyspeck 0.1 mm magnified as large as the observable universe, then inside that universe-sized flyspeck, place a 0.1 mm flyspeck—that's how small it is. Although conscious experience occurs over orders of magnitude longer in both time and space, it's within the interstices of these unimaginably miniscule timeframes—and micro spaces of electric neural analogs, where we accrue the likes, dislikes, the binaries of what we become, and the endless variegations of what we create. Yet, we have no access to these micro spaces, the black holes of neurons, impossible to map onto perceptions in the macro world, where we abstract thoughts, emotions, the concrete elements of touch and feel, the patterns that form our concepts of love, beauty and wonder.

But nevertheless, as new technology morphs into products that eradicate some and create new social structures, it undoubtedly changes what and who we are on the spectrum of Nature's species. Query: will *Homo futuro* be creative in the inventive and artistic ways that currently distinguish the modern human from one another and as a species from the animal kingdom? Will we remain autonomous, self-controlled, free-willed, if infused with semiconductor sensors, RFIDs, modified by CRISPR/Cas9 synthetic-DNA, or revised with molecular-computers affecting metabolic processes, our mind, its sensitivity, intuition, evaluation, and feelings? We must ask and answer these questions, while we can.

Humans have the capacity to assign meaning to practices in virtue of the purposes served. Machines do not, because they cannot sense fear, failure, or on the positive side, achievement and self-worth. Thus, a doctor takes the opportunity to make a statement, not necessarily in words, to palliate suffering or save a life. Her dedication to the medical arts fulfills a human purpose. A biologist projects a molecule into a framework not before imagined and sees a double helix. His ingenuity spawns a new brave world. A musician lays down a progression of chords that include intervals not previously combined and we hear the vestiges of a mournful wail that haunts our heart. A poet strings phrases that arrest an emotion that outlasts the lives first touched. We live for the moment that captures the essence of what we do to affirm our self-worth or we risk passing out of the world without experiencing, why we were put here in the first place.

What happens when humans incorporate these technological designs, not as humans as we know them, but in the specification of posthumans, who may have posthuman values, distinct from modern humans? Does the modern human stand to lose the preeminent position of assigning meaning to practices in virtue of the purposes served? What happens when the machine, imbedded in the remnants of what we call the "mind" supersedes the cognitive intellect of the human as we know it, individually, and in its collective societal manifestation? Humans are born to share the benefits and liabilities of nature and their cultures, free to adopt beliefs, values, and attitudes. But, in a *Homo futuro* era, through design, what we hold as values and attitudes, those acquired through natural contact and genetic affinity, may vanish.

Nick Bostrom and others speak to cognitive intelligence, a super intelligent machine, but few speak to attributes of an aesthetic intellect, and how, in the form of an algorithmic *Homo futuro*, one may acquire this genius as an APP that instills programmed values. Seneca said, "Nothing is terrible in things except fear itself." Should we fear this changing equation, one that comes about because we can't govern our curiosity to employ technology in all manner of our lives?

On the subject of culture in the age of modern technology, Neil Postman, writes: "… we are surrounded by the wondrous effects of machines and are encouraged to ignore the ideas embedded in them. Which means we become blind to the ideological meaning of our technologies" [1]. True enough, but does it also mean that technology can potentially blind us to the teleology of truth and beauty?

In a posthuman era, diverse structural and functional revisions of the human form will exist, where life-forms at each update improve physical or mental prowess, but conceivably change concepts of harmony, balance, and beauty in the classical sense. And if this is the case do these concepts take on new meaning in the psychological and aesthetic sense? It seems likely newer technological versions of posthuman would only drive differences among those exhibiting earlier versions further and further apart, that is newer paradigms would diverge from what an earlier generation may have regarded as aesthetic.

Needs, longings, and curiosity drove humankind's greatest achievements, beginning with stone tools, dating back between 3.3 and 1.8 million year ago. We cannot dismiss the caustic potential of genetic engineering, and how this wonderful tool may, in the long run, bring us to extinction. The model for evolutionary patterns varies because natural organisms, from the simplest to the most complex, survive because they adapt to their natural environments. On the horizon of those now being born, the patterns of human evolution will likewise fall into the recesses or opportunities that the future presents, not of a natural environment, but a technological environment, as it reengineers the DNA backbone, the one nature shaped over the course of billions of years [2]. We inexorably move toward a time when transhumans, cyborgs, humanoid robots, each in some profound way, threaten to rival humans for jobs, health care services, security, and soldiering. But, will this progression also threaten values heretofore represented by our humanities: art, literature, music, and philosophy? Harriet Tubman once said "Every great dream begins with a dreamer. Always remember, you have within you the strength, the patience, and the passion to reach for the stars to change the world." But, it's unlikely Ms. Tubman saw this far into the future, where the dreams of technologists now threaten to change the "*I*" in who we are—***agents of humanity***.

Notes

1. Postman, N. (1992). *Technopoly: The Surrender of Culture to Technology*. Alfred A. Knopf.
2. http://www.actionbioscience.org/newfrontiers/jeffares_poole.html (Last visited 8/7/2012).

CHAPTER 48

Timeless Borderless Creativity

We are what we think. All that we are arises with our thoughts. With our thoughts, we create the world.—The Buddha

Humans possess an astonishing ability to create across every conceivable domain, from arts and sciences to languages and performance. Significant contributions change the culture itself.[1] And, although we often associate creativity with particular individuals, more times than not, especially in the Pro-C category of creativity, rarely does the creative product represent the work of one individual, working in a vacuum, isolated, and alone [1, 2]. High levels of organizational creativity, according to one psychologist, require expertise, and people who approach problems flexibly and imaginatively, motivated, either by the organization itself, or intrinsically as within the nature of the specific individuals [3]. Throughout history new products and even art forms, such as bebop or impressionism, come from group efforts. In the book, Becoming Steve Jobs, Brent Schlender writes that Jobs took brainstorming walks with others [4]. Creativity and its close ally innovation appears periodically in societies, businesses, and academia as contagions spreading

[1]Jean-Baptiste Lamarck (1744–1829), held that an organism can pass on characteristics that it has acquired through use or disuse during its lifetime to its offspring, which, except for some untested hypotheses related to gene-expression, does not otherwise appear borne out by modern science, but his ideas strongly suggest that cultural inheritance does follow a model where cultural traits are passed on to successive generations.

J. R. Carvalko Jr., *Conserving Humanity at the Dawn of Posthuman Technology*, https://doi.org/10.1007/978-3-030-26407-9_48

within clusters of people, rising and falling, perhaps in the shape of a sine wave, slowly increasing, peaking, then slowing down. Graham Wallas, in his seminal work Art of Thought, considered creativity to be a legacy of the evolutionary process, which allowed humans to quickly adapt to rapidly changing environments [5, 6]. For some societies the period can range from a few to several hundred years.[2] The German economist Gerhard Mensch and others believed that innovation, that is the implementation of creative ideas, is cyclical, periodically manifesting as swarms having long time constants, which some have called super-waves or Kondratieff waves (K-waves), as posited by a Russian economist, Nikolai D. Kondratieff, who described the existence of 45–60 years cycles of economic development in the early part of the twentieth century.[3]

Moving down the world's timeline into the Renaissance, we see a phase of discovery and invention that kindled, then smoldered and rekindled like fire did a million years earlier, but this time millions of lumens brighter, into what began the modern era of progressive, rational thought. We ask what caused progress following the Dark Ages to self-sustain a three hundred year "rebirth" that sparked a six-hundred year continuous scientific and technological expansion? Was it a by-product of a collective mind accumulating sufficient information, amassing forces of imagination to accommodate a growing population with intensifying economic needs?[4] I believe so, and further believe that

[2] A rough timeline of historical progress discloses the Bronze Age with writings, philosophy, religion, art and music, burgeoning throughout the world, Egyptian pyramids 3000–1200 BCE; Greece, art, architecture, literature, philosophy 500–300 BCE; Rome art, language (Roman alphabet), literature, civil law, civil and architectural engineering, 100 BCE–350 CE; Mayan calendars, mathematics, astronomy, art 1000 BCE–900 CE; China art, literature, music, philosophy, science, papermaking, printing, gunpowder, compass, 200 CE–1100CE; Islamic Golden Age, agriculture, the arts, economics, industry, law, literature, navigation, philosophy, sciences, sociology, and technology, 800 CE–1400 CE; Renaissance 1300–1600 CE; Scientific Revolution, 1500–1600 CE; Industrial Age 1750–1850 CE; Computer Age; Information Age.

[3] Nikolay D. Kondratyev, Encyclopædia Britannica, see, https://www.britannica.com/biography/Nikolay-D-Kondratyev (Last visited 10/5/2019).

[4] Leonardo da Vinci, one of the most prolific inventor of the last Millennium made numerous machines, but his notebooks contained even greater numbers inventions that could not be realized due a lack of some essential knowledge, technology or material. Early Renaissance inventors were constrained to materials such as metals, wood or leather to shape and fashion their ideas. Although Venetians learned how to make colorless and transparent glass in the late 1200s, and which was used for a variety of utensils, they were largely unavailable to the experimenter dabbling in clocks, optical instruments, or precision tools.

dependent on the velocity of innovation and its accumulation that it may acquire a critical mass, where it destabilizes the very societies that it instantiates, self-sustaining, but subject to chaotic instability.[5]

General intellectual habits, such as openness, levels of ideation, autonomy, expertise, exploratory behavior do not favor one over another group, society or culture. Among humans, the trait for intelligence is ubiquitous. But, where the invention of language or simple tools originated was likely limited geographically in scope, and not the product of one person, but a pool of contributors who could rightly claim provenance. Yet, these important innovations came from somewhere, even if from some nondescript, combination or offshoot of some seeded event involving a deme lost in the annals of time.[6] As it stands, ancient important cultural artifacts still exist as cornerstones to civilization. We are surrounded by them, the wheel and axel, the hammer, the inclined plane, and the dam. We hardly pay them any mind. And even with most modern of devices, e.g., the smartphone, with which with we are surrounded as an innovative artifact driving social change, we hardly pay it any mind, either.

We mentioned more than once that a computer model can simulate analogs of natural phenomena, such as a biological cell, and additionally, both analogous to and in fact viewed, in a macro sense, as a social construct. Evolutionary computation takes this idea by modeling evolution itself. Societies modeled in this way, at least viewed from 50,000 feet, look like swarms of semi-autonomous units that adapt behaviors to meet goals within their environs [7]. This involves programming machine states and corresponding transforms subject to rules the designer wants to investigate. Also as posited earlier, self-referential systems and formal rules can describe the properties of physical systems, and according to some theorists, further explain how biological substrates, e.g., thoughts, can emerge from inanimate abstract patterns [8]. I'm not going in that

Although da Vinci must have understood that glass shaped in the certain ways made things larger when he proposed to "make glasses in order to see the moon large." See, Codice Atlantico, Leonardo da Vinci, Notebooks, II, 190 r.a.; and of course Galileo invented and reduced the telescope to practice in 1609.

[5]Thomas Friedman speaks to the acceleration of this sort to technology in Thank You for Being Late (2017). New York: Farrar, Straus and Giroux.

[6]A subdivision of a population consisting of closely related plants, animals, or people.

direction at this juncture, but I am pointing out that a predominant characteristic as among three types of behavior natural, artificial, and social, the first operates in accordance with laws of nature, and the latter two, in accordance with rules regarding social constructs. And, that if a formal system can assume properties of physical systems, then given rules and transformations that create self-sustaining artifices, theoretically, such artifices may behave in ways which contribute to its growth and expansion.

The evolution underway currently involves each of us, and the societies in which we live, in ways that it did not 50 or 100 years ago.[7] We are being swept up by the confluence of biological entities, at the phylogenic level, of human species at the ontogenic human level, society at the sociogenic level and technology at the technogenic level that is about to profoundly change civilization as we know it. Melded, these disparate information-carrying vehicles will form into technological complexes that from outside the cyber-sphere, appear as an electronic bubble, with signs that point to the beginning of a unitary supersized information colossus.[8]

[7] Two important developments on the world stage make the point how individuals, socially driven memes, and technology combined to affect political institutions, which in turn affected global security, economics and nature, as related to global warming. The Iranian revolution of 2009 was energized largely by Twitter and Facebook to protest what many Iranians considered a flawed presidential election. As demonstrations came alive via Twitter, the Iranian regime also used the Web, to identify protesters, via photos and associated personal information, and then widely disseminated propaganda, which when combined with shootings, tear gassing and arrests, put the restive population into a state of paranoia, which had the effect of tamping down the marches. See, Editorial: Iran's Twitter revolution. https://www.washingtontimes.com/news/2009/jun/16/irans-twitter-revolution (Last visited 3/18/2018). The 2016 elections in the U.S. and the U.K. were influenced by operatives, who used social media platforms to sway the election in two of the world's oldest democracies. In March 2018, *The Guardian* reported that a data analytics firm worked with Donald Trump's election team and the winning Brexit campaign to harvest millions of Facebook profiles of voters. Revealed: 50 million Facebook profiles harvested for Cambridge Analytica in major data breach, see, *The Guardian*. https://www.theguardian.com/news/2018/mar/17/cambridge-analytica-facebook-influence-us-election (Last visited 3/18/2018). See, https://grist.org/article/russian-trolls-shared-some-truly-terrible-climate-change-memes/.

[8] We see signs of this in the emergence of supersized organizations throughout the world: Amazon, Jingdong, Google, Facebook, Twitter, Instagram, Alibaba, Rakuten, B2W Companhia Digital, Groupon, Zalando, and eBay.

Notes

1. Woodman, R.W., Sawyer, J.E., and Griffin, R.W. (1993). "Toward a Theory of Organizational Creativity." *Academy of Management Review* 18 (2): 293–321. https://doi.org/10.5465/amr.1993.3997517.
2. Harvey, S. (2014). "Creative Synthesis: Exploring the Process of Extraordinary Group Creativity". *Academy of Management Review* 39 (3): 324–343. https://doi.org/10.5465/amr.2012.0224.
3. Amabile, T.M. (1998). "How to Kill Creativity." *Harvard Business Review*.
4. Schlender, B., et al. (2016). *Becoming Steve Jobs: The Evolution of a Reckless Upstart into a Visionary.* Crown Publishing Group.
5. Wallas, G. (1926). *Art of Thought*.
6. Simonton, D.K. (1999). *Origins of Genius: Darwinian Perspectives on Creativity.* Oxford University Press.
7. Fogel, L.J., et al. (1966). *Artificial Intelligence Through Simulated Evolution.* New York: John Wiley.
8. Hofstadter, D.R. (1999). *Gödel, Escher, Bach.* Basic Books, pp. P–2 (Twentieth-Anniversary Preface). ISBN 0-465-02656-7.

CHAPTER 49

Information Colossus

Why man, he doth bestride the narrow world, Like a Colossus, and we petty men, Walk under his huge legs and peep about.—Julius Caesar—Shakespeare

We have all looked up at one time or another and saw birds flying in a V formation, which, although beautiful to watch, raises the question: why? Same with fish that swim in schools, and so many other species that organize and move in herds, prides, and tribes. Evolutionary biologists explain, that at least with birds and fish, it's simply to minimize resistance to wind or water currents, but it's also related to genetic conditioning, which evolved to conserve energy as creatures moved through their natural mediums, searching for food and friendly climes [1].

Once understood, we come to appreciate better that many highly structured collective behaviors are the consequence of simple principles or rules, the purposes which help satisfy some important survival imperative, like successfully moving through natural mediums. Humans share survival imperatives similar to other species, especially as concerns adaptation to a changing climate, but also different imperatives as wrought by changing technology. Currently, technological progress advances exponentially, but human adaptability to technology, which includes its response to it, appears to progress linearly. The difference between these two phenomena causes cultural and political pressures, not the least of which affects everything from local jobs to mass migrations.

Society changes in fundamental ways when driven by population growth, wealth, poverty, war, and peace, and significantly, technology

J. R. Carvalko Jr., *Conserving Humanity at the Dawn of Posthuman Technology*, https://doi.org/10.1007/978-3-030-26407-9_49

and collective intelligence. Some believe that technological expansion will inevitably lead to a Singularity, the moment when superintelligence abruptly triggers runaway technological expansion, resulting in unfathomable changes to human civilization [2]. Powerful uncontrolled expansions are seen in a variety of phenomena, such as splitting of the atom, the organization of hurricanes and tornados, volcanic eruptions, and even cardiac arrhythmias. These occurrences dissipate from the lack of a sustained energy source. As a societal energy source, the effects of technology do not appear to be dissipative, and additionally contribute to such things as economic expansion at exponential rates.

I do not know if a Singularity will occur. Negative and positive feedback mechanisms exist in living systems, e.g., in ecosystems, and contribute to periods of societal expansion, contraction, and stagnation. The dynamics involved may follow linear or nonlinear patterns. Forces and negative feedback can be such that they influence a linear system's growth or output. In other cases, under similar forces, positive feedback acts upon a nonlinear system, such that it can, and often does swing widely, inducing major changes in output conditions.[1] An example, which most of us can relate to, is when a microphone is brought too close to a loudspeaker, producing a loud, highly irritating screech. Thus, feedback contributes at the boundary between order and chaos, and perhaps catastrophe. I do not intend to add more to the thesis than a Singularity is in the offing, and perhaps already underway according to scores, if not hundreds, of prognosticators. But, I do believe that intelligence and technology will combine in ways that will reconfigure society, and to address my thoughts on the subject requires a survey on what futurists say about the Singularity.

Techno-future observers for over half-century (e.g., Ulam 1958; Turchin 1977; Bostrom et al. 1998; Kurzweil 2005; Vinge 2008) have predicted a coming societal transformation, referred to a technological singularity, what Vinge suggested, as events "capable of rupturing the fabric of human history" [3, 4, 5, 6, 7].[2] These events are obviously related

[1]The feedback mechanisms may be such as friction, or stigmergy, which is an indirect coordination mechanism acting through an environment between agents or actions. The principle is that the trace left in the environment by an action stimulates the performance of a next action, by the same or a different agent.

[2]A wide range of opinions on singularity is found at: http://mason.gmu.edu/~rhanson/vc.html#bostrom.

to an unprecedented accelerative change in the capabilities of technology, cyber, AGI and the approaching era of human enhancement [8]. Combined, the future may be heading toward a catastrophic activation of a runaway cyber-technology, ironically increasing opportunity via a virtual world, while narrowing the options to operate outside the cybersphere, but in any case irreversibly changing civilization as we now know it [9].

In a 1993 essay, The Coming Technological Singularity, Vinge wrote that the human era would come to an end as superintelligence advanced at an incomprehensible rate [10]. The idea of its eventuality has been widely accepted by futurologists, although the mathematical models upon which this may come to pass, as well as the consequences of the Singularity, have been vigorously debated.

Bostrom writes: "'The singularity' has been taken to mean different things by different authors, and sometimes by the same author on different occasions. There are at least three clearly distinct theoretical entities that might be refered (sic) to by this term: A point in time at which the speed of technological development becomes extremely great. (Verticality) The creation of superhuman artificial intelligence. (Superintelligence) A point in time beyond which we can predict nothing, except maybe what we can deduce directly from physics (Unpredictability, aka 'prediction horizon')" [11]. At least 10 different hypothetical models have found currency, though in each case much is left to conjecture, to the point that no one theory has been widely accepted in the scientific community [12].

The various ideas about Singularity start from different assumptions about the likelihood of technological breakthroughs, their contribution, and timing. Importantly, no mathematical model appears to fully represent a social system, which by definition has thousands of independent inputs and degrees of freedom. In complex systems, features often only emerge as the system proliferates, especially when the underlying system puts effort into its own creation, as would an expanding societal model [13]. A simple analogy is how adding heat to an enclosed system with an expandable boundary, such as a balloon, undergoes internal thermodynamic changes that, in turn, influence the rate and extent of its deformation. Another analogy is how a current state of any cellular automata in and of itself makes it impossible to predict where it's headed, precisely because it is self-creating.

Other uncertainties are introduced in putting forth models that predict accelerating progress and corresponding human cognitive and cultural responses, and yet other uncertainties pertain to whether the effects are even quantifiable, so that we are able to develop theories having an empirical grounding.[3] Keep in mind that effects are not solely due to emergence or information, but additionally due to institutional change i.e., laws, norms, and rituals.

On the other hand, well established fields, such as econometrics, can dependably add to part of the solution, i.e., the emergent theories relative to the accumulation of technology or critical mass of information. For instance, the logistic curve has been empirically established as fitting the behavior of the diffusion of technology. Likewise, some ideas do not appear overly controversial, as the use of hyper-exponentials to model technological capability, or the accelerative exponential nature of technology growth, as proposed by Kurzweil's Law of Accelerated Return.[4]

As mentioned earlier, and which bears repeating, any theory starts by defining a working hypothesis, often the system's configuration and conditions of its operation. In fact, currently no Singularity-type model yet proffered can be considered sufficiently fleshed out so as to be tested, along the lines of the scientific method, where such a model might be disproved.[5] As said, much is left to conjecture to the point that no one theory stands out.

But despite these weaknesses, it hasn't prevented notable scientists, technologists, and philosophers from believing that current technological growth and progress will eventually lead to a transformation in our civilization, rearranging domains of knowledge and culture, and changing

[3]The information theorist, Claude Shannon, who quantified the notion of information, in Prediction and Entropy of Printed English, (1951) discussed upper and lower bounds of entropy of the English language. He showed that treating a whitespace, as between words, as the 27th letter of the alphabet actually lowers uncertainty, providing a quantifiable link between cultural practice and probabilistic cognition. This type of quantification is one example of what is needed for a theory of technological singularity to emerge, one that had the power of predictability.

[4]"An analysis of the history of technology shows that technological change is exponential … we won't experience 100 years of progress in the 21st century—it will be more like 20,000 years of progress." See, http://www.kurzweilai.net/the-law-of-accelerating-returns.

[5]Karl Popper demarcation refers falsifiability of a scientific theory, and generally the means by which we distinguish science from pseudo-science.

what it means to be a human, as we have been discussing throughout. Some take even a darker view than the dichotomization of the species. Before his death, Stephen Hawking expressed concern that AGI could result in human extinction [14].

Among a variety of mechanisms that supposedly will drive a technological singularity, several stand out as particularly noteworthy. One is copy-ability (i.e., the ability to self-replicate) of information and second, its progressive accumulation, either as realized as process and structure in physio-chemical forms, or as information itself. As noted by Anders Sandberg, of the Future of Humanity Institute, Oxford University, "If mental capital becomes copyable (such as would be the case for AI or brain emulation) extremely rapid growth would also become likely." But, copy-ability may occur in different mediums, biological, cultural, technological, and as Ander's suggests, within the conscious mind. Sandberg has what I consider a large "if," but I do believe that raising IQ alone gets us a long way into what he is prognosticating; as he states, "extremely rapid growth would also become likely."

Significantly, entities, whether animate, inanimate, or a combination of both, in the future would not only have the feature of copy-ability, but replicate entire processes, including respective cognates of cognition and behavior, ideas, memes, and cultural practices, which add to the accumulated sum and substance of succeeding generations.[6] We see process replication in online businesses, such as Amazon.com, which started by selling books and ended up putting thousands of book stores out of business. Amazon duplicated its underlying business method, entering retail and movie distribution, where it put established retail outlets into dissolution. It continues to try and replicate the process into the food distribution chain (e.g., by its acquisition of the grocery store Whole Foods), for example. This is the true impact of important inventions, generally. They are duplicated, refined, and integrated, to the point that they displace the old.

In some cases inventions have the effect, not of process duplication, but product duplication. Twitter, for example, achieves its power through the duplication of individual meme transfers. Another example

[6]Richard Dawkins defines as "...anything in the universe of which copies are made. Examples are a DNA molecule, and a sheet of paper that is xeroxed." See, Dawkins, R. (1983). *The Extended Phenotype: The Long Reach of the Gene*. Oxford: Oxford University Press.

is the wheel, which has powered civilization for about 3,000 years, and by today's estimates, is an essential part of over one and a quarter billion vehicles. Apropos to societal expansion, displacement, and duplication will depend on a combination of both process and product duplication, where entire processes are replicated, and in other instances the product simply multiplies.

In 2008, Vinge posited four scenarios by which the singularity will occur: (1) enhancement of human intelligence through human-to-computer interfaces—that is, intelligence amplification; (2) increased intelligence by improving the neurologicaloperation of our brains; (3) human, networks, computers, and databases becoming superhuman; and (4) networks of embedded microprocessors becoming sufficiently effective to be considered superhuman. Kurzweil includes other enabling technologies such as genetic-nanotech-robotics (GNR) mentioned earlier in connection with molecular computers [15]. But, many of the models fail to define a structure within which these forces would occur.

Importantly, few if any models of Singularity take into account DNA evolution of the kind that a world of genetic engineering of intelligence will yield. This includes a DNA evolution as may be realized through the Shulman and Bostrom IES proposal mentioned earlier, to screen embryos, or to employ CRISPR-like technologies to locate and directly modify genes, and finally the application of molecular computers imbedded into the anatomy to improve g-factors. Scientists who study DNA sequenced evolution develop models, such as Markov models, of which a variety differ regarding rates of nucleotide substitution, and which one nucleotide might replace another, as an evolution of a species occurs. We cannot begin to delve into such models' level of complexity here. With that in mind, it's interesting to conjecture what effect small increases in IQ bring, along with the current level of technological amplification, increased consolidation of both organizations (such as Amazon) and an increased meme traffic brought about through social media.

Notes

1. Hamilton, W.D. (1971). "Geometry for the Selfish Herd." *Journal of Theoretical Biology* 31 (2): 295–311. https://doi.org/10.1016/0022-5193(71)90189-5. PMID 5104951.

2. Eden, A.H., and Moor, J.H. (2012). *Singularity Hypotheses: A Scientific and Philosophical Assessment.* Dordrecht: Springer, pp. 1–2. ISBN 9783642325601.
3. Ulam, S. (1958). "Tribute to John von Neumann." *Bulletin of the American Mathematical Society* 64, nr 3, part 2, May, pp. 1–49.
4. Vinge, V. (2008). "Signs of the Singularity." *IEEE Spectrum* 45 (6), pp. 76–82.
5. Turchin, V. *The Phenomenon of Science: A Cybernetic Approach to Human Evolution.* Columbia University.
6. Kurzweil, R. (2005). *The Singularity Is Near: When Humans Transcend Biology.* Viking Penguin.
7. http://mason.gmu.edu/~rhanson/vc.html#bostrom (Last visited 02/26/2019).
8. Ulam, S. (1958). "Tributeto John von Neumann." *Bulletin of the American Mathematical Society* 64, nr 3, part 2, May, pp. 1–49.
9. Eden, A., et al. (2012). *Singularity Hypotheses: A Scientific and Philosophical Assessment.* Dordrecht: Springer, pp. 1–2. ISBN 9783642325601.
10. Vinge, V. (1993). "The Coming Technological Singularity: How to Survive in the Post-Human Era." In: G.A. Landis, ed., *Vision-21: Interdisciplinary Science and Engineering in the Era of Cyberspace.* NASA Publication CP-10129, pp. 11–22.
11. http://mason.gmu.edu/~rhanson/vc.html#bostrom.
12. Sandberg, A. (2010). *An Overview of Models of Technological Singularity.* https://doi.org/10.1002/9781118555927.ch36.
13. Physica, D. (1994). Special Issue on the Proceedings of the Oji International Seminar Complex Systems—From Complex Dynamics to Artificial Reality Held, 5–9 April 1993. Numazu, Japan.
14. "Hawking: AI Could End Human Race." *BBC. 2 December 2014.* Retrieved 11 November 2017.
15. Ibid., see note 4 above.

CHAPTER 50

Inevitable Integration

In Fig. 50.1, we hypothesize a model in line with Vinge's four scenarios and Kurzweil's GNR, the entirety of which influences development of four domains consisting of phylogenetic (P), ontogenetic (O), sociogenic (S) and technologic (T), spheres labeled POST. Furthermore, each domain is interconnected to exchange information and process material substance, e.g., such as nucleotide substitution. Vinge's theory is implied particularly as it considers the effect of an increased intelligence via improved neurological efficiency. This, of course, is a main thesis underpinning the model. Whether such a system actualizes, as indicated, is conjectural, as the intricacies of these types of models take more than a few pages to explore and fully flesh out. But, it does serve to illustrate one direction that a future society may follow as germline editing and AI bioengineered devices come to dominate the planet's intelligent form of life.

The model in Fig. 50.1 has an operational complexity related to its structure and underlying individual process mechanisms [1]. For simplicity, we might view it as a nonequilibrium Markov state-diagram, where each of the hubs, or POST elements represent a process that changes its state in accordance with an assigned probability. The elements that determine the rate of a change in any POST state are noted by the smaller circles labeled **a** through **i**. Note that each of the four POST elements remain in their individual states, as indicated by the arrows ***W*** through ***X***, until a change occurs. Positive feedback results, along the lines of Kurzweil' Law of Accelerating Returns, that urges the "evolutionary

J. R. Carvalko Jr., *Conserving Humanity at the Dawn of Posthuman Technology*, https://doi.org/10.1007/978-3-030-26407-9_50

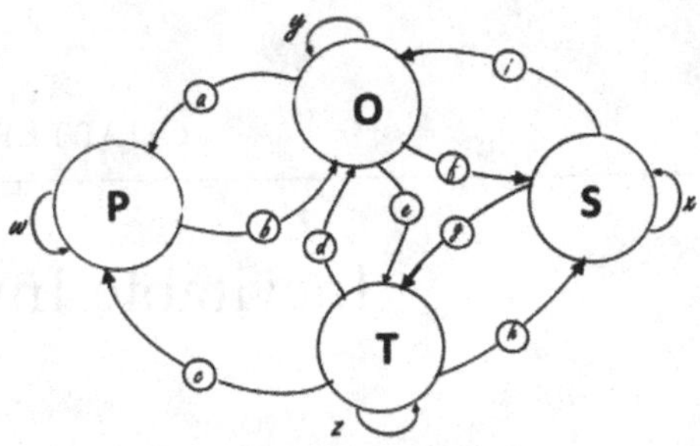

Fig. 50.1 State-diagram for a future society

progress" as an *ad seriatim* influence on its continuous development. The POST construction requires a distinctive method of computation, that is, its own rules or transfer function, relating inputs to outputs, referred to as its intrinsic computation.[1] For illustration, I have summarized one hypothetical computational approach, in the following note, where in its entirety, the model analogizes a pseudo-natural process.[2]

This is another way of saying that its operability depends on a transfer function, the relationship between inputs and outputs for each of the domains, and an interpreter that deals with or establishes the relationships between different variables and parameters unique to each of the domains. In the final analysis, what one domain communicates, the receiving domain must interpret.

The individual POST elements assume various states that emerge to create a future condition. The respective transfer functions of each of the components and system dynamics, in their totality, determine the condition or state the system assumes at any given time. Keep in mind that

[1] I am using the concept of transfer function (also known as system function or network function) in an engineering sense as the mathematical function which relates theoretically to a model's output resulting from each possible input.

[2] In one version, POST represents a four-state nonequilibrium Markov process. To generalize, the model designates each POST state by Ω, and the particular POST by the subscripts, w, x, y and z. The following state equations refer to the model depicted in Fig. 50.1:

A. $\mathbf{\Omega_{(w\text{-}z)}\ (t \le \dot{t})}$ = the state of a POST element during a time interval $\mathbf{T(0, \dot{t})}$.

B. $\mathbf{[(a\text{-}\ i)^{*}\Delta t]}$ = the rate of change (positive) during a time interval $\mathbf{T(t, \Delta t)}$.

C. $\mathbf{\Omega_{(w\text{-}z)}\ \{(t \le \dot{t}) + \Delta t\} = \Omega_{(w\text{-}z)}\ (t \le \dot{t}) + \Omega_{(w\text{-}z)}\ (t \le \dot{t})^{*}\ [(a\text{-}\ i)^{*}\Delta t] + \ldots}$ **[balance of inputs and outputs to the particular** $\mathbf{\Omega_{(w\text{-}z)}}$] represents the differential equation of the state of $\mathbf{\Omega_{(w\text{-}z)}}$.

as earlier shown, models can represent the real world, transforming and self-replicating what appears as meaningless symbols, to effectuate biological entities, and in this case memes, as well as phylogenic changes via genetic engineering, and technological transfer, which in totality alter the characteristics of social institutions and cultures. Daniel Dennett using artificial intelligence as a construct, claims that:

> Human consciousness is itself a huge complex of memes (or more exactly, meme-effects in brains) that can best be understood as the operation of a [serial] virtual machine implemented in the parallel architecture of a brain that was not designed for any such activities. The powers of this virtual machine vastly enhance the underlying powers of the organic hardware on which it runs, but at the same time many of its most curious features, and especially its limitations, can be explained as the byproducts of the kludges [i.e., ad hoc software repairs] that make possible this curious but effective reuse of an existing organ for novel purposes. [2]

As the POST output undergoes transformation we can anticipate one of the following occurs: (1) a disorganized social complexity evolves into one that is coherent and organized, (2) a social complexity assumes an indefinite state of oscillatory behavior, organized or disorganized, or (3) a social complexity that devolves into a self-extinguishing behavior. Note that these types of states are reflective of cellular automata.

Anders points out a notable lack of models that show "how an intelligence explosion could occur. … the most important and hardest problem to crack in the domain of singularity studies." I believe that, as discussed in the Introduction to this book, the intelligence explosion occurs due in part to genetic engineering. And in addition to, or in the alternative, as the Shulman and Bostrom study of IES suggests, in vitro embryo selection for increased intelligence traits could yield a 4.2 point IQ gain, and with increased iterations of 1 in 10, result in an 11.5 point gain. Over 5 generations, a 1 in 10 protocol would max out at a 65 point gain. Adding even 5 IQ points to any population would remarkably empower human capital.

The POST system has a large number of cross-linked sub-units encapsulating major elements implicated in a socio-technological system, including an intelligence explosion, as previously mentioned. Phylogenic or "P" refers to an evolutionary history and relationship among individuals or groups of organisms (e.g., species, or populations). These relationships generally have heritable traits, such as DNA sequences and a form, i.e., morphology under a model of gene evolution. The manner in which it used here focuses on phylogenetic mutability at the nucleotide or base pair level, which supports intelligence and memory. But, the ultimate goal is to state the relationship between the change at the biological level of an organism's genotype, as a consequence of genetic engineering, and the resultant physical manifestations in the organism.

Genomic organisms deliver one of two sets of function: genotypic (informational/inheritable instructions) and phenotypic (behavioral, physiology/morphology of the organism). As related to an organism, e.g., the brain, the entire assemblage of the genotypes may affect the phenotype *qua* neuronal cellular networks that carry out what is observed as psychological processes, and some, as we know, below the level of consciousness. For example, "P," the phylogenic component, stands for the probability that a single nucleotide polymorphism or SNP occurs, e.g., a C-G nucleotide pair switches to A-T due to some extraneous event. As pointed out, natural selection only operates on phenotypic expressions of the genotype. But circumstances exist, such as mutation, which can cause a phenotype to jump the gap as it were and cause a genotype to emerge.

Ontogeny is associated with the development of the larger organism of the human anatomy and relevant to our interests, where engrams form as conscious memory via mutability at the neuronal firing level.[3] Essentially, this is subject to genetic engineering for traits that express intelligence. The ontogenic element "O" plays a critical role in matters that require a physical presence in the world, which relates to intellectual and creative output.

Within the ontogenic element we include the molecular computer technologies that will eventually be implanted into the anatomy to improve various elements within the psychological g-factor complex: improved visuospatial processing (Gv), working memory (Gwm), and fluid intelligence (Gf).

Sociogenic "S" is associated with a social group, and is mutable at the idea or meme level, having a "memory" instantiated within its culture. A meme carries cultural ideas, symbols, or practices from mind to mind through writing, speech, gestures, rituals, advertising, and propaganda [3]. Memes are analogous to biological genes in that they replicate,

[3] In Hebbian engrams and cell assembly theory, neurons connect themselves to become engrams. When one cell repeatedly assists in firing another, the axon of the first cell develops synaptic knobs (or enlarges them if they already exist) in contact with the soma of the second cell. See, Hebb, D.O. (1949). *The Organization of Behavior*. New York: Wiley & Sons.

mutate, and respond to selective pressures. As ideas they penetrate institutions, industrial, commercial, political, religious, and social, using standard modes of communication, Internet, television, phone and all the ways in which the future of these technologies may evolve.

On this score, neurophysiologist Roger Sperry wrote: "Ideas cause ideas and help evolve new ideas. They interact with each other and with other mental forces in the same brain, in neighboring brains, and thanks to global communication, in far distant, foreign brains. And they also interact with the external surroundings to produce *in toto* a burstwise advance in evolution that is far beyond anything to hit the evolutionary scene yet, including the emergence of the living cell" [4]. The sociogenic component reflects culture, which obviously undergoes change at a faster rate than evolution, by many orders of magnitude.[4] Viral objects or patterns, such as memes, replicate themselves or convert other objects into copies of themselves. This has become a common way to describe how thoughts, information, and trends move into and through a human population. To this end, researchers can now characterize the frequency of meme patterns, such as peak value, percentage of lifetime elapsed before reaching the peak, meme lifetime, and ratio between rates of ascent and descent of the peak.

According to Susan Blackmore the transition from human to posthuman will occur in two stages: the first stage memetic, where the current technological culture fades into a society, loosely in control, but largely accepting and dependent on genetically engineered traits and in-the-body technology for well-being and enhanced abilities, security, and social mobility [5].[5] The second stage, moves toward a temetic state, where individual autonomy is lost and technology, as an autonomous agent takes over.[6] Max More supports Blackmore's premise:

> The transition from human to posthuman can be defined physically or memetically. Physically, we will have become posthuman only when we have made such fundamental and sweeping modifications to our inherited

[4]Danaylov, N. *ReWriting the Human Story: How Our Story Determines Our Future*, Chapter 2. https://www.singularityweblog.com/the-story-of-story/. 16 February 2019.

[5]Memetics as a concept arose in the 1990s to explore the transmission of memes as an evolutionary model.

[6]See, Blackmore, S. (2010). The Third Replicator. *New York Times*, August 22. https://opinionator.blogs.nytimes.com/2010/08/22/the-third-replicator/.

> genetics, physiology, neurophysiology and neurochemistry, that we can no longer be usefully classified with *Homo sapiens*. Memetically, we might expect posthumans to have a different motivational structure from humans, or at least the ability to make modifications if they choose. For example: transforming or controlling sexual orientation, intensity, and timing, or complete control over emotional responses through manipulation of neurochemistry. [6]

Technogenic "T" refers to technology changeable via induction of novel ideas coupled with the inventory of all technology that has preceded it. Susan Blackmore coined the word temes (short for technological memes), which represent "digital information stored, copied, varied and selected by machines." This incorporates the sum and substance of supercomputers to computers embedded in the anatomy. Currently, humans act as hosts through which information, via mediums of Internet, television, and phones, as obtained through conventional sources and social media is distributed under the influence of techno-memes.

Blackmore writes: "While (human) brains were having an advantage from being able to copy—lighting fires, keeping fires going, new techniques of hunting, these kinds of things—inevitably they were also copying, putting feathers in their hair, or wearing strange clothes, or painting their faces … is there a difference between the memes that we copy—the words we speak to each other, the gestures we copy, the human things—and all these technological things around us? … Let's call them techno-memes or temes" [7].

Blackmore's techno-memes parasitically bolt onto biological artifacts, such as the brain, serving to change thinking. Life controlled by techno-memes, might manifest in lives lived according to a deterministic, algorithmic logic. Over time, technological replication will surpass Neo-Darwinism, as the evolutionary paradigm for successive knowledge-based species.

According to Blackmore, a second level will be reached when the process moves the system to a temetic state, where individual autonomy is lost, and technology, as an autonomous agent weaves itself into the fabric of the human physiology, a physical transformation, where humans become part biological and part technological, including genetically engineered changes.

Let's digress into a thought experiment about what for some may represent techno-evolution, technological singularity or just Singularity. Anders identifies one kind of eventual phase transition, the kind supported by Valentin Turchin, Teilhard de Chardin, and Heylighen, where a shift to a new form of organization includes, by definition, *Homo futuro*, humanoid robot or both [8]. With that in mind, imagine that a phase transition occurs, which self-propels the system. An analogy is a way an engine, under the forces of an explosive gas undergoes a thermodynamic cycle, which uses the pressure difference given by a phase transition of the working fluid to move a piston connected to a crankshaft that turns a wheel. In our model, the engine is the societal system shown in Fig. 50.1, not one comprised of nuts, bolts, or even codons, and proteins, or semiconductors and software, but a product of humans, technology, and culture. Our hypothetical engine runs on the explosive force created from a mixture of rapidly circulating memes and techno-memes, the latter an agent capable of altering phylogenic traits.

Real-world systems are nearly all nonlinear dynamical systems, meaning their behavior undergoes a reaction that is not directly proportional to the forces that sets them in motion. If you pull a spring with an attached weight, it springs back into compression, oscillating up and down until it comes to rest. The physics of the operation is determinate, meaning that if we know enough about the spring's materials, the weight we apply, and the forces and distances we pull the weight, we can calculate how many times it will oscillate and for how long. The theoretical model I am proposing operates dynamically, but further exhibits nondeterministic behavior, meaning that even when we know the parameters and the input specification of the system, we cannot specify the output with certainty. It will behave ordered and disordered, oscillating stochastically, or according to a probabilistic pattern. One type of theoretical machine often used to consider these types of behaviors is a Markov processor, which receives input data and then outputs data dependent on transition probabilities inherent in the process. In Fig. 50.1, POST states are established via a Markov process.[7] It's important to note that

[7] "Markov chains, named after Andrey Markov, are mathematical systems that hop from one 'state' (a situation or set of values) to another. For example, if you made a Markov chain model of a baby's behavior, you might include 'playing,' 'eating,' 'sleeping,' and 'crying' as states, which together with other behaviors could form a 'state space': a list of all possible states. In addition, on top of the state space, a Markov chain tells you the probability of hopping, or 'transitioning,' from one state to any other state—e.g., the chance that a baby currently playing will fall asleep in the next five minutes without crying first." Powell, V., and Lehe, L., Markov Chains, Explained Visually, Tweet. http://setosa.io/ev/markov-chains/.

the Markov process is stochastic, which depends on a sequence of events in which the probability of each event depends only on the state attained in the previous one. However, as it pertains to POST particularly, a further consideration is that the probability itself may evolve i.e., change, dependent on the state of the system or its elements over time.

Briefly, the threshold between the disorder of system equilibrium and order is known as a phase transition. The conditions for a phase transition can be determined with the mathematical modeling of nonequilibrium dynamics. The initial amounts of reactants—in our case technology (rate of growth, processing power, speed, measure of AI sophistication), forces of human engagement, and cultural determinants (rate and quantity of meme transition)—determine the distance from equilibrium the system extends. The transition from order to disorder, in terms of a distance metric from equilibrium, is not usually a continuous function. The greater the initial collective concentrations of reactants, the further the system departs from equilibrium. As the initial concentration of reactants increase, an abrupt change in order occurs. This abrupt change is known as phase transition. At the phase transitions, fluctuations in macroscopic quantities increase as the system oscillates between a more ordered state (lower entropy) and the more disordered state (higher entropy). As the initial concentration increases further, the system settles into an ordered state in which fluctuations are again small.

In any case, if any of our terminal states were to actualize, society would experience an intense cultural change, although as we were to live through it, we would hardly acknowledge its influence [9].[8] As Crutchfield noted: "It is the observer or analyst who lends the teleological 'self' to processes which otherwise simply 'organize' according to the underlying dynamical constraints. Indeed, the detected patterns are often assumed implicitly by analysts via the statistics they select to confirm the patterns' existence in experimental data. The obvious consequence is that 'structure' goes unseen due to an observer's biases" [10].

[8] The Jesuit priest Pierre Teilhard de Chardin coined the term Omega Point, to describe an evolving universe which would eventually reaching a maximum level of complexity and consciousness.

Notes

1. Physica, D. (1994). Special Issue on the Proceedings of the Oji International Seminar Complex Systems—From Complex Dynamics to Artificial Reality Held, 5–9 April 1993. Numazu, Japan.
2. Dennett, D.C. (1993). *Consciousness Explained*. London: Penguin Books.
3. Blackmore, S. (2007). "Memes, Minds and Imagination." In: I. Roth, ed., *Imaginative Minds* (Proceedings of the British Academy). Oxford University Press, pp. 61–78.
4. Sperry, R. (1965). Mind, Brain and Humanist Values, the Information Philosopher. http://www.informationphilosopher.com/solutions/scientists/sperry/Mind_Brain_and_Humanist_Values.html (Last visited 05/02/19).
5. Heylighen, F., and Chielens, K. (2009). Meyers, B. (ed.). "Encyclopedia of Complexity and Systems Science: Evolution of Culture, Memetics." *Encyclopedia of Complexity and Systems Science* by Robert A. Meyers, Bibcode:2009ecss.book.....M. https://doi.org/10.1007/978-0-387-30440-3. ISBN 978-0-387-75888-6.
6. More, M. (1994). On Becoming Posthuman. http://www.maxmore.com/becoming.htm (Last visited 10/26/12).
7. Susan Blackmore: Memes and "temes."
8. Sandberg, A. (2010). An Overview of Models of Technological Singularity. https://doi.org/10.1002/9781118555927.ch36.
9. Magee, C.L., and Devezas, T.C. (2011). *Technological Forecasting & Social Change* 78, 1365–1378.
10. Ibid., see note 1 above.

PART VII

Policy and Ethics

CHAPTER 51

Wheels and Genes

Sarouk: Mensa, when you were a little girl, did you ever see a wheel?

Mensa: Oh yes, there were wheels every where you looked, but because they were so ever-present, no one took the time to really notice.

Sarouk: I have never seen one, what happened to them?

Mensa: As everything ascended into the cloud, we no longer needed cars, trucks; the few that were used for weapons of war, armored personal carries, eventually they too became obsolete.

Sarouk: And, did they just disintegrate. I mean could we find one if we knew where to dig?

Mensa: I suppose, they were simply steel cylinders with a rubber covering, so there might be steel relics, and even enough carbon left to analyze.

Sarouk: In the end just scraps of steel and rubber. Not even wheels at that point.

Mensa: That's right. Wheel was the word we applied to how we used those things. It only meant something in the scheme of human endeavors, made for human consumption. It had a function, to get us from point A to point B. Once it ceased to have any utility, it ceased having a function in society, so we scrapped it.

Sarouk: Do you think that because we are largely engineered that we in some way are like the wheel?

J. R. Carvalko Jr., *Conserving Humanity at the Dawn of Posthuman Technology*, https://doi.org/10.1007/978-3-030-26407-9_51

Mensa: You mean have we been modified to function in a certain way? To be a certain way according to someone's design or plan, or rather intent?

Sarouk: Yes, it seems all things that we produce, are produced for a reason, an inherent functionality at bottom.

Mensa: Unlike natural things,… trees, birds, oceans, they did not have a function. Humans were like that once, purely natural.

Sarouk: Once something has function it's been objectified.

Mensa: Yes, there were long periods when the powerful tried to use people as objects, we called it slavery. But, by and large, humans would fight to the death not to be objectified as such or in any manner for that matter.

Sarouk: Why was that, why would anyone die rather than live objectified?

Mensa: It had to do with autonomy and equality. We strove to be free from value judgments, who was better or worse, strong or weak, reliable and unreliable. We believed that at our core there was an essence about being human, one that we would defend to the death.

CHAPTER 52

Patenting the Transhuman

The question as to whether there is such a thing as divine right of kings is not settled in this book. It was found too difficult.—A Connecticut Yankee in King Arthur's Court—Mark Twain

When *Homo futuro* comes into existence, every novel genetically engineered innovation will represent yet another independent invention, each version having a provenance that traces back to a designer, or what we may characterize as an inventor. Regardless who that designer might be, in most cases the inventions will be owned by corporations. U.S. law does not categorically disallow patenting transhuman innovation, but the U.S. Supreme Court has weighed in on patentability when natural phenomena, laws of nature or the basic tools of scientific and technological work lie beyond the domain of patent protection.[1] There will

[1] 35 U.S.C. §101 sets out the subject matter that is patent-eligible. Section 101 reads as follows:

Whoever invents or discovers any new and useful process, machine, manufacture, or composition of matter, or any new and useful improvement thereof, may obtain a patent therefor, subject to the conditions and requirements of this title. The U.S. patent office, publishes its administration of the patentability of inventions through the Manual of Patent Examining Procedure (MPEP). MPEP 2106.04 deals with whether a subject is patent eligible under the U.S. statute: 35 USC 101. Determining that a claim falls within patentable subject matter recited in 35 U.S.C. 101 (i.e., process, machine, manufacture, or composition of matter) does not end the eligibility analysis, because claims directed to nothing more than abstract ideas (such as mathematical algorithms), natural phenomena, and laws of nature are not eligible for patent protection. Diamond v. Diehr, 450 U.S. 175, 185,

J. R. Carvalko Jr., *Conserving Humanity at the Dawn of Posthuman Technology*, https://doi.org/10.1007/978-3-030-26407-9_52

also continue to be differences around the world, where for example, the European Court of Justice may circumscribe some types of patents and the U.S. Supreme Court, other types, each based on similar subject matter.[2]

Europe codifies what is contrary to public order and morality regarding patentability. Article 6(2)(c) of Directive 98/44/CE deals with the legal notion of "human embryo" in patent law. This article is known as the ethical clause of the Directive, which at paragraph 1, establishes that "Inventions shall be considered unpatentable where their commercial exploitation would be contrary to ordre public or morality." Paragraph 2, then, enumerates a list of inventions that, cannot be patented if used for "human embryos for industrial or commercial purposes."

In 2004, the U.S. Congress failed to pass bills banning human cloning, but it did pass the Weldon Amendment, which made it illegal to grant a U.S. patent on human organisms, including fetuses and embryos.[3]

Private institutions, such as The Hastings Center in the U.S. or Nuffield Council on Bioethics in the UK have addressed patenting genomic technology [1]. In 2002, Nuffield Council issued its position

209 USPQ 1, 7 (1981). Alice Corp. Pty. Ltd. v. CLS Bank Int'l, 134 S. Ct. 2347, 2354, 110 USPQ2d 1976, 1980 (2014) (citing Association for Molecular Pathology v. Myriad Genetics, Inc., 133 S. Ct. 2107, 2116, 106 USPQ2d 1972, 1979 (2013)); Diamond v. Chakrabarty, 447 U.S. 303, 309, 206 USPQ 193, 197 (1980); Parker v. Flook, 437 U.S. 584, 589, 198 USPQ 193, 197 (1978). Supreme Court's decisions, regarding isolated DNA for example in Myriad, were deemed novel or newly discovered, but nonetheless were considered by the Court to be judicial exceptions because they were "'basic tools of scientific and technological work' that lie beyond the domain of patent protection." Myriad, 133 S. Ct. at 2112, 2116, 106 USPQ2d at 1976, 1978 (noting that Myriad discovered the BRCA1 and BRCA1 genes.

[2] In October 2011, the European Court of Justice issued a landmark decision on the patentability of biotech inventions in the case "Oliver Brüstle vs Greenpeace." In a nutshell, the issue was whether a patent should be granted for the neural precursor cells (stem cells) and the processes for their production from embryonic stem cells and their use for therapy in diseases such as Parkinson's, Huntington's, and Alzheimer's. Europe's top court decision effectively banned patenting any stem cell process that involved destroying a human embryo.

[3] See, Congress Bans Patents on Human Embryos, Washington (February 4, 2004). https://www.nrlc.org/archive/news/2004/NRL02/congress_bans_patents_on_human_e.htm.

in "The Ethics of Patenting DNA," outlining the ethical concerns it had with patenting human DNA [2]. The document focused on the special nature of human genomes as well as their mutations or variations, and for numerous reasons tended to oppose patenting, except within the confines of strict limits.

Even before Nuffield's cautionary vision, much water already had passed under the proverbial bridge. On the patent front, depending on interests, we find ourselves in a rapids occupied by competitors, institutions and individual researchers. What an invention vis-à-vis a patent represents is often misunderstood, so let's take a brief look at it through both a legal and ontological lens.

In theory intellectual property bifurcates into two categories. The first category (ontological) delineates the existence and the properties that attach to classes of material object; for example, a doorstop that comprises: a shaft, rubber bumper, and a screw, the kinds of things with which inventions and their protection, patents typically cover. The second category (jurisprudential) comprises a legal system concerned with classes of rights, powers, and interests that attach to the material object.[4] But, because this second category deals with objects in a juridical way, it implies that social aims such as ethics are part of the blend.[5] I will address the social element in a subsequent chapter.

Over the past 40 years, thousands of patents have issued on inventions that are incontestably beneficial to health and general well-being.[6] But, not all attempts at patenting products of gene engineering succeed, such as Dolly the Sheep. In 2014, the Federal Circuit for the Court of Appeals, denied patentability for Dolly because the cloned sheep was "an

[4]In the U.S. once a patent is issued, it protects the invention in the U.S. and its possessions for a term of 20 years from the date of filing. Anyone can make, use or sell the invention in countries where it's not patented.

[5]Martin P. Golding referring to H.L.A. Hart notes: "…Hart points our various facts about human nature that make necessary some of the rules of social morality and law: men are vulnerable and liable to harm; they are approximately equal in intelligence and physical abilities; they are not completely selfish but have limited good will toward others; and they are limited in their powers of foresight and self control…" Golding, M.P. (1975). *Philosophy of Law*. Prentice-Hall.

[6]The identification, alteration and cloning of genes that produce proteins has led to new medicines and genetic mutations that cause disease has been employed for instruments and diagnostic. Patents that assert property rights over DNA sequences have been granted in both these areas.

exact replica of another sheep" and thus "does not possess 'markedly different characteristics than any [farm animals] found in nature'" [3].[7] In this decision we read into the proposition that one can not claim a subject, if the claims read on something that is already found in nature [4].

Genetically modified organisms, as mentioned earlier, include plants, mammals and microorganisms, such as bacteria. In the U.S., most of these products are patentable, if they are deemed novel and nonobvious. Europe, however looks askance at GMO products. But, in the U.S. the food and chemical industry has fought vigorously to keep GMO products on grocer's shelves. In fact, agricultural technology progresses along the traditional lines of improving the quality of produce, but biotechnology has now turned plants into hosts to manufacture mammalian proteins. In the trade, such plants are referred to as plant bioreactors. Bacteria with genes from foreign sources are routinely integrated into plant genomes to replicate and propagate. Developments in synthetic biology only accelerate this type of progress.

As mentioned in an earlier chapter, the first significant case involving the patenting of life forms, went to the Supreme court, as *Diamond v. Chakrabarty*, which dealt with a bacterium that had markedly different biological features from anything found in nature.[8]

Chief Justice Warren Burger, writing for a majority of the Court, called attention to the concerns of eminent scientists who opined that "genetic research and related technological developments may spread pollution and disease, that it may result in a loss of genetic diversity, and that its practice may tend to depreciate the value of human life" [5]. In any case, in 1980, the decision handed down in Chakrabarty was a watershed moment in history, setting the stage for a revolution in science, making it profitable to develop and monopolize for a period of time, an unending stream of bioengineered products.

[7] In 1996, removing a nucleus of an unmatured sheep egg, and then using a second sheep nucleus having a genetic blueprint, was fused that into the enucleated egg cell, which developed into an embryo, implanted into a surrogate producing a clone. See, Dolly the Sheep is Cloned, BBC (February 22, 1997). http://news.bbc.co.uk/onthisday/hi/dates/stories/february/22/newsid_4245000/4245877.stm[http://perma.cc/2RTX-RYG6].

[8] *Diamond v. Chakrabarty*, 447 U.S. 303, 206 USPQ 193 (1980), Court held that microorganisms produced by genetic engineering are not excluded from patent protection in a case involving the creation of an organism not found in nature.

As also mentioned earlier, in 1988, U.S. Patent 4,736,866, issued for Transgenic Non-human Mammals, a mouse having an inherent susceptibility to certain forms of cancer. The '866 patent covered a eukaryotic animal whose germ and somatic cells contained an activated oncogene sequence introduced into the animal, or an ancestor of the animal, at an embryonic stage. At the time, the '866 patent, again raised ethical issues about whether patents should be granted for genetically engineered animals. Never before had the patent office allowed patentability for a complete, living mammal. The OncoMouse® development was a turning point, as to what kinds of genetic alternations would be permitted under U.S. patent law, Since the late 1980s, more than 17,000 patents have issued, referring to transgenic nonhuman mammals.

But, not to exaggerate the point, discovering a single new gene or the protein that subsequently could be produced in a "mouse" may not suffice for patentability, as it must be shown that the gene or protein has a use in some medical treatment or research procedure, e.g., target for a drug to treat a particular disease. These type of guard rails are under constant pressure from researchers and vast commercial enterprises. Thus, the idea that anything remains unchanged in this field simply is not borne out by history.

In March 1998, the patent office allowed Thomson, et al., a patent covering human embryonic pluripotent stem cell lines, meaning the lines are capable of differentiating along each of the three germ layers, ectoderm (skin, nerves, brain), the mesoderm (bone, muscle), and the endoderm (lungs, digestive system) of cells in the embryo, as well as producing within the germ line (sperm and eggs).[9] Although other embryonic stem cell lines had been patented earlier, this was a breakthrough of enormous magnitude, because it dealt with isolating and maintaining in culture human stem cells. At the time this raised and continues to raise controversy, because of the destruction of embryos, which occurs when stem cells are derived. In other instances, opposition only relates

[9]U.S. Patent Nos. 5,843,780 (1998), 6,200,806 (2001), and 7,029,913 (2006). The '780 patent claimed: 1. A replicating in vitro cell culture of human embryonic stem cells comprising cells which (i) are capable of proliferation in in vitro culture for over one year without the application of exogenous leukemia inhibitory factor, (ii) maintain a karyotype in which the chromosomes are euploid through prolonged culture, (iii) maintain the potential to differentiate to derivatives of endoderm, mesoderm, and ectoderm tissues throughout the culture, and (iv) are inhibited from differentiation when cultured on a fibroblast feeder layer.

to whether we are dealing with pluripotent stem cells, which can develop into specialized cells or totipotent stem cells, which can give rise to both the placenta and the embryo [6].[10]

In March 2012, the patent office issued Patent No. 10,227,625, a method for engineering and utilizing large DNA vectors to target, via homologous recombination that modifies, in any desirable fashion, endogenous genes and chromosomal loci in eukaryotic cells. The field also encompasses the use of these cells to generate organisms bearing the genetic modification, the organisms, themselves.

In the U.S., a landmark, 2013, Supreme Court case, captioned Association for Molecular Pathology v. Myriad Genetics, Inc., again challenged the validity of gene patents, which in this instance covered isolated DNA sequences, cDNA a synthesized gene, and related methods to diagnose the susceptibility to breast and ovarian cancers [7]. The litigants testing the legality of the patent were women who claimed to have been harmed by Myriad Genetics, the company that sought to monopolize the means for testing the BRCA1 and BRCA2 genes, which correlate with a heightened risk for cancer. It should be noted that the patent income contributed substantially to Myriad's revenues, which in 2011 was approximately $US 400 million. The company would not permit competition, so that women had to pay whatever Myriad charged, and thereafter were barred from seeking second opinions. Before the case was heard in the Supreme Court, it was heard in 2012, by the U.S. Court of Appeals for the Federal Circuit (CAFC), which ruled in that DNA sequences extracted from the body were patentable. Although the case involved testing for breast cancer, one legal issue devolved into what was meant by "information," on the theory that DNA represented a code and therefore was not protectable. Skirting the hard definitional issues of what is meant by such things as "information" the CAFC ruled in Myriad's favor, taking into consideration the potential impact that invalidating gene patenting would have on the life sciences industry.

The case bounced between the appeals court and the Supreme Court on issues that were not dispositive, but finally the U.S. Supreme Court

[10]The first stem cell patents were directed to hematopoietic stem cells (e.g., U.S. Patent Nos. 5,436,151 [1995] and 5,670,147), fetal/neonatal cells (e.g., U.S. Pat. 5,004,681), and mesenchymal cells (e.g., U.S. Pat. 5,827,740). Embryonic stem cells were first patented from animals (e.g., birds [U.S. Pat. 5,340,740 and U.S. Pat. 5,656,479] and mice [U.S. Pat. 5,453,357 and]).

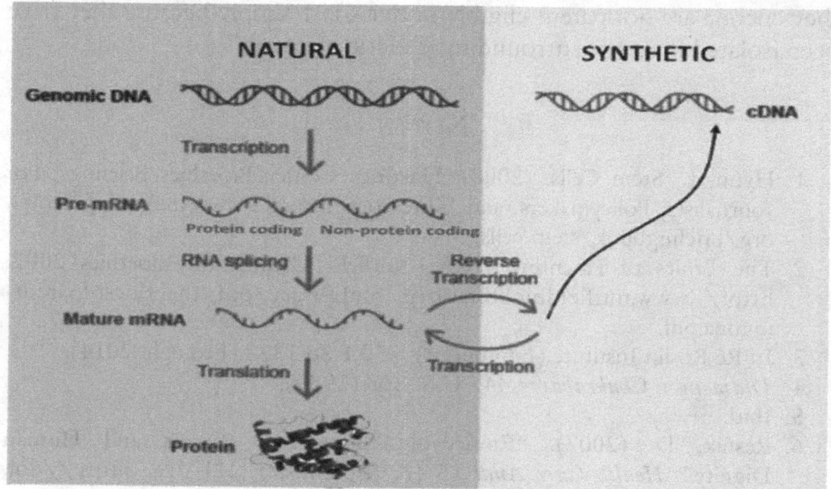

Fig. 52.1 Natural/synthetic DNA

heard arguments from all sides, and in 2013, Justice Clarence Thomas delivered a majority opinion holding: "A naturally occurring DNA segment is a product of nature and not patent eligible merely because it has been isolated, but cDNA is patent eligible, because it is not naturally occurring" (See, Fig. 52.1).[11]

Allowing cDNA to be patented unfortunately does not prevent utilizing it in a genetic engineered operation, and depending on the particular cDNA, claiming patent protection. Thus the patent allowed a monopolization on any process utilizing that particular embodiment of the cDNA gene. The decision went on to indicate that: "Nor do we consider the patentability of DNA in which the order of the naturally occurring nucleotides has been altered. Scientific alteration of the genetic code presents a different inquiry, and we express no opinion about the application of §101 to such endeavors. We merely hold that genes and the information

[11] In genetics, complementary DNA (cDNA) is DNA synthesized from a single-stranded RNA (e.g., mRNA) template in a reaction catalyzed by the enzyme reverse transcriptase. cDNA may be used to clone eukaryotic genes in prokaryotes.

they encode are not patent eligible under §101 simply because they have been isolated from the surrounding genetic material."[12]

Notes

1. Hyun, I., Stem Cells. (2008). Hastings Center Bioethics Briefings, For Journalists, Policymakers, and Educators. https://www.thehastingscenter.org/briefingbook/stem-cells/.
2. The Ethics of Patenting DNA, Nuffield Council on Bioethics 2002. http://www.nuffieldbioethics.org/fileLibrary/pdf/theethicsofpatentingdna.pdf.
3. In Re Roslin Institute (Edinburgh), 750 F.3d 1333 (Fed. Cir. 2014).
4. *Diamond v. Chakrabarty*, 447 U.S. 303 (1980).
5. Ibid.
6. Resnik, D. (2007). "Embryonic Stem Cell Patents and Human Dignity." *Health Care Anal* 15 (3, September): 211–222. https://doi.org/10.1007/s10728-007-0045-9.
7. Association for Molecular Pathology v. Myriad Genetics, Inc., 133 S. Ct. 2107 (2013), 569 U.S. 576 (2013).

[12] 35 U.S.C. §101 refers to the section of the patent statute that states the subject matter eligible for patenting: any new and useful process, machine, manufacture, or composition of matter, or any new and useful improvement thereof.

CHAPTER 53

Virtuous Deliberations

***Can there be anything more splendid than to put the whole world into commotion by a few arguments?*—Voltaire**

The long history of technology began some 3.3 million years ago, referred to as the Stone Age, an epoch divided into periods: Paleolithic (2.5 Mya), Mesolithic (2.5 Mya-10,700 ya), and Neolithic (12,200 ya-4000 ya). The Bronze Age followed the Neolithic Period signaling the end of prehistory with the introduction of written records and advanced metalworking, such as smelting copper and tin and combining them into bronze products. As emphasized throughout, we see that the constant push for innovation serves as one among many strategies humans excercise to survive the milleniums of planetary transformation. But now technology has gone well beyond its purpose to assist in the species' basic survival, engaging in projects that foster increased living standards, better health, economic enrichment of peoples and institutions, and that support political power via bot propaganda technology and the latest weapon systems. Yet, the compounding advancement of our technological proficiency also jeopardizes the survival of our species.

In a world driven by political and economic interests, government policy often appears rule-based or consequence-based, either which discounts the ethical consideration. And, although normative issues go hand-in-hand with technology, it's less than clear how *Homo futuro* will affect the world intellectually and spiritually, i.e., change the manner in which we consider how things ought to be, how to value them, which

J. R. Carvalko Jr., *Conserving Humanity at the Dawn of Posthuman Technology*, https://doi.org/10.1007/978-3-030-26407-9_53

things will be deemed good or bad, and which actions will be deemed right or wrong. Perhaps we will find a kinder, more enlightened world. But, no time in history has been free from the exploitation of human beings.[1] And on a broader scale, consider the rise of physics in the twentieth century, which led to the expansion of knowledge-about how the Universe works, an age of unparalleled technological progress, which also brought us the atomic bomb and chemical warfare. In a *Homo futuro* world, superintelligence will undoubtedly be put to the task of inventing all manner of matter and apparatuses, and likely will advance along the lines that satisfy a particular need, one which may diverge from what modern humans consider utilitarian or agreeably aesthetic.

To mollify the effects of transhumans behaving contrary to civilized notions of conduct, we may be moved to imprint our respective cultural values (e.g., equality under the law) onto *Homo futuro*, and likely for a significant period as they assimilate. But, time will come when *Homo futuro* is recognizably dissimilar, for example, a decidedly quicker thinking species. At that juncture, it will likely exercise powers to achieve its own particular social, political, and economic ends. As Wendell Wallach, writes "technological development is at the risk of becoming a juggernaut beyond human control" [1].

No consensus has been forthcoming among scientists regarding the imposition of prohibitions or moratoriums to germline editing. That said, in March 2019, eighteen scientists and ethicists from seven countries called for an international governance framework, including a global moratorium on changing heritable DNA (in sperm, eggs or embryos) to make genetically modified children [2]. They went on to propose the "establishment of an international framework in which nations, while retaining the right to make their own decisions, voluntarily commit to not approve any use of clinical germline editing unless certain conditions are met."

The New England Journal of Medicine, published two articles, both strongly in favor of germline editing. This position supports what's occurring in the field, where Dieter Egli at Columbia University experiments with gene-editing human embryos "for research purposes" and Werner Neuhausser of Harvard's Stem Cell Institute announced plans

[1] The rationalization to use weapons systems on civilization populations makes the point. Genetic engineering has the potential to ameliorate suffering, but what rationalizations will accompany its use to further a political or economic agenda?

to edit sperm cell DNA. In the UK, Robin Lovell-Badge of London's Francis Crick Institute published a personal perspective, calling for public debate, leaning in favor of expanding the current British limits on gene intervention.

In 2017, the U.S. National Academy of Medicine and the National Academy of Sciences published an exhaustive report on human genome editing, but in light of He Janaki' germline editing in 2018, have also called for an international commission on the most controversial use of that technology [3]. The report focused on principles of "do no harm," autonomy, and the application of responsible science to gene-editing technologies.

In November 2018, the U.S. National Academy of Sciences and U.S. National Academy of Medicine, the Royal Society of the UK, and the Academy of Sciences of Hong Kong convened the Second International Summit on Human Genome Editing at the University of Hong Kong.[2] The summit brought together more than 500 researchers, ethicists, policymakers, representatives from scientific and medical academies, patient group representatives, and others from around the world to consider the scientific, medical, and ethical requirements relative to germline editing.[3] Following the summit, the International Commission on the Clinical Use of Human Germline Genome Editing (ICC) was created by the U.S. National Academy of Medicine, the U.S. National Academy of Sciences, and the Royal Society of the U.K., with the participation of science and medical academies around the world. The aim of the ICC is to identify a number of scientific, medical, and ethical requirements that should be deliberated, and could inform the development of a potential pathway from research to clinical use, and whether the society concludes that a particular heritable human genome editing application is acceptable, according to their standards and practices.[4] The first meeting of the

[2] https://www.nap.edu/catalog/25343/second-international-summit-on-human-genome-editing-continuing-the-global-discussion.

[3] A November 27, 2018, article in Nature, International Journal of Science, How the genome-edited babies revelation will affect research, gives a summary of what was concerning to many of the attendees. See, https://www.nature.com/articles/d41586-018-07559-8.

[4] In early 2019, acknowledging that clearer guidelines are needed, a commission was being organized by the U.S. National Academy of Medicine to deal specifically with gene editing projects involving human embryos.

commission was held at the National Academy of Sciences in Washington, DC, August, 2019.[5]

By the end of 2018, forty countries had banned germline engineering, which included most of Europe, with the exception of the UK. Notably some of the largest nations did not, notably China, India, U.S., Russia, and Brazil. Gene editing is banned in at least 40 countries. For example, the UK allows the creation and use genome-edited human embryos, sperm, or eggs in research, under strict licensing conditions, but it's illegal to use them in assisted reproduction. Recognize that historically, scientists intent on conducting unethical research will find ways to justify their work, and further what one country does or does not stop another country from going its own way. Currently, it is underway in countries such as the U.S.—though not with federal funds, because it's against the law. But it can be practiced with private and state funding. Japan appears poised to allow gene editing set to allow the gene editing of early-stage embryos. For example, guidelines released in 2018 would restrict the manipulation of human embryos for reproduction, but they are not legally binding. However positive sentiment prevails for allowing the repair of genetic mutations that cause inherited diseases.

A multitude of influences work their will on science and technology policy. Inevitably, economic interests, sway public policy advantage. In the U.S. we see this in policies regarding global warming, universal health care, gun legislation and taxation. As between nations, the policies may well differ, which concerning *Homo futuro*, because if even one country were to openly practice germline editing, without regulation or restriction, the entire world would be affected. Clearly, political, cultural, and religious ideologies will drive policy, as it has in the ongoing debates over abortion rights, cloning and embryonic stem cell use throughout the U.S. and Europe.[6] Legislators and court decisions have weighed in

[5] International Commission Launched on Clinical Use of Heritable Human Genome Editing. See, http://www.nationalacademies.org/gene-editing/international-commission/index.htm (last visited 05/22/2019).

[6] In 1994, over thirty organizations representing indigenous peoples passed formal declarations against patents on life forms and indigenous knowledge. In 2003, the U.S. Congress passed an amendment which made it illegal for the U.S. Patent and Trademark Office grant patents on human organisms, including fetuses and embryos. European Parliament and Council. European Directive on the Legal Protection of Biotechnological Inventions. 1998. Accessed 5 June 2006. Available: http://europa.eu/eur-lex/pri/en/oj/dat/1998/l_213/l_21319980730en00130021.pdf.

on patenting human organisms, including totipotent embryonic stem cells, embryos, and fetuses, but no government agency or legislation currently directly regulates the production of transhumans, cyborgs, and humanoid robots.

To deal objectively and effectively with genetic engineering vis-a-vis the production of transhumans, requires an appreciation that the norms, conventions, and limits of human behavior will change forever. Germline engineering affects the entire species. However even somatic cell genome editing, albeit contained with the anatomy of a single individual and not passed on to future generations, if done to effectuate increases in IQ, would nonetheless be significant if widely practiced. As for now, few policymakers well understand, either at the theoretical or the policy level, the potential impact of the new technologies. After all, politicians and business executives are not philosophers, scientists, mathematicians, or ethicists, and it will take a fair amount of effort before we see movement in the policymaking sector.

Notes

1. Wallach, W. (2015). *A Dangerous Master*. Basis Books.
2. Lander, E., et al., Adopt a moratorium on heritable genome editing, 14 March 2019, Vol. 567 Nature 165.
3. National Academies of Sciences, Engineering, and Medicine. (2017). *Human Genome Editing: Science, Ethics, and Governance*. Washington, DC: The National Academies Press. https://doi.org/10.17226/24623.

CHAPTER 54

The Ethical Claim

***In short, to assert that a certain line of conduct is, at a given time, absolutely right or obligatory, is obviously to assert that more good or less evil will exist in the world, if it be adopted than if anything else be done instead.*—G. E. Moore**

I may have house plans and the materials needed to erect a fine structure, but the plans say nothing about whom will someday live inside or whether it will ever be elevated to that hallowed place called "home." Analogously, a genome lays out a plan in a four-letter alphabet that composes three letter words into sentences that confer life. Transfixed by this marvel, its not surprising that we might overlook the deeper meaning of how in profound ways it reveals who we are. And, the idea of monopolizing elements of the human anatomy, for example through a system of patenting, ignores this deeper significance. We have waged fierce debates about whether animals should be the subject of patents, and now we face whether forms of humans may become subjects of patents on many of the same grounds.

The ideal enshrined in the 1948 Universal Declaration on Human Rights, acknowledged the "inherent dignity" and "equal and inalienable rights of all members of the human family."[1] In this preamble, dignity implies autonomy and the right to self-determination, both which put at risk personhood itself if the idea of patenting were to entangle the transhuman in any significant way. Perversely, because novel synthetic genes and

[1] UNESCO's Universal Declaration on the Human Genome and Human Rights (1997) refers to the human genome as the 'common heritage of humanity.'

J. R. Carvalko Jr., *Conserving Humanity at the Dawn of Posthuman Technology*, https://doi.org/10.1007/978-3-030-26407-9_54

cDNA genes, particularly are patentable, there does not appear any logical reason why when engineered into the human anatomy, the imprimatur of patentability would somehow collapse.[2] In fact, it is likely the case that the combination of the novel gene and the anatomical element to which it attaches can form the basis for a combined patent claim. As indicated earlier, U.S. law bans patents on human organisms, fetuses and embryos, which likely includes totipotent stem cells separate and apart from their constituents to which they are attached, e.g., as compositions of matter *products*, but nothing categorically restricts patenting genetic engineered process or products, which include them as elements in a claim.[3] As mentioned earlier, in the Association for Molecular Pathology v. Myriad Genetics, Inc., the U.S. Supreme Court held, in 2013, that human genes cannot be patented, because DNA is a "product of nature." It begs the question as to whether core elements of our genetic makeup are patentable, because the Court found that cDNA was patentable, and therefore if it were essential to the novelty in an inventive product or process, there otherwise would appear no restriction in allowinig the patent to move forward. This type of patent would monopolize making a particular genetic engineered modification, for example, one made for to carry out some medical therapy.[4]

Margaret Farley writes: "At the heart of tradition, however, is a conviction that creation is itself revelatory, and knowledge of the requirements of respect for created beings is accessible at least in part to human reason. This is what is at stake in the tradition's understanding of natural law… the concrete reality of the world around us, the basic needs and possibilities of human persons in relation to one another, and to the world as a whole" [1].

[2] Synthetic DNA preparations are eligible for patents because their purified state is different from the naturally occurring compound, whereas cDNA is known to be synthesized, or manufactured from an mRNA or messenger RNA template.

[3] Legislation that specifically addresses this is found in the Leahy–Smith America Invents Act (AIA), Pub. L. 112–29, sec. 33(a), 125 Stat. 284, which states: Notwithstanding any other provision of law, no patent may issue on a claim directed to or encompassing a human organism.

[4] For example, U.S. patent, 10,266,582 is typical of hundreds of cDNA patents, in this case one that couples to a human receptor. It claims: cDNA comprising a nucleic acid sequence encoding at least a portion of a **human** G protein-coupled receptor 156 (GPR156) protein, wherein the portion comprises an aspartic acid at a position corresponding to position 533 according to SEQ ID NO:4, and wherein the cDNA comprises at least 30 contiguous nucleotides of the nucleotide sequence encoding the portion of the **human** GPR156.

Immanuel Kant observed that science stays within the empirical domain, and matters such as morality and virtue are not experiences, as they deal with purpose, which in the application of science and technology, we do not generally consider. Alasdair MacIntyre writes:

> Abstract these conceptions of truth and reality from [their] teleological [final causes, goals, purposes, and aims] framework, and you will thereby deprive them of the only context by reference to which they can be made fully intelligible and rationally defensible. Yet the widespread rejection of Aristotelian teleology and of a whole family of cognate notions in the sixteenth and seventeenth centuries resulted in just such a deprivation. In consequence, conceptions of truth and rationality became, as it were, free-floating. [2]

Institutions, commercial, academic, and government protect interests, such as technology, through the application of monopolistic devices, when given the opportunity. Patents provide legitimate economic reasons for granting monopolies.[5] The main issue has been the advisability of allowing a monopoly on a genetic engineering technology, which in turn claims a relationship to the potential creation of a transhuman.[6]

On one dimension patents granted for transhumans might be considered prima facie "good," when, for instance, the monopoly serves to incentivize inventors to advance medical science, similar to the rationale allowing the patenting of drugs. Although this practice itself has a storied history of producing systems of inequitable distribution. Nevertheless, a novel, nonobvious advancement to a technology satisfies the basic patentability requirement for purposes of U.S. patent law. The invention may be one thing, but in a second dimension, there is the matter of an essential feature of the invention, which would actually embrace a form of life. This second consideration takes the matter beyond utility under patent law theory, it moves the matter into moral significance. Patent

[5]As a reward for inventions and to encourage their disclosure, the United States offers a seventeen-year monopoly to an inventor who refrains from keeping his invention a trade secret. Universal Oil Prods. Co. v. Globe Oil & Ref. Co., 322 U.S. 471, 484, 64 S.Ct. 1110.

[6]U.S. law has not considered this, but it's likely it would be prohibited by most European nations, which have a "morality clause," which allow respective governments to refuse to issue patents on moral grounds, European Patent Convention policies reflect this as well.

monopolies are obviously not prima facie "bad," but how we use and exploit forms of life (i.e., ownership) may be. To complete the thought we move to a third level to consider whether an invention as used in its intended application, advances the community's traditions and practices, i.e., its deeper humanistic values.

For example, genetically engineered embryos are unqualified, a form of life, and thus owed respect and dignity in accordance with deontological teachings.[7] It seems that ownership of these beings, in the property sense, at first blush falls outside the boundaries of respect and dignity. I may own a bird, and use it for my amusement, companionship or as a commodity for sale, but ownership of human embodiments seems counter to morality. It is not a matter of whether this third dimension trumps the lower dimensions, for the other considerations remain extant in our reckoning. I propose that the third dimension defines something about who we are.

What I suggest here may lack originality, except in the manner I propose for getting to an objective with definitive elements. When a new subject matter of invention comes before the patent office, it often winds up in litigation, where the debate centers on utility. Such has been the case of countless new technologies: telegraph, software, life forms, gene technology, and business methods, to name but a few. In deliberating whether these inventions serve the utility requirement under patent law, the larger teleological interest may not be vetted. Once, the law has been established to allow or not allow a technology within the scope of patentable subject matter, the inquiry ceases to be a factor in the patent allowability analysis. The patent office and the courts are prone to set a course and even though it may reflect a conservative directional change, overtime a technological change creeps in and we move from patenting plants, to bacteria, to animal cells, to human cells, to the genetically engineered *process* by which we create transgenic species, and perhaps a posthuman (as bizarre as this may sound).

In prosecuting a patent, the participants hardly consider teleological considerations that ought to accompany each incremental technological change. One marker that might be used to signal a deeper policy consideration for patentability is when inventions are ends in themselves,

[7] An action can only be good if the principle behind it is a duty to the moral law. Central to Kant's construction of the moral law is that it applies to everyone, regardless of their interests or desires.

having the potential to directly implicate universal human values. We will return to this issue later.

Identity requires us to focus on distinguishing a specific instance of a particular class from other instances of the class by identifying certain characteristic features. I refer to these as essential features, for if they are missing, identity can not be established. Having seen the elements a, b, and c on a shaft, the inventor exclaims: "That is my patented door stop!" This illustrates a simple example of an identity statement. But, underlying this statement a unity stands for the elements or parts that serve to make the identification, the patent claims unique and thus a jurisprudential object.

When formulating an intellectual property object, one needs to capture the underlying structural and functional components. Some of these are extraneous and others serve as the essential features of an object. The analogy is "I shall remain who I am throughout my life, regardless my age or appearance." Those who know me, identify me by name because they can perceive me in some invariant way. What essential feature about me does not change, so that I remain who I am? Clearly, certain physical attributes incrementally change, but my personality and other essential features combine to give uniqueness to my identity.

In an analogous way essential features of a patented invention, circumscribed by claims, does not change. The law grants favored protection to this invariant quality of an intellectual property. I propose that essential features go beyond physical structural features. As in life, it would seem plausible to consider features of an object of our ingenuity, as both part of our world, but at the same time being a part of something else, separate. It projects us into a separate dimension where the domain of the telos becomes distinctive. For instance, neither commercial enterprise nor patent law reflects on objective ontological features of life, such as conception, gestation, birthing, aging, and dying or what MacIntyre refers to as "internal good."

As we consider the ethical soundness of patenting life forms or its processes, we need to accept three premises. First, we must accept as self-evident that life forms are biological. Biological forms divide into those that have neurological systems and those that do not. Those that do not have neurological systems know nothing about assigned function. Said in another way, nature produces no facts beyond its intrinsic existence. The second premise is that only life forms having neurological systems can assign functionality to natural and constructed things. Third, once a life form bearing a neurological system defines, explicitly or implicitly, the function of something through how its to be used, whether a natural phenomenon or artifact, such as another life form, it introduces a new set of facts that relate to teleology. In a stubborn way, nature seems to impose a greater responsibility for those having knowledge of teleological considerations, those richer in synaptic connection, as we humans demonstrate. Biological forms such as the stem cell or the embryo would seem helplessly at the mercy of species higher in the biological phylum for a determination of its fate, its dignity and respect.

I use the broad notion of neurological systems, because not only human and higher animals seem to have the capacity to appreciate when something works or does not work, or operates more or less efficiently. One might argue that the seagull "appreciates" a hard macadam road more than a softer dirt road. It senses the concept of "better" or "worse," "reliable," or "unreliable." How far down the phylum do we travel? What about bees, ants, and spiders all builders, do they "appreciate" when something has "malfunctioned?" These are objective qualities. But at the risk of ignoring G.E. Moore's concerns in Principia Ethica (1903), that assessing what's good or bad may be indefinable, I nonetheless believe that at times there exists irrefutable criteria for what is "good," or "bad."

Spinoza observed that we learn good or bad by our experience with them in the first instance. If I invent an ashtray, I need to show to the satisfaction of the patent office that it is novel, nonobvious and useful. The last requirement would require that I prove to the satisfaction of the patent office that the ashtray usefully contains cigarettes and their ashen by-products. This does not set a value, good or bad about the product. Patent law generally disregards purpose, although in bio-patents, as mentioned, they are given favorable treatment if they are for research or discovering and curing disease. In the course of history, certain items were once deemed ipso facto not patentable (e.g., gambling machines, atomic bombs), for public policy reasons.

In furthering a policy to make the environment safe from fire, the ashtray can be regarded in the first instance as prima facie "good." Most people should regard this as an objective statement of value.

However, something inherent in an ashtray directly implicates a dangerous habit. In the second dimension of analysis, we find that the ashtray serves to facilitate smoking, which causes cancer. Cancer destroys the quality of life and life itself. Cigarette smoking, being destructive of life, must be bad. The ashtray in this calculus would be as an objective statement of worth determined as serving a negative purpose.

Let me explain how I employ the phrase "objective statement of worth." Cigarette smoking, being destructive of human life is "bad" because it unites the factual judgment that cigarettes destroy life with the moral judgment that smoking is wrong. But, as MacIntyre says: "… the moral element in such a judgment is always to be sharply distinguished from the factual. Factual judgments are true or false…" I wish to dwell on the factual judgment, because it stands independent of what I may think or feel.

We can go beyond this second dimension into yet a higher dimension to consider whether the use to which we put the ashtray comports with our ideals for moral virtue (value life rather than death) or supports our "practices" as MacIntyre depicted them. Does the commercial value, community interest, or even scientific purpose outweigh the objective statement of worth (being destructive of human life)? The commercial objective leads to jobs, wealth, and industrial progress. The community effort leads to safer homes for our families. In the case of the ashtray, our assessment of the "good" and "bad" may not be black and white. But, as Sartre observed, "man makes himself." The choice remains ultimately and unavoidably ours.

In matters of technology, utility deals with those qualities having to do with how we use something. A compass leads the mariner to a chosen destination, but does not offer guidance about the desirability of the destination. It may lead the mariner to a place where he or she finds purpose, but in and of itself, the device called a compass has done all that its form allows. Nature and science are agnostic. But technology is not, as it represents the product of human intention. Its design and embodiment carry within it, a moral component, and essentially embodies the practice of ethics itself [3]. This applies to all technology whether embodied in artificial intelligence or biology. If life forms comprise technology then life forms becomes, not essentially means and ends, but a statement of our humanity. I am not finished with this, as I think we need to look at utility from a slightly different direction, as well.

Notes

1. Holland, S., Lebacqz, K., and Zoloth, L. (eds.). (2001). *The Human Embryonic Stem Cell Debate, Science, Ethics, and Public Policy*, M.A. Farley, Roman Catholic Views on hES Cell Research. MIT Press.

2. Alasdair MacIntyre Phi Sigma Tau Lectures (published in 1990 as First Principles, Final Ends, and Contemporary Philosophical Issues).
3. Verbeek, P-P. (2006). "Materializing Morality: Design Ethics and Technological Mediation." *Science, Technology & Human Values* 31 (3).

CHAPTER 55

Runaway Utility

[I]t takes a human being to generate a human being.—Aristotle

Aristotle wrote that the ultimate aim in life is finding happiness [1]. In our quest to locate this often elusive emotional state, we each take different paths, each of us driven by disparate conceptions of what is good, each making choices in its satisfaction along life's journey. To an extent, what we strive toward would seem to align with our potentialities, existential, physical, and social. I regard these as constituting the "form" of what and who we are, our inner core or essence [2]. This holds true for all living things.[1] It would be pure fancy for me to imagine that I could swim across the Pacific Ocean, but for a whale, the possibility exists. A whale after all was made to swim and the ability to swim vast distances derives from its natural form. In the first instance then our potentiality and inescapably utility depends upon form, which in turn determines those life functions we best serve. To the degree living things exploit the full measure of their form, they can be said to demonstrate a certain utility, and importantly fulfill their purpose.[2]

This idea that function follows form applies to all things and when we acknowledge that something performs a function, we invariably

[1]Perhaps the most salient example of a change in form are genetically engineered changes in anyone of the 46 chromosome complement.

[2]Aristotle's *eudaimonia* as a final cause or telos, the ultimate purpose or good that we seek.

J. R. Carvalko Jr., *Conserving Humanity at the Dawn of Posthuman Technology*, https://doi.org/10.1007/978-3-030-26407-9_55

attach valuation. Valuation itself remains contextual, limited by the scope of its formation. A screwdriver exhibits excellent performance driving screws, but poor performance driving nails. A software business method accurately calculates a poverty index by determining the cost of living adjusted for inflation, but fails in palliating poverty. A human liver grown in a pig egg, produces a transgenic pig that enters into a world as an alien due to its extraordinary biological origin.[3] Words such as "poor," "fails," "alien," and "extraordinary," in connection with life, are not ones that can be easily steered clear of as we travel deeper and deeper into the technology of genetic engineering.

Each of the aforementioned technologies has efficiency or utility in virtue of their particular physical form or attribute. Each assumes a social position depending upon the interpretation we assign to its function. In this context, the word "utility" is a general word that relates to form and further links how well something meets its intended function or virtue. Although the word "utility" implies how well something works in accordance with its form and function, that fact does not necessarily lead to the assumption that it will in the future function the same way.

In matters of patenting, the concept of "utility" maintains a conspicuous position. Here it deals with whether an invention meets the qualifications of an article, machine, composition of matter, or a process, but again does not resolve the wisdom of monopolizing the technology. Because of the narrowness of how we employ the concept of "utility" in the marketplace, we have essentially disengaged the teleological considerations that often lie dormant in the totality of its form. As such, we consider only the means to an objective and again not the ends. In this case inventing *Homo futuro*.

Obtaining exclusive rights in one's invention dates back to the Greeks, in the era of Aristotle, who noted that Hippodamus of Miletus proposed a law "to the effect that all who made discoveries advantageous to their

[3] The idea of for growing human organs inside pigs or sheep for human transplant is not theoretical. In 2015, the National Institutes of Health placed a moratorium on federal funding of this area, but nine months later announced its intention to lift the ban—Stanford University, put in motion plans to produce human organs—by taking pluripotent stem cells generated from a person's skin or blood cells and implanting them into pre-embryos of pigs and sheep. If it doesn't integrate well into a pig's innards, causing discomfort, is it ethical to turn pigs into organ factories, for human purposes? Chen, I. (2018). "How Far Should Science Go to Create Lifesaving Replacement Organs?" *Undark-Truth, Beauty, Science*. https://undark.org/article/dilemma-science-ethics-organ-farming/.

country should receive honours [3]. The operative word is "advantageous," which through the centuries comes down as a watchword circumscribing that humankind and their institutions lay claim. Words such as advantageous and utility are impliedly foundational conditions for granting patents.

In the early 1800s, U.S. Supreme Court Justice, Joseph Story made the point that a "useful" invention is one "which may be applied to a beneficial use in society, in contradistinction to an invention injurious to the morals, health, or good order of society, or frivolous and insignificant…For instance, a new invention to poison people, or to promote debauchery, or to facilitate private assassination, is not a patentable invention. But if the invention steers wide of these objections, whether it be more or less useful is a circumstance very material to the interests of the patentee, but of no importance to the public. If it be not extensively useful, it will silently sink into contempt and disregard" [4].

When a government contemplates absorbing new subject matter into the patent statute (e.g. inorganic chemistry, organic chemistry, biochemistry, and now life forms), it first inquires directly into the concept "use" and its cognates (useful, utility, utilization), and obliquely invokes the philosophical notion of utilitarianism. When the U.S. patent office and the courts grappled over whether patent protection should be extended to cover software and life forms, the statutory requirement for "utility" was spiritedly contested. This standard of patentability reemerges repeatedly in analyzing the justification of patenting because the U.S. Constitution and the patent statute refer to the word "use." Now that we have opened the discussion to biological forms, let us look at "utility" from yet another angle, utilitarianism.

The subject of utility has had a distinguished history, having been considered in depth by the philosophical giants of the Enlightenment period, and continuing in that tradition to vitally influencing current philosophical dialogue. "Utility" combines the ascription of actual use and the assignment of social beneficence. Its seminal exponent Jeremy Bentham (1748–1832) believed that actions are motivated from considerations of least pain and maximum pleasure.[4] This simple idea implies

[4] "By utility is meant that property in any object, whereby it tends to produce benefit, advantage, pleasure, good, or happiness (all this in the present case comes to the same thing), or (what comes again to the same thing) to prevent the happening of mischief, pain, evil or unhappiness to the party whose interest is considered: if the party be the

a moral purpose or telos, which had been readily understood by social and political reformers of the late eighteenth and nineteenth century [5]. Unquestionably, utilitarianism has had a profound impact on the development of criminal law and social welfare.

Much of the advocacy for utilitarianism spawned from the good works of Bentham's godchild, John Stuart Mills who communicated widely through his famous treatises On Liberty and Utilitarianism. Mills sought to enlarge and refine the theory of utilitarianism by both distinguishing higher levels of "pleasure" and pointing out that creativity itself may be the ends to happiness. This of course plays directly into the idea of innovation. But, in the main, utilitarianism is more a sociological theory and has had a large influence on the development of a variety of social institutions; and incidentally not the worse fate that history conceivably has bestowed. Its affect can be traced to Anglo-American views on the suffrage movement, education, labor, health care and from time to time, prison reform. It stands for the proposition that the goodness or wickedness of a thing should be measured in terms of its consequences to the well-being of an individual and the greater community. In this respect, its precepts managed to affect our notion of what should and should not be patented for the greater good. However, as technology changes and as a consequence our social reality changes, this kind of calculation no longer seems to lead us to an ethically satisfying end point for patent purposes. It all seems rather arbitrary.

Utilitarianism supports restrictions on one's liberties where the balance of the competing interests ultimately weighs in favor of one or another collective good. In this way restricting someone from capitalizing on anything they did not invent can be justified by a patent system that operates in the final analysis to motivate inventors to disclose their technology—for the good of all. This kind of calculus supports outcomes where patented drugs command higher prices, only the affluent can afford, but denying them to individuals who cannot afford monopoly prices.[5] Such free-market forces may work well for producing

community in general, then the happiness of the community: if the particular individual, then the happiness of that individual." Bentham, J. *An Introduction to the Principles of Moral and Legislation* (ed. J.H. Burns and H.L.A. Hart); Chapter 1, §3, p. 12.

[5] For a discussion on the extent to which profiting from patents attract questions on the moral right to use the patented invention, see, Jeremy Waldron, From Authors to Copiers: Individual Rights and social Values in Intellectual Property, 68 Chi.-Kent l. Rev. 841 (1993).

better automobiles, rocket ships, and washing machines, but whether the underlying social policy can be justified under all circumstances, the distribution of life saving drugs for example, is a question begging an answer.[6]

Much of the ethical discussion surrounding stem cell research focuses on the sanctity of the embryo and even the blastocyst. Is it moral to use them and then destroy them for medical research? Other questions will inevitably follow when and if these kinds of biological hosts are used in the mass production of therapies. Many thoughtful voices provide differing points of view, but in the final analysis, many of the arguments are grounded in the ideals of utilitarianism as framed by the American political, economic and social experience. When President Clinton tasked the National Bioethics Advisory Commission (NBAC) in 1998 to review the propriety of stem cell research, he pointedly asked them to proceed by "balancing all ethical and medical considerations." The following year the NBAC report issued amidst controversy, but clearly its moral intention sounded in a "benefit: harm" rationalization. Despite Clinton's charge and the plain language of the report in terms of cost–benefit analysis the commission nonetheless insisted that it did not appeal to any "particular school such as deontology or utilitarianism."

We must separate what is patentable from what is permitted in research laboratories, where in the former case a governmental imprimatur may denied, but in the latter case, no legislation exists that would deny the right of institutions to move ahead with research that may well destroy the biological products labeled embryos or stem cells. So, although there is no actual ban on the use of human embryo research in the U.S., there is a prohibition of using government funding to support research on human embryos. The 1994 Human Embryo Research Panel (HERP) refused to find that embryos were in fact legally protectable life forms, based upon a finding that they lacked developmental individuation in the preimplantation embryo, lacked sentience and experienced a high rate of natural mortality. Twenty-five years later the debate has not abated, and will not, until Americans reach a consensus on what constitutes "life." Given U.S. history vis-a-vis the abortion controversy, a consensus does not appear on the horizon. Utilitarianism as a guide fails in

[6]In its more modern incarnation Mill's brand of utilitarianism (act-utilitarianism) has been invoked to judge the utility of broader social policies rather than particular actions.

those circumstances where an irreconcilable difference in the nature of the very matter exists, or we find the value of the matter inextricably tangled in a community of disparate values.

Notes

1. *Aristotle, Nichomachean Ethics*, Book, I.
2. Ibid.
3. Aristotle, Politics II (1268a6) (Penguin Classics 1981).
4. Lowell v. Lewis 1 Mason 182, No. 8568, 1 Robb.Pat.Cas. 131 (1817) (C.C.D.Mass., Also see, Brenner v. Manson, 383 U.S. 519 [1966]).
5. See, MacIntyre, A. (1984). *After Virtue* (2nd edition). University of Notre Dame Press.

CHAPTER 56

Middle Fields of Moral Force

Everything in this world improves: Swedish matches, operettas, locomotives, French wines, and human relations.—Anton Chekhov

A question raised throughout this book is whether we ought place limits on what we permit to materialize by way of our supposed anatomical improvements, that is the innovations and inventions discussed, and the degree to which we allow ownership vis-à-vis a patent monopoly to lay claim to elements of *Homo futuro.* MacIntyre deals with the problem of relativism in modern moral theory by what he claims the consequences of the failure of the Enlightenment project [1]. In support of his argument, he advances the proposition that when we detach conceptions of truth and reality from final causes, goals, and higher purposes, we deprive ourselves of rationally defensible arguments [2]. He presents as support for his assertion that when the individual moral agent eschewed pre modern teleology, a new position of moral authority had to be assumed. In attempting to establish a new grounding, new sets of rules with assigned moral meaning were created. The first project to fail was utilitarianism followed by the deontological Kantian-centered "practical reason."

One could argue that genetic engineering has a strong utilitarian value. As earlier mentioned, utilitarianism evidences qualities associated with liberal social improvements, but it also fails as a standard against

J. R. Carvalko Jr., *Conserving Humanity at the Dawn of Posthuman Technology*, https://doi.org/10.1007/978-3-030-26407-9_56

which we judge *moral purpose* as it applies to the things we invent.[1] Part of the failure is due to looking merely at the greater numbers served and the consequences of an action. The first may not always lead to the moral high ground, and the second may not objectively account for all consequences.[2] As if these weaknesses were not enough, a more powerful reason for its failure is that utilitarianism depends on two disjoint categories in the pursuit of the execution of its moral injunctions: the general happiness and the individual happiness. One may not be compatible or practicably attainable within the ambit of the other. Certainly in a democracy, many voices can easily silence what is or is not a preferred or right course of action and where that action may lead. Some voices are heard, but who ultimately serves to judge the rightness of a course of action and an ultimate destination in the stream of competing values?[3]

Neither general nor individual happiness is susceptible of independent verification, since these originate in one's personal belief. Therefore, who qualifies to judge the proper ends except relative to one's own predilections? The answer may lie in our potential to locate objectivity in what we prescribe as "good." But is this possible? G. E. Moore would probably answer "no," it's not possible. Quine says it another way, that if we were to eliminate the subjective component to our judgments, we must first eliminate intentions, purposes, and reason for our actions.[4] If we

[1]What follows is essentially adopted from MacIntyre's argument that the postmodern sundering of moral value was replaced by other self-centered moral rules that for one or another reason do not return us to an Aristotelian account of human value.

[2]Perhaps the general fear that the law of unintended consequences should be the mantra of the pharmacological and food science. We need only to be reminded of Thalidomide, which lead to profound birth defects or the revelation that Creutfeldt-Jacob disease (a/k/a "mad cow disease") was linked to how we fed cows to reduce costs to in the productions of beef.

[3]In Ethics, vi, 6, Aristotle writes: "Regarding practical wisdom we shall get at the truth by considering who are the persons we credit with it. …"

[4]Quine's general thesis is even more striking. "The totality of our so-called knowledge or beliefs, from the most casual matters of geography and history to the profoundest laws of atomic physics or even of pure mathematics and logic, is a man-made fabric which impinges on experience only along the edges. Or, to change the figure, total science is like a field of force whose boundary conditions are experience. A conflict with experience at the periphery occasions readjustments in the interior of the field. Truth values have to be redistributed over some of our statements. Re-evaluation of some statements entails re-evaluation of others, because of their logical interconnections—the logical laws being in turn simply certain further statements of the system, certain further elements of the field. Having

aim to find the right purpose, then must moral maxims have independent and objective authority? Quine suggests that we seek out rational means in finding answers, but counsels us to separate ourselves from the very objects of our pursuit: to find purpose. But, how is this possible, especially, in view of the conundrum raised by Olya Kudina, where she points out, "... beyond mediating the moral habits of and behaviors of people, technologies also mediate the meaning of the value frameworks themselves"? [3].

If we were to turn to the Kantians for guidance, we would be informed that: "any rational agent is logically committed to the rules of morality in virtue of his or her rationality." In the sphere of human endeavors, rationality distinguishes itself as a classic "form" of existence. As much as we find comfort in the premise, it offers little direction as to what final purpose we should chose to pursue. An appeal to reason always has an air of objectivity, but as we will shortly see, our 'rational agent' presents us with other problems. However, such problems actually illuminate the nub of the teleological quandary as it relates to an analysis of "utility," both as a social means to an end, such as well-being or supposed happiness, when it comes to genetic engineering. Especially, in genetic engineering, where the product itself folds back into the anatomies and lives generally of the individuals deciding what limits we should impose on development.

Kant asked why science and mathematics succeeded in yeilding objective knowledge, but that knowledge-about values did not [4]. He concluded, as did Plato 2000 years earlier, that we see things as they appear and not the way they essentially are. The very forms or categories we impose on our perception, such as causality, substance, action, reaction, time, and space do not allow us to know in a metaphysical sense. We need to find answers beyond the limits of ordinary experience.

re-evaluated one statement we must re-evaluate some others, whether they be statements logically connected with the first or whether they be the statements of logical connections themselves. But the total field is so undetermined by its boundary conditions, experience, that there is much latitude of choice as to what statements to re-evaluate in the light of any single contrary experience. No particular experiences are linked with any particular statements in the interior of the field, except indirectly through considerations of equilibrium affecting the field as a whole." See, Quine, W. The Verification Theory and Reductionism (1951).

In Kant's subsequent major work, Critique of Practical Reason, he proposed a process of looking beyond individual desires and purposes. As such, he formulated the unconditional imperative or "ought," which moves the matter of discretion and consequence outside the purview of relative morals and ethics. In Kant's famous counsel, he says, "Act only on that maxim whereby you can, at the same time, will that it should become a universal law" [5]. The proposition that humans should be never treated as means, but as ends serves as such an example [6].

At the margins of a clear moral constraint or conversely the power to act upon one's desires, wants and needs without economic or political opposition, Kant's categorical imperative may have merit, because it invokes conscionable limitations. A matter reaching the public forum, as to what is or is not good social policy, typically does not repose at the margin [7]. The more difficult questions posed by whether life forms ought be accorded patent protection tends toward the middle fields of moral force, where advocates for both sides line up ready to do political and social battle.[5] Only time will tell how far society will venture in allowing commercial interests to lay claim to elements or the entirety of what will become the *Homo futuro*.

Notes

1. MacIntyre, A. *On Virtue*. Oxford University Press.
2. MacIntyre, A. Phi Sigma Tau Lectures (published in 1990 as First Principles, Final Ends, and Contemporary Philosophical Issues).
3. Kudina, O. (2019). "The Technological Mediation of Morality, Value Dynamism and the Complex Interaction Between Ethics and Technology." Dissertation, University of Twente.
4. Kant, I. (1781). *The Critique of Pure Reason*.
5. Kant, I. (1788). *Critique of Practical Reason*.
6. Kant, I. (1959). *Foundations of the Metaphysics of Morals*, trans. L.W. Beck. Indianapolis: Bobbs-Merrill, p. 9.
7. Hegel, "Adding Ethical Substance to Kant's Empty Formalism." In: Singer, Ethics, pp. 132–134.

[5] Furor Over Cross-Species Cloning, WSJ, March 19, 2002 reported that fusing human DNA and the egg of a cow to create embryos for harvesting embryonic stem cells and possible transplant treatments has stirred a world wide debate.

CHAPTER 57

Crossing Point: Respect for Form

When the discussion moves from ethics to law and science, we are reminded to define and specify the context in which terms are used [1]. In the matters we have been addressing, "personhood," "human," and "life," differ in their legal, social, scientific, religious, and philosophical etymologies. For example, "personhood" stands in both the juridical and moral categories. These distinctions are crucial in discussions about legal or moral rights, each of which in turn require definition and context. I prefer not to engage in defining "personhood," but to simply refer to "life forms." But as Wynn Schwartz, reminds us, "Whether there are persons other than ourselves is neither a trivial nor a purely academic question. We should remember that, from time to time, successful attempts have been made to strip the status of person from some of us."[1]

I have chosen the term "life form" rather than person, personhood, or human to encompass all the various stages of human development, as well as so-called organoids, tiny clumps of organ-like tissue that can self-assemble from human stem cells in a petri dish, and life that may be chimeric in form, i.e., which include human organs and other instantiated human genetic material. In arguing the prudence of patenting life forms, we must not only consider positive law or legislation, but natural law as well. The natural law of which I refer is far less expansive than that theorized by Thomas Aquinas. My narrow view of natural law is that

[1] Schwartz, W.R. (1982). "The Problem of Other Possible Persons: Dolphins, Primates, and Aliens." *Advances in Descriptive Psychology*, Vol. 2.

J. R. Carvalko Jr., *Conserving Humanity at the Dawn of Posthuman Technology*, https://doi.org/10.1007/978-3-030-26407-9_57

which harmonizes nature in the scientific sense with both moral codes of conduct and legislation.

Nowhere does the notion of life form, function and utility come together in more stark terms than where the greater good spills over into the debate regarding the ethics of such issues as embryonic stem cell research, genetic engineering of human embryos, or patenting the processes leading to transhumans. Two factors make this subject the center of ethical attention. The first is that embryos can be used to alter the germline. This may not be the only technology or biological entity, where this can be accomplished. The second reason is that these practices often result in the destruction of embryos. Other medical and scientific procedures may also result in the destruction of the embryo. Therefore, neither of the issues is necessarily unique to stem cell research or genetic engineering of human embryos. Nevertheless, both the destruction of the embryo and the potential for alteration of the germline, suggests that arguably human existence is at issue. As a starting point in analyzing the controversy over the wisdom of a liberal research policy, it would seem reasonable to draw boundaries around what does and does not pass for its essence. These boundaries, although not sufficient for a consensus on what is or is not "life," may well serve to establish the limits to patenting inventions that have the potential to alter human germlines.

If we limit our discussion to bestowing a moral status to life forms, then one such separation leads to the proposition: If a biological form maintains the potential to maturate into a human life form (not necessarily a person, but at the most rudimentary level) it may be accorded moral status and if it cannot reach that potential it may simply be accorded an amoral status. If we accord an amoral status, relative to at least embryos that will never mature into life forms, we ask, what is wrong with patenting the *life forms* that cannot mature into "personhood" by any definition of that term?[2]

I follow Ronald Dworkin's idea that distinct individual liberties exist, not derived from some abstract right to liberty, but from the right to equal concern and respect [2].[3] In this regard, I find that the concern

[2]A central question is: are persons different from microscopic human life forms in kind or only different as to degrees of complexity? A material difference in kind would strongly suggest warranting a different modicum of respect and dignity due.

[3]This forms the central thesis in his influential work on jurisprudence.

and respect attaches to the feature of *form* extant in human biological entities. If I follow this logic, I reach the conclusion that we should not permit the patenting of a feature of *form* that envelops incidents of human essence. This means whether the biological specimen can or cannot maturate into a viable living form is irrelevant. Viable would mean the potential to maturate into personhood. I am advocating that our threshold be less than what might be considered viable by this definition. There are possibilities that must be avoided as we may create life forms that although never viable, nonetheless experience sentience or pain.[4] To be clear about this, the form of a stem cell or modified post-human is manifest in that it properly works to achieve an end (recall the whale metaphor earlier). We should not permit the patenting of any invention that alters such form as may potentially affect the ends to which the organism naturally aspires. I offer no corresponding opinion about whether the research or inventive activities ought or ought not to proceed.

According to Thomas Aquinas, natural law governs us in our journey toward our earthly goal and human law governs us as members of our communities [3]. If in respect of the law's function in society, a positive law were not to support natural law and the common good, Aquinas believed it would be a perversion of what we were to consider law.[5] In some sense, Aquinas enlarges the scope of human law so as to embrace natural law. When a law affects values going to the core of one's beliefs (e.g. religious faith, free speech, sanctity of life), it must move away from the precepts of positivist jurisprudence toward considerations that assert the moral rights against the state that exist prior to a consideration of the majority interests. I believe that reasonable minds will differ on the very point at issue. It would appear that if we permit patenting altered life *forms*, no possible standard for gauging what is or is not permissible

[4]Natalie Kofler molecular biologist, founder of Editing Nature and scholar at Yale's Interdisciplinary Center for Bioethics, points out that altering the germline goes beyond a single "human" or "life form" as to its impact on humanity vis-a-vis future generations. One example is culturing brain organoids from a person's pluripotent stem cells. What do we owe the cultured organ, if it can experience pain, memory or emotions.

[5]As Augustine says, that which is not just seems to be no law at all…Every human law has just so much of the nature of law as it is derived from the law of nature. But if any point of it departs from the law of nature, it is no longer a law but a perversion of law (Summa Theologia, I–II, q. 95).

can be rationally developed. We would only encourage a relativistic ethic subject to the whim of political and economic forces that may not always work toward the goal of concern and respect for the integrity of *form*. We return full circle to the narrative, and that moral pronouncement is relative. If the matter would be left in a political maelstrom of uncertainty, the policy ought to tip in favor of moral interests that support the "good" that we may not be morally entitled to remove from the sphere of humanity. As stated earlier, these objects are not goods at all, but virtues.

Evolution constitutes a scientific fact. This fact represents a human, that is an objectively observed independent phenomenon of nature. Evolution performs or causes only that which we observe as a fact of nature. Any further attribution, such as the function of evolution was to cause humans to become toolmakers, depends on what we opine or more accurately interpret to be the effects of evolution. It would be subjective. I have repeatedly argued that no causal or functional relationship exists between nature, such as evolution and human assignment. Perhaps not only functions are figments of our consciousness, but the artifacts, to which we apply the quality of function, are as well. If our assignment of meaning and consequent purpose is relative, what distinction, if any remains relevant to an ethical inquiry into the use to which anything is put?

One distinction recognizes that certain artifacts are made agentive in that an agent constructs them, defines their use and employs them in a particular way. In contradistinction, RNA represents a naturally occurring artifact, the purpose of which is to manufacture proteins. In a manner of speaking it is formative and not agentive, because it involves no human agency in forming its final product. A normative direction does not apply: neither good nor bad, better nor worse. It also follows its form with precision. It "properly works." In other words, function follows form exactly in virtue of its physical nature. When non-agentive artifacts cross the threshold through human tampering and become agentive artifacts, we carry forward normative considerations.

Agentive processes, unlike non-agentive formative natural processes and artifacts, always situate themselves in a teleological framework. A hammer is agentive. It constitutes a non-naturally occurring device that we use to drive nails into pieces of wood to join them in some semi-permanent way. It can drive them for better or worse depending on the degree to which it conforms to the essence of its form. If I take a naturally occurring artifact and use it as a tool, it acquires a teleological characteristic. For example, an expressed sequence tag (EST), which is a naturally occurring DNA artifact taken from an RNA strand, is used in the creation of a marker for some disease.[6] It can be a good or bad marker. Contrast this to a hypothetical final purpose where we can use the EST for the dual purpose of advancing human well-being or destroying life, as if we were to use it to construct a biological weapon.

We employ agentive artifacts and processes ultimately to conform the environment to our worldview. In conforming things to our worldview, we actualize the ultimate end. Human inventions are employed by virtue of choices we make about how the world will support and exist within a framework of efficient and final causes. We bring our reasoning and our character to bear on how we use the tools at our disposal to find or fulfill some ultimate purpose.

MacIntyre restates the argument as originally formulated by Alan Gerwith: "Every rational agent has to recognize a certain measure of freedom and well being as prerequisites for his exercise of rational agency. Therefore each rational agent must will, if he is to will at all, that he possess that measure of these goods" [4]. We find in this statement the conjunction of "rights" and "telos." One has the "right" to a patent and the patent must be "good."[7] If one or the other is absent, we can say that we have not exercised a rational agency. When an agent fits the world to his or her mind's eye, the matter not only presupposes a right, but a telos. A double predicate exists that requires two separate truth-values settled: a right to *X* and a "good" served by *X*. However, only "a right to do *X*," one-half of the proposition, and double predicate needs be proffered before patent issues. Once congress legislates or a court judges a subject matter as within the ambit of the statute, that is having a right to do *X*, the matter of *X* being a "good" course of action, defaults for the world, for all time to a truth-value of one.

[6] An expressed sequence tag is a short sub-sequence of a cDNA sequence. ESTs may be used to identify gene transcripts, and are instrumental in gene discovery and in gene-sequence determination.

[7] Francis Fukuyama makes the point in Our Posthuman Future that "Classical political philosophers like Plato and Aristotle did not use the language of rights—they spoke of the human good and human happiness, and the virtues of duties that were required to achieve them. The modern use of the term rights is more impoverished, because it does not encompass the range of higher human ends envisioned by the classical philosophers."

This last point respecting "reasoning" raises an important issue that needs to be confronted if we hope to find support for our actions in the sphere of final causes, goals, and higher purposes. The central issue speaks to whether the agent finds a legal system supportive of his or her ambition to change the world to conform to his or her mind's eye view. After all, truth and reality must be objective, to be rationally defensible. We begin by observing that in a utilitarianism-based regime, freedom in many forms constitutes the "good." Certainly, the patent statute is upheld as constitutional and although those interests that test their rights under the statute from time to time may be disappointed with the outcome, in most industrialized nations a just process seems evident. We do not find fault with the justice of the patent process. This may not be true for the teleological process, where it is far from clear how we find the ultimate purpose served by patenting forms of artificial intelligence or life forms. A patent always lays claim to the form of the idea. When the form implicates the pattern within which the idea lives, such as DNA, we must be especially circumspect and insist that its teleological considerations at least be addressed in a reasonable fashion. We must ask, by what moral authority does one command the right to do *X*, that is, conforms the world to one's vision?

In respect to the current bioethical debate, there exists many issues and many visions that cannot all be accommodated or treated under one regime or with the same prescription. It may be that no theory or justification universally applies to dealing by way of example with stem cells, cloning, biological computers or transhuman engineering. Stem cells, transgenic species, sequencing of expressed genes and artificial intelligence will find applications in the joining of biological language and artificial languages. This would seem an unavoidable fact. And, each of these developments change social reality when viewed through the patent process, or the closest form of ownership relevant to diagnosing possible moral problems and suggested solutions. I believe that among the foregoing list, we will find some technology that seems prudent to patent, and other of which would be unwise. I believe that those that should be elevated to a "good" in the Aristotelian tradition will be found in *forms* that have been extant from the beginning of organic life as we have come to understand it. In this form, neither should we tamper nor attempt to exercise dominion and control.[8]

My purpose here does not include weighing in on what is life, or the related abortion debate. The focus throughout this book has been on the nature distinctively common to human "life forms" in terms of its structural features and the formative power to engage in certain activities. Humans as rational, logical beings have discovered that these powers can be employed to ends not designed into the organ by nature. In this revision of nature, we find beneficence in the avoidance of human suffering. We also find we can pervert nature into services that satisfy our pleasures. However, in the matter of patenting life forms, it is not only the failure of utilitarianism to rescue us from moral uncertainty that contributes to an incessant colloquy, it is also our failure in recognizing that the issue of what constitutes "life" may be a non sequitur. I do not believe that identifying "life" by some specification propounded by political edict, science, or philosophy helps move forward the gravamen of the debate—whether anyone should monopolize life forms—because, "life" does not serve as the appropriate criterion. *Human form* from which *human essence* flowers does serve as the basic criterion. If we affect form, we run the risk of affecting human essence.

Many different views surround what constitutes human nature and its essence [5]. Clearly, it would be useful if we could find its meaning in objective, absolute terms. As I previously indicated, such terms typically invoke differing legal, social, scientific, religious and philosophical etymologies. To fully appreciate what we mean by the term human essence in moral terms would for instance require a thorough inquiry into the history of cultures. Perhaps, as I imply here, that despite a reference *to essence*, that the matter itself has no universal basis and may be only a relative or even subjective term. I prefer to use the word in its scientific connotation. By way of example, the essential nature of DNA resides in the intrinsic explicitness of its code. It speaks irrefutably in a language that is both epistemically objective and ontologically absolute.

[8]A satisfactory resolution of this question will lead to an answer to the central question in this book: do we use the patent system to encourage and reward those who endeavor to do so? The answer does not turn on the analysis of what any individual "ought to do" about patenting or not patenting, for the matter of patents is one of law. As a matter of law, it is not a question of what private individuals or corporations want, or ought to do, but what under law they have a duty to do or no right to do.

Notes

1. Mahowald, M.B., and Mahowald, A.P. (2002). "Embryonic Stem Cell Retrieval and Possible Ethical Bypass." *American Journal of Bioethics* 2 (1), Winter.
2. Dworkin, R.M. (1977). *Taking Rights Seriously.* Harvard University Press.
3. Basic Writing of (1954). St. Thomas Aquinas and Anton C. Pegis, ed. (vol. II). New York: Random House, p. 784.
4. MacIntyre, A. (1984). *After Virtue* (2nd edition). University of Notre Dame Press; see also, Gerwith, A. *Reason and Morality* (1978).
5. Erlich, P. (2000). *Human Natures: Genes, Cultures, and the Human Prospect.* Island Press/Shearwater Books.

PART VIII

Final Thoughts

CHAPTER 58

Lost in Time

Sarouk: We've covered lots of ground, but how do you think our humanity has been altered, you know changed during your lifetime?

Mensa: Without getting too deep into my philosophy of life, the human form, for millions of years, experienced first hand, causal associations where reflexively we expressed intentions and desires. One to one, our sensations directly interfaced with the world. We heard music, saw color, tasted the bitter and the sweet. But, engineers knew that at bottom, much of what we felt, what we considered our humanity, was reducible to physics or chemistry. Take music, it's the compression of waves, color's a packet size of energy. Technology intercepted nature reprocessing and repackaging sensations to conform to some corporate specification. To what some engineer decided appealed to our tastes.

Sarouk: That's a somewhat cynical characterization of what has been a concerted effort to make the world safer, one in many ways less hostile then the one you were born into. One where competition for the most outlandish beliefs consumed the world in bloodbaths.

Mensa: Yes, but we lost music inspired by musicians, art inspired by artists … we've lost all that.

Sarouk: So, are you saying we have surrendered some part of what you'd once regarded as your humanity, because it's all artificial, all produced by computers? I'd say a small price to pay for what we gained in return.

J. R. Carvalko Jr., *Conserving Humanity at the Dawn of Posthuman Technology*, https://doi.org/10.1007/978-3-030-26407-9_58

Mensa: The concept of humanity includes more than the arts, music and literature. Those were simply ways of expressing the deeper essence of what all living beings once sought out. What we used to call autonomy and a unique identity.

Sarouk: When you say identity, autonomy, what do you mean?

Mensa: I mean a sense of freedom, the freedom to be. Free, free to have independent thought, free to believe in a supernatural being. This has been replaced by algorithmic-centric rules built into our anatomical computers. Real identity flows, not just from our rational side, but our emotional side, too. Now, hearts no longer throb, because we don't fall in love, because we no longer cry, because we've lost the means to purify the soul.

Sarouk: But, wouldn't you agree that we're now unburdened? We've eliminated ways of thinking, negative thinking, we've improved intelligence by orders of magnitude.

Mensa: Yes, but if post-humankind is unmoored from its identity, its autonomy, how can it know when it goes too far, reaches the abyss, when it takes that final step from the last remnant of the Homo sapiens, transitioning into the soul of some creature devoid of any remembrance of what came before it, of the emotions that once made us human?

CHAPTER 59

The Inner Eye

The way to truth is long.... I must keep on pursuing..., Qu Yuan (340–278 B.C.)

Unlike machines, creative people share an enlightenment, which American philosopher, Ruth Nanda Ashen describes as: "…the in-sight of this inner Eye that purifies and makes sacred our understanding of the nature of things; for that which was shut fast has been opened by the command of the inner Eye. And we become aware that to believe is to see." Can technology develop any algorithm that matches Ashen's inner eye: "To believe is to see."

It takes an inner eye to see what the artist or inventor sees, standing at the edge of originality or novelty, an unfolding horizon, which lacks an antecedent or remains hidden in an obscure mountain of complexity. The unfolding represents the transition between a *moment* and an *instant*, moments which pass in the life of a creative quest, but and then the instant when all that passed is concretized.

The philosopher Kierkegaard put it this way: "It is brief and temporal indeed, like every moment; it is transient as all moments are; it is past, like every moment in the next moment. And, yet it is decisive, and filled with the eternal. Such a moment ought to have a distinctive name; let us call it the *Fullness of Time*" [1].

Such is the experience of the solitary artist, perhaps Georgia O'Keeffe, first witnessing a New Mexico sunset, or the Thomas Hardy startled by the intricacies of Ramanujan's proofs, or occurred when Charley Parker riffed on something Dizzy Gillespie blew in *Koko*, for the first time.

J. R. Carvalko Jr., *Conserving Humanity at the Dawn of Posthuman Technology*, https://doi.org/10.1007/978-3-030-26407-9_59

These "wow" instants transform art and in some cases civilization, to wit: Einstein's $E=Mc^2$, that instant of truth about the light, energy, and mass, or the instant when an artist imagined the eternal woman and sculpted *Venus de Milo*, or heard a timeless polyphony and composed *The Requiem*, or dreamt of that undying journey, the one that presaged Homer and penned *Ulysses.* Euclid's elucidation of geometry, Galileo's idea of focusing light through a glass, or Fresnel's idea that the imposition of glass between two light bearing mediums could be described with geometrical certainty, and in and of each case worthy of eminence, but to only then discover successor geniuses, LaPlace and Fourier whose equations of complex wave fronts let us imagine light beyond its physical brilliance. The love of the feel of the cold glass that to its progenitor symbolizes not a piece of glass, but what one has come to love, the poetry of engineering, the laws of science, the language of mathematics.

Mary Oliver in *Ode to Spring* wrote: "For me, the poem opens into mystery. How could it not, since it's about the 'dazzling darkness' that's forever coming down the mountain toward us?"

Indeed! Down the mountain toward us, but what's concealed in this deeper darkness—? At the level of consciousness we appropriate rationale, but beneath it lives the expression claimed beautiful, a complex of creative roots, nutrients, where unobserved nature frolics, wildly naked.

Science only hints at the "what" and "why" of a phenomenology, which taps into perception and conception. Our states of mind flip-flop neurons that chart the unfathomable abstract caverns of the mind, imperceptible, objects of thought through brains-folds, teasing us into seeing universal truths in poetry, art, a piece of music. Truths embedded in the DNA common to our species, that which paradoxically we refer to as our sense of self, the essence of each of us, our "*I?*"

And, now comes technology: an algorithmic calculator deducing, inducing, and mimicking creative geometry, a plagiarized Rorschach model, analogized primordial arrangement of genes that survived the 100s of millennia, since the first hominid turned skyward and screamed.[1] If we surmise that such may come to pass, what does that conceptual space look like, how would we go about gathering the raw materials, the

[1]The Rorschach test is a psychological test in which subjects' perceptions of inkblots are recorded and then analyzed using psychological interpretation, complex algorithms, or both. Some psychologists use this test to examine a person's personality characteristics and emotional functioning.—Wikipedia

information or qualia of a person's personality, her emotion, and lay it bare on a silicon chip?

The extant psychological blot, our essence, plainly before us in our art, music, crafts, writings, and inventions. But, the neural triggering and its concealed symbology, the isomorphic linkages to the real-world we inhabit remain mysteries. Self-expression reveals that which we are, or at least our potential. It does not easily surrender what takes place in that 0.1 mm flyspeck of a universe-sized flyspeck space of the neuron that paints the canvas of my creative reality.

Whether we internalize life as some grand plan that moves us cosmically, harmoniously and synchronously or internalize it as an unconnected minute-by-minute discrete series of events, also depends on our personality. We each experience events differently, providing the potential to be singularly innovative as we approach our experiences from different psychological zones. So, we find that how we choose to view our world does not affect our potential to soar creatively nor does it affect the medium upon which we travel, it only determines where we land.

Nature determines the biological factors that bear on our lives, but as conscious beings, we remain free to create our reality, the world we come to know through individual perceptions, judgments, choices, the endless ways in which we exercise our creative urges. No other earthly creature appears as prolific as humans, for better or worse. We live in a malleable world, where, within realistic limitations, we influence it and vice versa. A good deal of these impacts have to do with the values set by those of the society into which we are born and raised, the families, mentors and peers that teach us right from wrong, good from bad, likes from dislikes.

To reveal the essence of whom we are, i.e., not simply some biological machine that on the face of it contains more information than we are aware of beneath the level of symbols. No, we are destined to manifest those creations at a macro level, where they are observed by others, and through this power, to analyze, to psychoanalyze and to thus peer into the interstices of those unimaginably miniscule timeframes—the micro spaces, of our mind, where our lives take on meaning. But, the complexities of the mind are such that we can resolve these only to a certain degree.

We have successfully constructed computers that at a high level represent a form of thought. For example when I speak into Microsoft's Cortana program, it analyzes and interprets the sounds I make to represent the word Dog into a picture or definition of that four-legged soul that sits at my feet every night. The program does not understand what a dog is, but it creates a one-to-one relationship between my vocal cords' utterances, which forces air to vibrate a membrane that transduces mechanical energy into electrical energy, which comprises a signal, which amplitude can be analyzed as to strength and frequency and converted into a series of numbers, which serve as an address to another series of numbers, which when processed by a digital analyzer executes a program to display on my computer screen a picture of a dog.

Currently, it's impossible to deconstruct the brain such that its neural networks can be simulated by computer circuits so as to produce an image, say of the dog. We have invented a way of representing, at a symbolic level, the air pressure leaving my vocal chords, to produce a picture of a dog, but what comes before the expression of the word dog are the psychological stages of what motivates the vocal expression: desire, fear, hope, love, or doubt. This aspect of creating a computer that mimics humans, their communications with the outer world, for whatever motivates the expression, gets to the heart of the question, whether a computer will ever create in the way humans do.

Note

1. Kierkgaard, S. (1946). *Philosophical Fragments, A Kierkegaard Anthology*, ed. Bretall. NY: The Modern Library, Random House.

CHAPTER 60

Time to Return to the Sea

I sit on the dock watching the tide roll in, wave after wave. I came from the depths of the sea about one billion years ago to marvel at the sizzle of sunrises and muted sunsets, the lovers, who scattered themselves on the beach, the fisherman's wife, who watched new ships come, old ships sail, watching for her fisherman, who left before dawn. I choose to remain here, in a calm that lingers over the angler on the jetty, hoping to catch the big one—, reflecting how, as a droplet in the scheme of things, I've watched this ocean through star-filled nights and nor'easters. I've seen the lives bent by the gale, and I too, have felt the buffeting of the uncertain storms, winters when the shore slurries ice, when the tides go out, when the gulls leave, wondering now if its my time to return to the sea.

Life altering technology comes at us at bewildering speed, poised to create a posthuman future of transhumans, cyborgs, and humanoid robots, who will exhibit new levels of cognitive and creativity abilities, while slowly integrating into the mainstream of what we presently consider the modern human.[1] Joel Garreau, author of Radical Evolution argues that we are at an inflection point in history; humanity as we know it today will be decidedly different on the other side of the inflection [1]. Our species risks the riptides of a new alignment, obliging us to consider

[1] Aristotle called attention to the four "causes," the idea that we know something when we know its origin, what it's made of, formal cause and its teleology or purpose. The philosopher John Locke in An Essay Concerning Human Understanding (1689) refers to 'the being of any thing, whereby it is, what it is'.

J. R. Carvalko Jr., *Conserving Humanity at the Dawn of Posthuman Technology*, https://doi.org/10.1007/978-3-030-26407-9_60

if we should welcome or erect barriers to buffet against the deluge of change. We can no longer ignore the evidence, it portends the inevitable, perhaps returning us to a sea from which we shall not, as before, again materialize.

Genetic engineering cuts two ways. It offers to cure intractable diseases, but also advances a posthuman future. In the hands of the irresponsible, technological revolutions can rip through the fabric of delicate ecological and social patterns—driving species to extinction. We have witnessed this with climate change caused by our voracious appetite for fossil fuel. Human history reflects its true achievement, not based on technology, but by its passion to embody the virtues of responsible stewardship of the planet, which includes preserving life in all forms and engaging in universal social justice. If, through common neglect we irreversibly apply harmful technology, without regard to ethical boundaries, then we will destroy any semblance of a moral legacy for the future of humanity. Those in positions of power, whether that power comes from the pen, pulpit, purse or politics, must organize for the future welfare of the planet and humanity as we know it, otherwise the existential threat of the irrevocable alteration of *Homo sapiens* stand a good chance of coming to pass.

We live in an era rife in the vices of commercialization, competition, politics, and war, each pushing the boundaries of technological conquest. However, with advances in know-how comes responsibility, for tools in the hands of irresponsible agents threaten the fabric of civilization. In modern life values often spill over into ecclesiastical and political vessels, raising questions in the minds of many, as to the suitable level of institutional engagement, leading to government regulation to control the extent to which a society will allow certain established conventions or modifications thereof to exist, or how far an institution, such as a corporation should be allowed to control the operation, distribution and use of technology, which have deleterious effects on society. Except in rare instances, for example nuclear and biological weapons, governments and international conventions, governments deal ineffectively with phenomena such as global warming, war and ideologies that lead to starvation, mass migrations, death, and the destruction of societies. We now face the alteration of *Homo sapiens* as we transition to *Homo futuro*. Query: will governments heed the warning, and work alongside one another to provide for a controlled development of how humans will evolve?

In the past quarter-century, we have experienced a new social connectivity through the Internet and smartphones, better medicines and medical technology, such as imaging devices and electronic prosthetics. Progress comes in many forms and on many fronts. In some cases technology wreaks havoc. And, we need not go far before we see signs that technology fails to reach many throughout the world, the many who suffer from technologically driven wars, or from starvation, lack of clean water, health care, and poor education. Without regulation, existential threats, posed by genetic engineering, put at risk humans and many other species with which we share this good earth.

That having been said, we can imagine that when we change elements affecting our genotype and phenotype, inevitably we risk changing our present essence into a new essence, the intangible feature of life that reflects who and what we are in the world as we have come to be. The core of the idea that I have advanced considers that humanity constitutes a cycle of natural patterns that forms our very existence.

Paleontologist, Stephen Jay Gould, in *Wonderful Life*, reminds us that evolution has been fashioned by unpredictable sequences of antecedent states [2]. The evolutionary paradigm suggested is one where organisms bring about changes in their environments, many of which are evolutionarily and ecologically consequential. But, as we transition from *Homo sapiens*, a species constrained by nature, toward *Homo futuro*, a being that will function within a new set of constraints, largely bio-cybernetic, this new race will enter a world where accident may be removed from evolution, and more directly nature's randomness, the part of life that adds to life's wonder, its challenges, the wellspring of wisdom, aesthetics, discovery, and innovation.[2]

James Watson, co-discoverer of the structure of DNA and former head of the Human Genome Project said, "We used to think that our fate was in our stars. Now we know, in large part, that our fate is in our genes" [3]. I take issue with Watson to the extent that our fate is more likely as recited in a Greek parable, where a student tried to trick his wise master. The student planned to conceal a bird in his hands. He planned to ask the old man to guess what he was holding and, if he guessed a bird, the boy would ask whether it was dead or alive. If the old man

[2]Evolutionary processes widely acknowledged are natural selection, including sexual selection, genetic drift, mutation, and migration.

guessed the bird was dead, the boy would let the bird fly away. But, if the wise man guessed the bird was alive, the boy would crush its life and open his hand to reveal a dead bird. The boy finally asked, "Is the bird alive or dead?" And, the old man replied, "My son, the answer to that question is in your hands."

Notes

1. Garreau, J. (2005). *Radical Evolution: The Promise and Peril of Enhancing Our Minds, Our Bodies—And What It Means to Be Human.* Doubleday.
2. Gould, S.J. (1989). *Wonderful Life: The Burgess Shale and the Nature of History.* New York: Norton.
3. Time, March 15, 1993.

Patents Cited

1	Pat. 3,106,698
2	Pat. 4,060,713
3	Pat. 4,259,444 A
4	Pat. 5,004,681
3	Pat. 5,340,740
4	Pat. 5,453,357
5	Pat. 5,656,479
6	Pat. 5,827,740
7	Pat. 5,985,659

Index

J. R. Carvalko Jr., *Conserving Humanity at the Dawn of Posthuman Technology*, https://doi.org/10.1007/978-3-030-26407-9

C

D

E

Printed in the United States
By Bookmasters

Printed in the United States
By Bookmasters